AUGUST
SEPTEMBER
JULY
MARCH 1855
FEBRUARY
JANUARY 1855
BALANCE
AGAINST ENGLAND
Exports
NATIVE STOCK
3206

AF888633

Data by Design

The MIT Press
Massachusetts Institute of Technology
77 Massachusetts Avenue
Cambridge, MA 02139
mitpress.mit.edu

© 2026 Lauren F. Klein, Tanvi Sharma, Jay Varner, Margy Adams, Shiyao Li, Nicholas Yang, Dan Jutan, Jianing Fu, Anna Mola, Zhou Fang, Yang Li, and Silas Munro

All rights reserved. No part of this book may be used to train artificial intelligence systems or reproduced in any form by any electronic or mechanical means (including photocopying, recording, or information storage and retrieval) without permission in writing from the publisher.

This work is licensed under a CC BY-NC-SA 4.0 International license. To view a copy of this license, visit https://creativecommons.org/licenses/by-sa/4.0/legalcode.en for the full legal text. Note to users: A Creative Commons license is only valid when applied by the person or entity that holds rights to the licensed work. This work may contain components to which the rightsholder in the work cannot apply the license. It is ultimately your responsibility to independently evaluate the copyright status of any work or component part of a work you use in light of your intended use.

(cc) BY-NC-SA

The MIT Press would like to thank the anonymous peer reviewers who provided comments on drafts of this book. The generous work of academic experts is essential for establishing the authority and quality of our publications. We acknowledge with gratitude the contributions of these otherwise uncredited readers.

This book was made possible by generous funding from Furthermore: a program of the J. M. Kaplan Fund. Additional funding was provided by a 2018–2019 National Endowment for the Humanities (NEH) and Andrew W. Mellon Foundation Fellowship for Digital Publication (FEL-257658-18), a 2021–2022 NEH Office of Digital Humanities Advancement Grant (HAA-281011-21), and a 2023–2024 Emory College of Arts and Sciences Chronos Fellowship, funded by the Abraham J. & Phyllis Katz Foundation. This book is freely available in an open access edition thanks to the generous support of Emory University and the Andrew W. Mellon Foundation.

This book was set in VTC Du Bois by Vocal Type (Tré Seals) and PP Neue Montreal by Mat Desjardins by Polymode (Tanvi Sharma and Silas Munro). Printed and bound in the United States of America.

Library of Congress Cataloging-in-Publication Data is available.
ISBN: 978-0-262-05618-2

10 9 8 7 6 5 4 3 2 1

FRONT COVER, detail from "The Amalgamation of the White and Black Elements of the Population in the United States." Courtesy of the Library of Congress, Prints & Photographs Division, LC-DIG-ppmsca-33916. **BACK COVER,** Elizabeth Palmer Peabody, from *Chronological History of the United States*. Courtesy of the Internet Archive. Digitized by the Library of Congress.

EU Authorised Representative: Easy Access System Europe, Mustamäe tee 50, 10621 Tallinn, Estonia
Email: gpsr.requests@easproject.com

Emory College of Arts and Sciences | Digital Publishing in the Humanities
Emory University | Emory Center for Digital Scholarship
Mellon Foundation
National Endowment for the Humanities
Furthermore: a program of the J. M. Kaplan Fund

Data by Design
Visualization and Power from Abolition to the Dawn of Data Science

Lauren F. Klein with

Tanvi Sharma, Jay Varner, Margy Adams, Shiyao Li, Nicholas Yang, Dan Jutan, Jianing Fu, Anna Mola, Zhou Fang, Yang Li, and Silas Munro

The MIT Press
Cambridge, Massachusetts
London, England

For my family

PREFACE 8

INTRODUCTION — A History of Visualization and Power 14

CHAPTER 1
DATA — The Society for Effecting the Abolition of the Slave Trade and Responsible Design
Before there is data, there are people. How can the lives behind the data guide our visualization work? 48

CHAPTER 2
IMAGE — William Playfair and the Politics of the Simple View
Visualization has never been neutral or objective. How can we learn to recognize the decisions that shape each design? 76

CHAPTER 3
PEOPLE — Shanawdithit and Our Place in the Colonial Frame
Maps, like all visualizations, tell stories. How can we learn to see the lines of power that they draw? 102

CHAPTER 4
KNOWLEDGE — Elizabeth Palmer Peabody and the Value of Epistemology
Visualization can produce knowledge in many ways. What perspectives do we gain when we expand our view of the past? 136

CHAPTER 5

CHANGE

Lula Iola Mack, W. E. B. Du Bois, and Visualization for Liberation

Visualization can bear powerful witness to oppression. How can we use charts to make change in the world?

168

CHAPTER 6

LABOR

Data by Design and What We Learn from Visualization Work

Any visualization project is the work of many hands. How can we surface the meaning of this labor?

206

EPILOGUE 228

ACKNOWLEDGMENTS 234

NOTES 240

WORKS CITED 272

INDEX 284

PREFACE

I've always loved charts. I love their endless variety. I love their combination of logic and creativity. I love the feeling—you've likely had it too—when a vague inclination, a hunch you have about the data, is suddenly *right there in front of your eyes*. Which is why, when thinking back to the origins of this project, it was surprising—to no one more than myself—to realize that the idea to write a new history of data visualization was prompted not by any particular chart, or even by a dataset, but instead by a paragraph of dense historical prose.

It was the fall of 2012, and I had been teaching at Georgia Tech for just over a year. I had arrived in Atlanta with a newly minted PhD in early American literature. But even in my interdisciplinary humanities department, my students were mostly aspiring engineers. Their goals were actually quite familiar to me. Before going to graduate school, I'd worked as a software developer. After eight years of research and writing, however, I thought I'd left that life of code sprints and client demos far behind. But there I was in my Georgia Tech classroom, looking out from my podium into the faces of 35 students who wanted nothing more than a chance at a job like the one I'd once had. I decided to honor that. I'd teach a course on data visualization. After all, I still loved charts.

I had a fair amount of experience with visualization design from my time in the tech world. But creating charts for websites hadn't taught me quite as much about visualization's history. In the industry, Edward Tufte's *The Visual Display of Quantitative Information* had been our definitive guide. The book included a short history, centered on a man named William Playfair. He was the "remarkable" Scotsman, born in 1759, whom Tufte credits with improving upon the then-emergent line and bar chart forms, and with inventing the pie chart outright. In Tufte's telling, Playfair's contributions seemed nothing short of revolutionary. So I figured I should start my own research there.

I tracked down a facsimile edition of Playfair's *Commercial and Political Atlas*, first published in 1786 and revised and expanded in 1801.[1] It was a surprisingly hefty hundred-page tome that contained many of the earliest instances of Playfair's field-defining charts. From Tufte, I knew the visual beauty that awaited me inside the book—vivid red and yellow data lines, delicate hand-tinted shading, all set against a stark black grid. But I decided to begin with the editors' introduction, pencil in hand, like the literature professor I'd recently become.

Only a few pages in, however—the bottom of page 3, to be precise—my pencil slowed. There was that paragraph, the one that would alter the shape of my academic career. It started with a mention of Playfair's childhood tutor, a man by the name of William Small. I felt a nagging sensation. I circled the name and jotted a question mark in the margin. But I still couldn't place it, so I kept on reading. "In 1758 he [Small] joined the faculty of William and Mary College in Williamsburg, Virginia," the next sentence began.[2] Now I knew I'd seen this name before. Sure enough, as the edition's editors, Howard Wainer and Ian Spence, went on to explain, Small had been a professor at William and Mary for six years, where he had taught none other than Thomas Jefferson. Hence that nagging feeling. I'd been researching Thomas Jefferson for years.

William Small—Jefferson's professor and Playfair's tutor—was Scottish, and the other thinkers associated with his intellectual milieu, known as the Scottish Enlightenment, have long been credited with providing the philosophical foundation for the establishment of the United States. The more famous men in this group included Adam Smith, David Hume, and Henry Home, Lord Kames. But Jefferson credited Small, in particular, for seeding his own philosophical thinking. "From his conversation," Jefferson wrote in his autobiography, "I got my first views of the expansion of science and of the system of things in which we are all placed."[3]

In one of my earliest published essays, I'd analyzed this line as evidence of Jefferson's emerging empiricism: his belief that a universal "system of things" could be determined through sensory experience and direct observation of the world.[4] I'd connected this belief to Jefferson's meticulous record-keeping, which included spreadsheet-like tables that he used to document and draw conclusions about not only plants and animals but also people—including the people he enslaved. Among the most difficult truths of this early form of empiricism—demonstrated here by Jefferson, but practiced by many more—is that the big-picture conclusions of these men, drawn from their own firsthand observations, produced seemingly generalizable knowledge that was, in reality, deeply limited by the exclusionary worldview that so many of them shared.[5]

When I realized that the very same man had taught both Thomas Jefferson and William Playfair, and that both had been inculcated in the empiricism—and the exclusions—of the Scottish Enlightenment, I knew there was more to the story of data visualization than had been told to that point. There was more, even, to the story of William Playfair. Tufte, after all, had focused only on Playfair's charts. Wainer and Spence had focused only on Playfair the person. But the expanded story—the one that would place Playfair alongside figures like Thomas Jefferson in their full historical, philosophical, and political complexity—had yet to be told. And at that moment, I also realized something else: that I wanted to be the one to tell it.

***

Fifteen years later, you are reading the end result of that realization. But I begin with an account of the origins of this project to answer two important questions: why does *Data by Design* begin when it does, and why does it begin where it does? There are, after all, a vast array of examples of historical data visualization—many of which date back much farther than the European colonial era, and which emanate from all over the world. But my interest in this particular time and place is a direct consequence of what I realized while reading about Playfair: that there is a more complicated story about this particular strain of data visualization that has gone untold up to this point. We need additional stories, of course. But we also still need to understand how the standard story of modern data visualization—mostly European and North American, and mostly undertaken by white professional men—is entirely inseparable from the histories of colonialism, capitalism, and slavery that produced it.[6]

***

I still have the facsimile edition of Playfair's *Atlas* on my desk. I keep it there (or so I tell myself) as a reminder of where this project began, and how much it has evolved since. Thinking back to my first read-through of the editors' introduction, it's as if that passing mention of Playfair's childhood tutor was the key to unlocking my own scholarly expertise. I know from personal experience how tempting it can be to want to celebrate "founders" and "fathers." I study the founding era of the United States. But if I've learned anything at all from my time with Thomas Jefferson, it's that we must also look at the shadows that he and those like him cast. Because in those shadows are often unseen people and untold stories, stories that help to bring a more complete picture to light. Who were the people standing in the shadows of William Playfair? What stories about the emergence of modern data visualization could they tell?

Of course the path from initial idea to completed project is never a straight line, and *Data by Design* is no exception. I set it aside to write two other books.[7] I birthed two children. I moved from Georgia Tech down the road to Emory, where, since 2019, I've worked to build programs and curricula at the intersection of the humanities and data science writ large. But just as important, the kind of stories I sought to tell, and the ways I sought to tell them, continued to evolve.

The same dot-com boom and then bust that had sent me to grad school had also resulted in a host of early web technologies entering academic research. These projects coalesced into a field called digital humanities, defined by the use of computational methods (and related digital technologies) to explore questions about literature, culture, and other areas of humanistic concern.[8] At Georgia Tech, I'd followed my colleagues in the bench sciences and established my own research group, which I'd called—after the field—the Digital Humanities (DH) Lab. But my lab wasn't filled with test tubes or mass spectrometers; it was filled with people: a fantastic group of computer science and interaction design students eager to put their classroom skills to use.

My DH Lab students were this book's first interlocutors. Together, we explored its central formal and methodological questions: could a history of data visualization also *include* data visualization? Could we tell an expanded story about the emergence of modern data visualization using visualization itself? The answers to these questions turned out to be self-evident—even if they took an incredible amount of time to build out: yes and yes. Across three cohorts of students, first at Georgia Tech and then at Emory, we together designed the custom visualizations, page-level interactions, and underlying platform that constitute the web version of this book.[9] Along the way, we were joined by members of Emory's Center for Digital Scholarship as well as of the graphic design firm Polymode, each of whom brought additional expertise in software engineering, data visualization, art, and design.

When the MIT Press offered to publish a physical version of *Data by Design*, we returned once again to this project's earliest reference points: Tufte's history and Playfair's charts. In the end, it was viewing them from the perspective of a team that allowed us to identify our own story's main plot. If the narrative we'd inherited from Tufte was one of singular achievement, ours was one of a project that could not be completed alone. While Playfair was focused on the impact of his visualizations as they struck his viewers' eyes, we drew just as much insight from our own design process as we did from our charts' final form. Above all, if we wanted to communicate visualization's transformative power—the reason *why* William Playfair, Edward Tufte, and so many others before and since have found themselves captivated by charts—we realized we would need to write about the processes, and the *people*, who enable each visualization to come to be.

So now a note about these people. While I am the author of these actual words, there are eleven additional authors of *Data by Design*: Tanvi Sharma, Jay Varner, Margy Adams, Shiyao Li, Nicholas Yang, Dan Jutan, Jianing Fu, Anna Mola, Zhou Fang, Yang Li, and Silas Munro. Each of them has worked with in (or with) the DH Lab at some point over the past ten-plus years. Each has also contributed their own unique perspective and forms of expertise. Several are visualization designers, others are visualization researchers, yet others are humanities researchers, and still others are software engineers. As this book unfolds, you will see how we have come together to forge our collective authorial voice. You will also see how this book has been directly shaped by each member of this capacious team. I say this both as a practical observation and as an implicit claim of the book: for visualization to be its most impactful, it must leverage not only the skill sets but also the meaning-making capacities of each of its allied fields.

This is the reason that I tend to write as a collective "we" in these pages. At times, when discussing specific design decisions, this "we" is a strict one; it includes only the members of the project team. I have attempted to make these instances explicit. But more often than not, this "we" also includes you, the readers of this book. It does so not because I assume that we share all aspects of our identities, or our disciplinary and professional perspectives. I certainly do not intend to erase our differences, many of which may be quite profound. Rather, I write as a "we" to invite you into this project: to join us in working toward a more informed and more intentional practice of visualization viewing and design.

With that said, different perspectives—especially disciplinary ones—do carry different expectations, and for this reason it seems worth addressing them head on. To readers from the humanities: you will likely find the book's historical and theoretical frameworks quite familiar; my hope is that you will be energized by seeing how these frameworks can be applied to the history and practice of visualization design. For readers coming from the digital humanities in particular: my hope is that this book will provide you with a model of a critical-creative practice that expands the ethos of your work. For those in the fields of visualization research and visualization design: you will likely recognize some of the historical examples and technical reference points, but you may find the way I've chosen to tell this story a bit unfamiliar—or even opaque.

For this reason and more, it's worth being explicit about the form of the history offered here. This history is not intended as different in content; rather, it's different in approach. By this I mean that, in the humanities—and in particular, in the mode of literary-historical scholarship in which I was trained—our primary evidence comes from detailed analysis. In this case, it consists of a set of core examples of early data visualization, the processes that led to their creation, and the writing that accompanies them in their finished form. Interwoven with this analysis is relevant historical, philosophical, political, and cultural context. This context is included with a calculated goal: to tell a story about each chart that is so seamless, and therefore so convincing, that you cannot help but see even the most familiar chart in ways you never have before.

This kind of story is what, in the humanities, we call an "argument." But to be clear on this point as well: it's not the same kind of argument as the one you might have with your kid about bedtime, or the one that you have in your head when you yell at a politician through a screen. Rather, an argument in the humanities is intended to be absorbed—and, crucially, further questioned—so that additional insight can emerge. The gist of the argument presented in this book is that the history of data visualization is a great deal darker, and more complex, than has been acknowledged to this point. But there is a second core tenet of current humanities scholarship that this project rests upon, which is also worth stating out loud. This is the belief that generative new paths forward can be revealed through a better understanding of the complexities of the past.

A refrain of this project, which will hopefully soon become a familiar tune, is that we, the authors of this project, want the ideas we express on these pages to become action in the world. Consistent across our project team is a belief in the power of data visualization. We believe that if used intentionally, data visualization can help to create new knowledge. In the right circumstances, visualization can wield a uniquely transformative force. With that said, this book will also introduce you to some difficult historical truths: modern data visualization has always been entangled with capitalism and colonialism; seemingly neutral visual forms have always been shaped by politics and ideology; visualization has been wielded consistently, over centuries, to consolidate power and control. But this same history also offers counterexamples—of how visualization has also been used, over centuries, to reveal the workings of oppressive power and bring about enduring change. These counterexamples are what inspire our work in *Data by Design*.

In this book we argue, in both charts and words, that data can be visualized with intention, with ethical clarity, and for the purpose of uplifting us all. To achieve this goal, it will require far more than revising our best practices (although, as you will see in the "takeaways" that conclude each chapter, we do provide our own version of those). It will require a commitment to acknowledging the power of visualization as well as its perils—to admitting how visualization can prompt the insights that lead to new knowledge just as easily as it can encourage viewers not to further question what they see.

At a time when our collective knowledge, and our common humanity, are both under attack, the stakes of this endeavor—indeed, of all scholarly work—could not be higher. We are constantly being told to distrust—to distrust each other, to distrust our institutions, to distrust the very foundations of democracy that Thomas Jefferson, for all of his flaws, helped to create. Against these forces of algorithmic preferences and authoritarian agendas, and the rapacious capitalism that underlies both, we must continue to insist upon the value of greater knowledge and understanding. What *Data by Design* seeks to contribute to this larger project, then, is increased clarity about how we have arrived at the present moment, a sharper sense of the shape of our current knowledge, and inspiration for how we can envision a better future to come.

INTRODUCTIO

A History of Visualization and Power

Even from the highest point in town—the bell tower of the church of San Pedro Teozacoalco—it was impossible to see farther than a few kilometers in any direction. The two major mountain ranges that run along each of Mexico's coasts converge in the region—what is today the Mexican state of Oaxaca—resulting in a "geological jumble" of tree-covered peaks and scrub-filled valleys, punctuated only by the occasional vertiginous cliff.[1] The year was 1580, according to the Spanish.[2] But the Ñudzahui (Mixtec) people had been navigating this terrain for millennia. The ridgelines that separated the main town of Teozacoalco from its 13 smaller dependencies were as familiar to the Ñudzahui as the backs of their hands. Even the growing number of Spanish colonial officials and other citizens of the Crown settling in the area had at least a general sense of the lay of the land. While this terrain could be learned through experience, there was still no way to see it all at once—not by those who climbed to the top of the bell tower, and certainly not by those like King Philip II of Spain, who would never set foot on the North American continent during his lifetime. Which was why the king—or, more accurately, his royal cosmographer—had requested a map (FIGURE 0.1).

The cosmographer was hoping the king would benefit from something most people today take for granted: that by visualizing the region's complex terrain, he would gain immediate insight into the land over which he sought to rule.[3] The map he requested would take more than a year to research and design, and another three years to travel across the Atlantic. And now, nearly four hundred and fifty years after its creation, the "Mapa de Teozacoalco," as it has come to be called, exemplifies the central argument of this book: that visualization is a tremendous form of power. This is true both because of the knowledge that visualization helps to surface, and because of the people who decide how to put that knowledge to use.

In today's world, we experience the power of visualization in ways large and small. It's what pulls us into colorful charts of election returns, which show us quickly and clearly which candidate has prevailed. It's what fills us with despair—or alternately, determination—when we encounter a graph of global temperatures' upward rise. It's what tells us to evacuate from the path of a hurricane when we see our town or city inside a shaded danger zone; or simply to move to another room of the house when we see that our cell coverage has shrunk to a single bar.

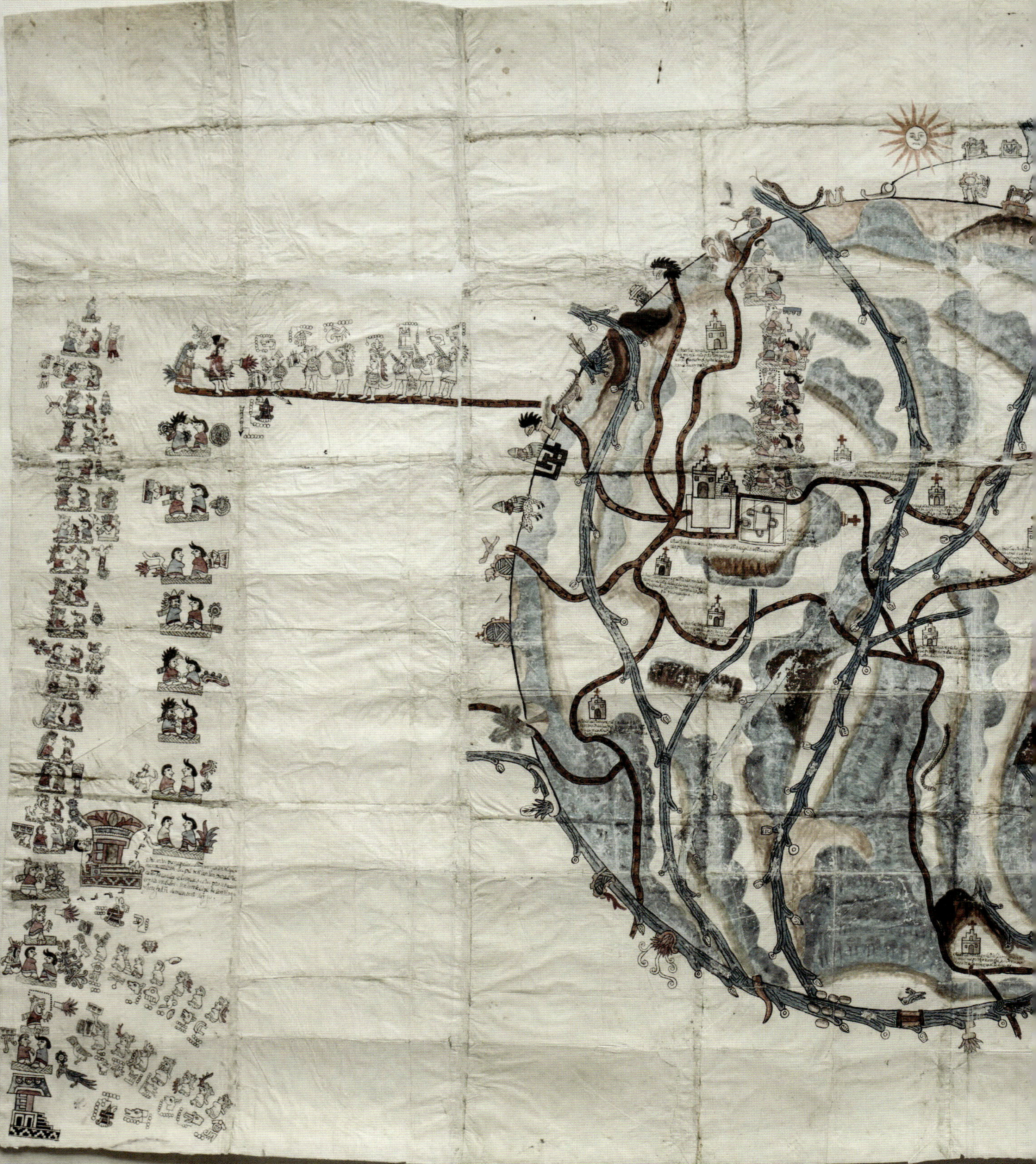

FIGURE 0.1
The “Mapa de Teozacoalco” produced by the *Relaciones geográficas* survey (1580). Courtesy of the Joaquín García Icazbalceta Manuscript Collection, Benson Latin American Collection, LLILAS Benson Latin American Studies and Collections, The University of Texas at Austin.

FIGURE 0.2
Fernanda Bertini Viégas and Martin Wattenberg's "Wind Map" (2012), viewable at http://hint.fm/wind/. Image courtesy of the artists.

FIGURE 0.3
The "Wind Map" as installed in the exhibition "Energy," October 21, 2019–January 26, 2020. IN2430.17. Photograph by Heidi Bohnenkamp. Digital image © The Museum of Modern Art / Licensed by SCALA / Art Resource, NY.

This power is real, and it is remarkably effective. Data visualization communicates complex information easily and efficiently. It enables meaningful patterns to emerge from data that is too complicated, too diffuse, or too subtle to otherwise see. At its best, this power is deployed in the service of new knowledge. Computer science researchers who study visualization often describe it as a tool to prompt insight, by which they mean the sparks of recognition that lead to a deeper understanding of the topic or issue at hand.[4] In the present moment, awash in data of myriad forms, this insight is immensely valuable. Visualization helps us cut through the noise, directing us to what truly matters within an otherwise overwhelming data sea. It also doesn't hurt that so many examples are, quite simply, beautiful. Some are literally works of art. To wit, Fernanda Bertini Viégas and Martin Wattenberg's "Wind Map," first released on the web in 2012, has been exhibited at New York's Museum of Modern Art (FIGURES 0.2 AND 0.3).

But the allure of visualization is inseparable from its danger. Because maps and charts (and their interactive analogues) all promise ease and efficiency, they mask their own complexity. And as data visualization has become a primary means by which we gain insight about the world, we have come to see these images simply as *the way things are*, rather than what they actually represent, which is a set of specific design decisions made by specific people, in specific places and times, and for specific purposes.[5] Put another way: we too often assume that visualization is transparent, a method that lets us see *through* to the significance of the data underneath. But the reality is that the choices made when visualizing data are nearly as significant as the data itself.

Which brings us back to the "Mapa de Teozacoalco." The creator of this particular map was an Indigenous painter whose name was surely known to his family and friends, although it went unrecorded on the map itself. As a result, the Ñudzahui man's name has been lost to time. But his work secures his place in a long lineage of visualization designers that also includes British antislavery activists, Scottish political economists, US women educators, and many more. Each of these figures, as we'll learn over the course of this book, recognized the tremendous power of data visualization and wielded it to their deliberate aims.[6]

As we journey from the era of the mapmaker across two hundred years of political, philosophical, and scientific thought, in order to arrive at visualization's so-called "modern age," we will also learn how the power of visualization is, always, a double-edged sword.[7] By placing each map or chart that we encounter in its particular context, we will absorb a fundamental truth: that visualization is only as powerful as how and by whom it is put to use. And as we meet designers whose own life experiences taught them just how easily data visualization could effect harm—that same Ñudzahui painter, along with a Beothuk captive and a man you might have heard of by the name of W. E. B. Du Bois—we will come to recognize the power, and responsibility, that we hold as designers and viewers of data visualizations ourselves. My sincerest hope is that you will marvel at the beauty and originality of these charts, created centuries before our time, while also coming to see the additional significance—and the additional power—that enters into each chart from outside its frame.

Visualization and Power

We know from a 1583 inventory that the royal cosmographer personally received the "Mapa de Teozacoalco," evidently one of dozens enclosed in a folder of "maps and plans" returned from the land he knew as New Spain.[8] But there is no equivalent record that can attest to whether the king himself ever saw the map, let alone what insight he might have gained. Because of his place in history, however, we do know a great deal about his worldview: shaped by a literal Renaissance education; focused on questions of land and resources; and fixated on financial gain.[9] This gives us some sense of the parts of the map that likely struck his eye. Perhaps the bold brown footpaths that indicate the region's major roads? Or the vinelike turquoise lines that reflect its many rivers? (Lest they be confused as mere embellishment, note that they are populated every so often by a scaly white fish.) The king would also likely have noted the numerous mountain ranges, and the churchlike icons that represent each of the towns in the larger community, or *yuhuiayu*.[10] These icons were plotted so precisely in relation to the map's natural features that they have been used by present-day archaeologists to pinpoint the locations of these historical towns.[11]

But would the king have known how to interpret the map's circular projection, intended to signify the *yuhuiayu*'s unity and wholeness, largely unfamiliar to European viewers, but typical of Mesoamerican mapping techniques of the time? And what of the bell tower of San Pedro Teozacoalco, depicted inside the circle on the upper left, atop which our own story began? Examining the map from a continent away, would the king have seen how this icon was intended to indicate the elevated status of the town of Teozacoalco, both as a colonial center and an enduring site of Native rule?[12] Or would this very symbol of vision and power have remained, ironically, opaque to the Spanish king, its significance as obscured as the horizon to those who climbed the actual bell tower's steps?

In truth, we cannot know precisely how the king might have interpreted the map. But with some knowledge of Ñudzahui history and culture, we can intuit at least a little about what the mapmaker *wanted* him to see. Consider the pairs of figures that rise up from the

FIGURE 0.4A
The ruling lineage of Teozacoalco, with the third dynasty of Teozacoalco on the bottom and the current ruler at the top.

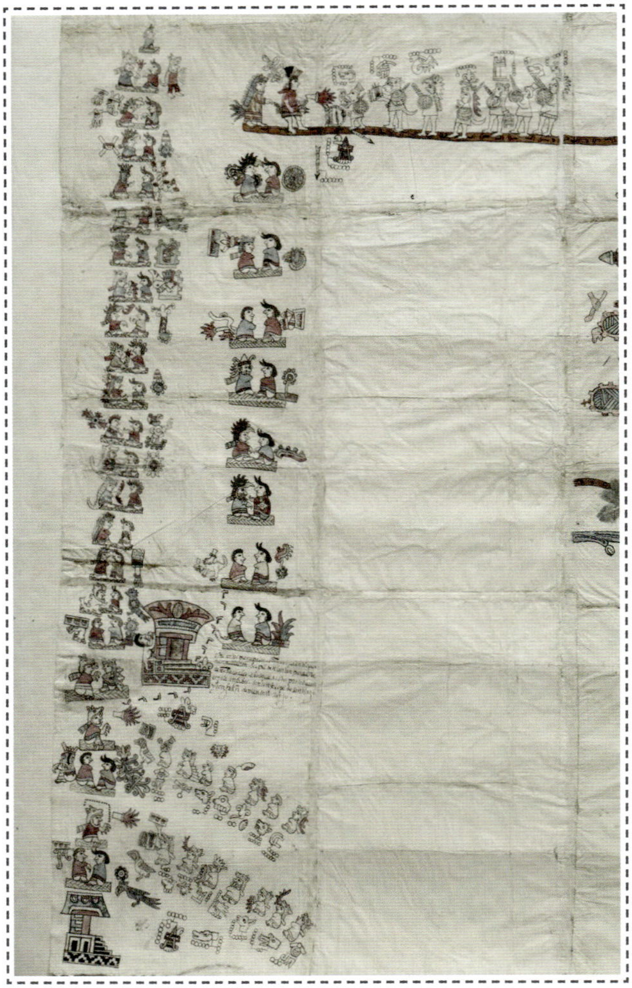

FIGURE 0.4B
The genealogies of Teozacoalco (center-left) and Tilantongo (far left) intentionally configured in two vertical lines.

main town, plotted just to the right of the bell tower icon. These pairs represent the ruling lineage of the *yuhuiayu*. The bottom, we know from extensive scholarship on the subject, is the start of the third dynasty of Teozacoalco. The dynasties continue upward through six generations to the unmarried male heir, pictured alone at the top of the line, who was likely the Native ruler of Teozacoalco at the time (FIGURE 0.4A).[13]

Scholars of Ñudzahui history have verified the accuracy of this genealogical diagram, as they have the longer family lines that appear on the left side of the page (FIGURE 0.4B).[14] But they have also documented something more: that there was not only significance to the information encoded in these diagrams; there was also significance to the manner in which the figures were drawn. For the vertical genealogical structure that would have been legible to the Spanish king was not how the Ñudzahui people themselves understood their own history, as historian Yanna Yannakakis explains. In a culture in which marriages across ruling families were employed to forge political alliances and to build collective strength, the Ñudzahui conceived of the relationships among the ruling families of the region more as a network than as the simpler parallel lines depicted on the map. By drawing these lineages as visibly distinct from each other, even as their genealogies were known to be deeply intertwined, the mapmaker was very deliberately representing the history of his people in a visual language that the king of Spain—his intended viewer—could easily understand.[15]

***

This is just one example of what can be gained by placing a map or chart in the specific context of its creation, and over the course of this book we will encounter many more. Alongside the Ñudzahui mapmaker, we'll consider everyone from the eighteenth-century Londoners—"men of active business, or high rank"—for whom some of the very first line graphs and bar charts were designed, to the members of the Atlanta University class of 1900 who found themselves, in the winter term of their senior year, sprinting to create a set of 63 (or 64) charts that would travel to the Exposition Universelle in Paris in several months' time.[16] Again and again, we will come to see how additional context serves only to enhance the knowledge that visualization brings to light. Among my greatest hopes for this book is that you, the reader, come away with unassailable evidence of how this context makes our interpretation of any particular visualization more accurate, and our knowledge of its subject matter more complete. For increased accuracy of interpretation, and greater overall knowledge, are—I believe—goals we all share.

The "Mapa de Teozacoalco" also underscores this book's argument, which has been stated once already, and will be elaborated as this book unfolds. This has to do with the power that data visualization commands. As we have already begun to see, this power derives both from the practice of visualization and from the people who put that practice to use. The Ñudzahui mapmaker was not merely employing his knowledge of European representational strategies to make the map easier for the king to interpret, after all. For the mapmaker, the king's interpretation carried profound personal and political stakes. By 1580, when he put his final flourishes on the map, his people had already endured several generations of colonial rule. King Philip II's ascent to the throne, in 1556, had ushered in a proliferation of new laws increasing Spanish control over the Native population, and Spanish profit from Native land. More recently, the effects of colonization had become visceral, with a deadly epidemic that raged across the region for several years. This environment would have confirmed the urgency of the mapmaker's charge. It was not merely to respond to a royal decree; it was to create an unequivocal attestation to his people's sovereignty and their enduring claims to the land.[17]

Here visualization, power, and context come together, illustrating how the intrinsic capabilities of data visualization are inflected by specific design decisions—decisions that are, in turn, influenced by the worlds in which those decisions were made. It can be very tempting to believe that it is possible to employ data visualization in a neutral capacity, just as it can be to believe in the complete objectivity of the information on display. But from the charts in this book will emerge evidence of the opposite claim: that complete objectivity is an illusion; that context and power always enter in. This argument does not undermine the value or

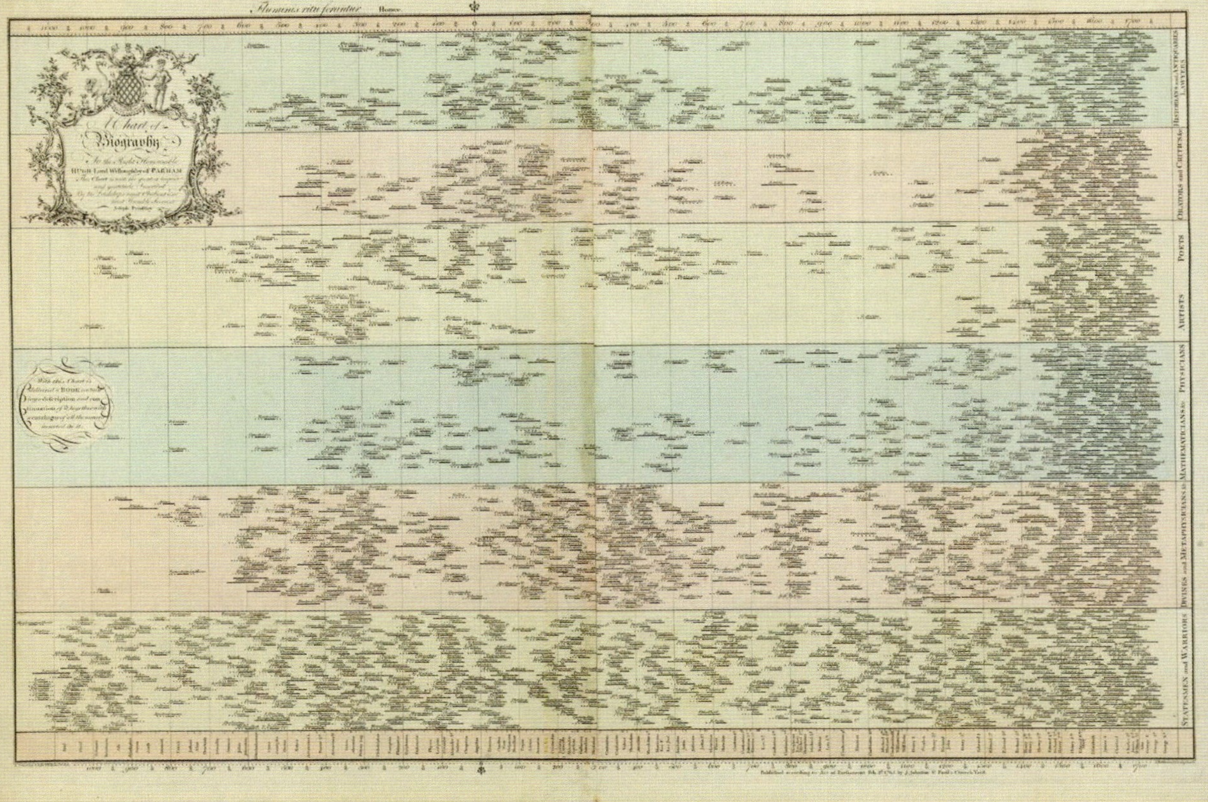

Flumina ritu feruntur
Homer
A Chart of Biography
HISTORIANS and ANTIQUARIES LAWYERS
ORATORS and CRITICS
POETS
ARTISTS
MATHEMATICIANS & PHYSICIANS
DIVINES and METAPHYSICIANS
STATESMEN and WARRIORS

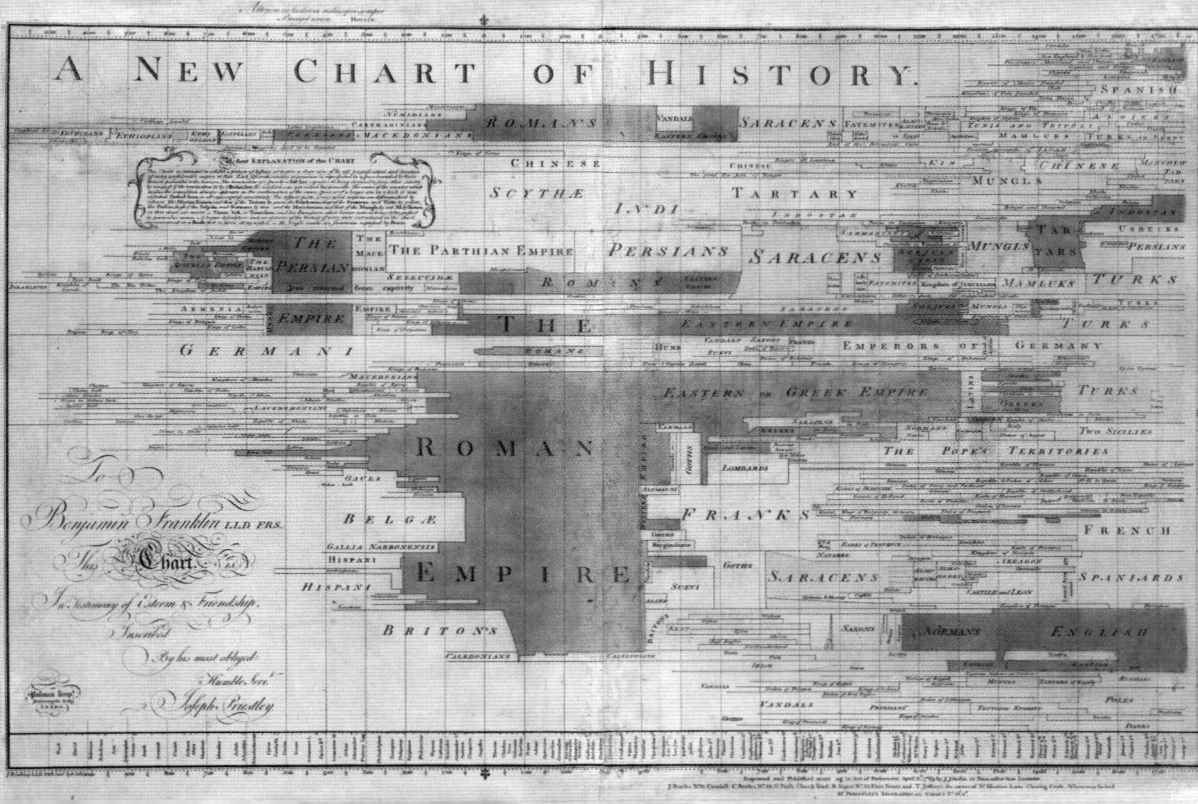
A NEW CHART OF HISTORY.
ROMAN'S
SARACENS
CHINESE
SCYTHÆ
INDI
TARTARY
MUNGLS
THE PERSIAN EMPIRE
THE PARTHIAN EMPIRE
PERSIANS
SARACENS
TARTARS
MAMLUKS
TURKS
THE EASTERN EMPIRE
GERMANI
EMPERORS OF GERMANY
EASTERN OR GREEK EMPIRE
ROMAN EMPIRE
THE POPE'S TERRITORIES
TWO SICILIES
FRANKS
FRENCH
BELGÆ
GALLIA NARBONENSIS
HISPANI
SARACENS
SPANIARDS
BRITONS
CALEDONIANS
NORMANS
ENGLISH
VANDALS
POLES
AMERICA
AFRICA
CHINA
INDIA
PERSIA
TURKY IN ASIA
GERMANY
TURKY IN EUROPE
ITALY
FRANCE
SPAIN
GREAT BRITAIN
To Benjamin Franklin LLD FRS.
This Chart is In Testimony of Esteem & Friendship, Inscribed By his most obliged Humble Serv.t Joseph Priestley

function of data visualization. On the contrary, it shows how much more we can do with data visualization when, like the Ñudzahui mapmaker, we recognize its power. Only then can we make informed decisions—about both design and interpretation—with our knowledge of this power in clear view.

What Is Visualization?

Before moving forward, let's pause to consider the concept at the core of this project: visualization. What does this term even mean? One of the most interesting things to note about the term "visualization" is that, despite its present ubiquity, it has only recently entered into widespread use.[18] Well through the nineteenth century, most visual representations of data were described, at least in English, simply as "charts." A chart could refer to a maritime map—indeed, this was where the term originated, in the late sixteenth century, via the Latin *charta* and the French *charte*.[19] Or a chart could describe a historical timeline, as in British polymath Joseph Priestley's "A Chart of Biography" (1765) or "A New Chart of History" (1769), which circulated widely across England and America in the second half of the eighteenth century (FIGURES 0.5 AND 0.6). We know this because numerous copies of the charts have been found folded up between the pages of other books.[20] Still further, "chart" could reference the genre of statistical diagrams that proliferated throughout the nineteenth century, in England and France and especially in the United States, as government officials and private citizens alike attempted to make visual sense of the vast troves of data generated by the decennial census.[21]

Some people still prefer the term "chart" for its clarity and ease of use. This is the main reason that I, too, use the word "chart" interchangeably with "visualization" throughout this book.[22] Other terms I use, mainly for variety's sake, include variations on the term "graphics." This nomenclature derives primarily from an 1878 treatise, *La méthode graphique,* by French polymath Étienne-Jules Marey, which inspired a host of twentieth-century historians and theorists.[23] These include the American mathematician Howard Grey Funkhouser, whose 1936 essay on the "Historical Development of the Graphical Representation of Statistical Data" represents the first modern scholarly account of the history of data visualization; the French cartographer Jacques Bertin, whose structuralist approach to data visualization, as expounded in his landmark book *Sémiologie graphique* (1967), continues to influence the teaching and practice of data visualization today; and the American statistician Edward Tufte, whose canonical *Visual Display of Quantitative Information* (1983) employs the term "data graphics" to distinguish his particular purview, to name just a few.[24]

Notably, none of these early chroniclers of the field use the term "visualization" at all. It was only in the 1990s, when visualization research began to emerge as a subfield of computer science, that this term began

Definitions.

External Cognition	Use of the *external world* to accomplish cognition.
Information design	Design of *external representations* to amplify cognition.
Data graphics	Use of *abstract, nonrepresentational* visual representations of data to amplify cognition.
Visualization	Use of *computer-based, interactive* visual representations of data to amplify cognition.
Scientific visualization	Use of interactive visual representations of *scientific data,* typically *physically based,* to amplify cognition.
Information visualization	Use of interactive visual representations of *abstract, nonphysically based data* to amplify cognition.

FIGURE 0.7
The hierarchical definition of visualization provided in Stuart Card, Jock Mackinlay, and Ben Shneiderman's influential visualization textbook, *Readings in Information Visualization: Using Vision to Think* (Morgan Kaufman, 1999).

FIGURE 0.8
Mona Chalabi, "US Workers on Strike" (2023) as it appears on Instagram. Used with permission of the artist.

to enter into conversation—first only in academic circles, and later in the mainstream.[25] For at least a decade, the term denoted something quite specific: "the use of computer-supported, interactive, visual representations of data to amplify cognition," as Stuart Card, Jock Mackinlay, and Ben Shneiderman explain in their field-defining textbook, *Readings in Information Visualization: Using Vision to Think* (1999).[26] They also emphasized the distinctions among visualizations of data of different forms (FIGURE 0.7).[27] But over the years, these distinctions have come to matter less and less, as visualizations have increasingly made use of multiple sources of data, and multiple visual and interactive forms.

Perhaps as a reflection of this evolution of the field, Tamara Munzner, in the definition that she offers in her also now-canonical textbook, *Visualization Analysis and Design* (2014), shifts the emphasis of the term from the *type* of data being visualized to the *task* for which any particular visualization is designed. "Computer-based visualization systems provide visual representations of datasets designed to help people carry out tasks more effectively," she explains.[28] Over the past decade, visualization researchers have expanded the scope of visualization further still. The field has seen the introduction of practices such as *data physicalization*, which involves the creation of three-dimensional representations of data that can be explored with the whole body; tasks such as externalizing "hunches" about errors in a dataset, involving the creation of a web toolkit that enables users to draw over existing visualizations; and frameworks such as "visualization vibes," which open space for viewers and researchers alike to acknowledge the feelings (or, if you want to be academic about it, "affective responses") they have about particular visual aesthetics.[29] As these examples suggest, it has become increasingly clear that visualization research, as a field, is now capacious enough to include a wide range of forms and approaches.

FIGURE 0.9
ABOVE: Shirley Wu, "Hong Kong Artists, Women" (2020), commissioned by M+, courtesy of the artist.

FIGURE 0.10A AND B
BELOW: Stefanie Posavec, "A Week of Clocks," front and back, from Giorgia Lupi and Stefanie Posavec, *Dear Data* (2016), courtesy of the artist.

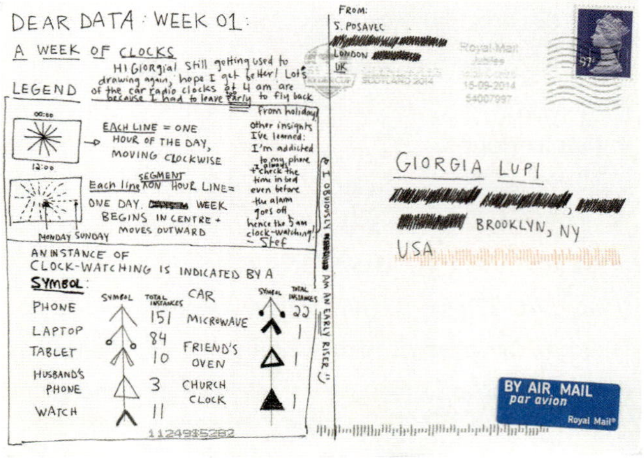

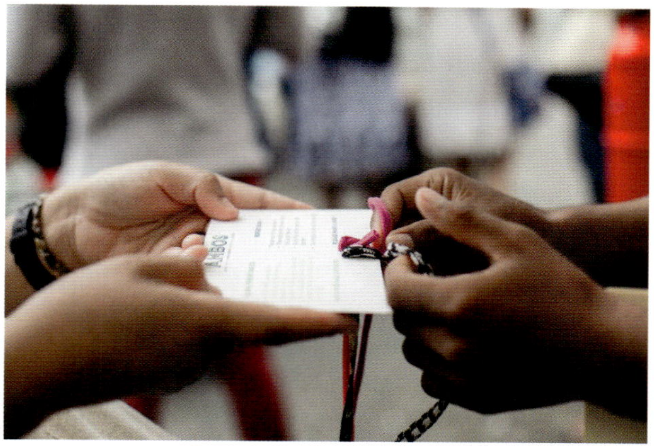

FIGURES 0.11 AND 0.12
AMBOS Project, *Border Quipu / Quipu Fronterizo* (2016–2018).
ABOVE: Digital image © 2026 Museum Associates / LACMA, licensed by Art Resource, NY.
RIGHT: Photo by Gina Clyne, courtesy of Tanya Aguiñiga and AMBOS.

FIGURE 0.13
Annina Ruest, "A Piece of the Pie Chart" (2017). Image courtesy of LACMA.

And yet it matters to me that these numerous forms and approaches are not just included in the ever-widening category of "visualization," but are specifically recognized as forms of and approaches to *data* visualization. I will readily admit that my motivation is, in large part, a reflection of my feminist ethos, and my knowledge of how narrow definitions have long been used to exclude women and people of color from technical fields.[30] But I am also motivated by what I see all around me in the world today: an expanding array of visual representations that, creatively and compellingly, make use of *data* of a variety of forms. Pulitzer Prize-winning data journalist Mona Chalabi is now sharing—on her Instagram feed—hand-drawn visualizations on topics as far-ranging as the history of labor strikes or the number of journalists killed in Gaza (FIGURE 0.8). Professional (and also award-winning) visualization designers Giorgia Lupi and Stefanie Posavec famously published a book of a year's worth of visualizations of the data of their daily lives, which they drew on post-cards with pencils and markers and sent to each other in the mail (FIGURE 0.10A AND B). Data scientist and designer Shirley Wu created a website to draw attention to the underrepresentation of women artists from Hong Kong on Wikipedia. The site looks like a stylized ink painting, but in its overlapping pastel peaks in fact encodes an abundance of information: about the length of each artist's Wikipedia entry, the number of references the article contains, and its citation count, all of which the user can interact with and explore (FIGURE 0.9). A team of artists and activists known as AMBOS (Art Made Between Opposite Sides) visited all 18 of the official border crossings between the US and Mexico and created an 18-strand khipu, with each knot representing a commuter crossing the border that day (FIGURES 0.11 AND 0.12). The artist-technologist Annina Ruest went so far as to create a machine to print pie charts with data about the gender gap in STEM fields on actual edible pies (FIGURE 0.13). The list is near-infinite, but the question remains the same: are these not data visualizations, all?

This is the reason that I use the terms "visualization" and "data visualization" interchangeably in this book. With that said, I am far less interested with delineating the precise boundaries of what a data visualization *is* than I am with asking how we—as viewers, designers, and researchers alike—understand the work that data visualization actually *does*. I therefore define visualization, which is also data visualization, which is also data *communication*, as any visual representation of data that contributes to knowledge.[31]

This definition is intentionally broad. It includes representations of data that are both static and dynamic, both images and interactions. They can be made by computers or by hand, and presented in a wide range of mediums and platforms—from rag paper to phone screens to Ruest's edible pies. They do, ideally, "amplify cognition," if the latter word is taken to mean "the acquisition or use of knowledge," as Card et al. helpfully specify. They might also "help people carry out tasks more effectively," as Munzner pointedly states. But as we will learn over the course of this book, additional knowledge can also sometimes lead to additional uncertainty, or to an increased awareness of the knowledge that will forever remain out of reach. A visualization might prompt a user to refuse—rather than carry out—their initial task. Or it might be designed for personal reflection, without a specific task in mind. In the case of a visualization that appears in nature, like the rings of a tree, it might not even be designed by a person at all![32]

Which is all to say that if we care about data visualization, and especially about innovations of form and content and process that push forward our ideas about what visualization can do, then we should embrace a definition of data visualization that includes rather than excludes, and that does not seek to erect barriers based on the type or size of data being visualized, the particular methods employed, the visualization's medium or platform, its intended purpose or audience, the backgrounds or credentials of its designers, or any other seemingly sharp but in reality quite fuzzy lines.[33] Put another way: I intentionally use the term "data visualization" as capaciously as those in the eighteenth century used the term "chart." In my view, this is an affordance of the term: a feature, not a bug.

This discussion brings us back, once again, to the "Mapa de Teozacoalco." In spite of the welcome trend in the field of visualization to include rather than exclude, I can still see how easy it may be for some to dismiss the idea that this particular map has much to teach those making visualizations today. Cochineal—the iconic red Mesoamerican pigment used on the map—is, after all, a far cry from Claude Code.[34] But technological and temporal differences such as these matter much less to me than what we learn by viewing historical and contemporary visualizations as conceptually linked.[35] Indeed, what we learn from the "Mapa de Teozacoalco," as we do from any other map or chart that we encounter in the world, is how visualization is indeed a powerful method of producing insight, but that insight is rarely limited to what is captured by the data alone. From the map's trails of footprints to its genealogical diagrams, from the bell tower of the church to the context that enters from outside of its hand-painted walls, it seems to me that the most important thing we can do is to continue to ask questions: about how the map's information is encoded, about the uses and limits of the insights it prompts, and about the forces of power that shape any decision to give data visual form.

FIGURES 0.14 AND 0.15
Two prominent examples of proto-data visualizations surfaced by Howard Grey Funkhouser in 1936. Little is known about the image on the left, an anonymous tenth-century diagram of the changing locations of constellations in the night sky. On the right is a set of fourteenth-century charts of the relationships between quantities created by French philosopher Nicole Oresme. Published in Funkhouser, "A Note on a Tenth Century Graph," *Osiris* 1 (1936).

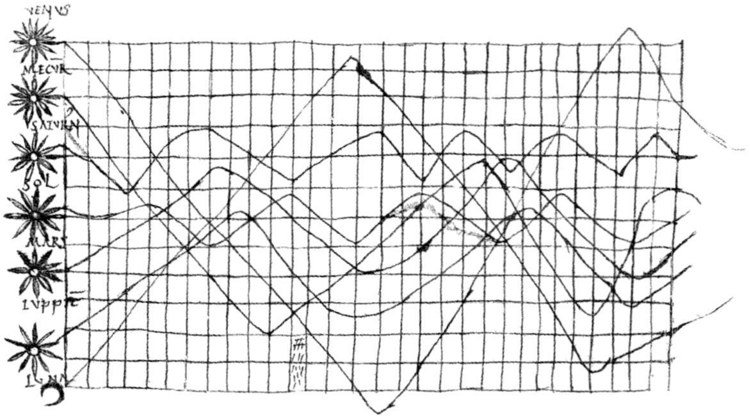

An Expanded Origin Story

In the very same year that the Ñudzahui mapmaker was weighing his power as the creator of a map for a king, Galileo Galilei, whose famed observations about the moon and the planets would soon help to usher in the modern scientific age, had just entered his first year of medical school at the University of Pisa. Already bored by his coursework, but—in a relatable moment—not yet prepared to tell his parents that he no longer wanted to become a doctor, Galileo turned his attention from the front of the lecture hall to the ceiling, contemplating a swinging chandelier.[36] His subsequent observations about the pendulum form would, like his astronomical experiments, go on to inspire an entire area of scientific inquiry. But considering these contributions in relation to the "Mapa de Teozacoalco" underscores a point no less profound: that Galileo's mathematical musings were far more closely connected to the mapmaker's project than he could then know.

When the story of the emergence of modern data visualization is told, it often eases in through Galileo and the other key figures of the Scientific Revolution.[37] The vast scientific advances of the sixteenth and seventeenth centuries, across astronomy, physics, mathematics, biology, chemistry, and more, were bound together by their belief in the value of knowledge gained through empirical evidence. Never before in Western thought had there been so much interest in evidence gained through experimentation, observation, and other forms of direct sensory experience. Put another way, the "revolution" of the Scientific Revolution was this: knowledge was no longer just handed down from heaven, divinely received, but was something that people—real people, not God—could create themselves.[38]

While this empirical evidence could be registered by any of the senses, the sense of sight was the most valued.[39] After all, it was by observing the night sky that Galileo, having successfully negotiated with his parents to shift his studies from medicine to astronomy and math, was able to confirm the phases of Venus and the existence of mountains on the moon. Fifty years later, in England, in an experiment canonized by historians of science (albeit less memorable to students of high school chemistry), Robert Boyle developed his law describing the relationship between the pressure and volume of air by recording his observations about his self-fashioned "air pump" and inviting members of the public to do the same.[40] The profundity of these

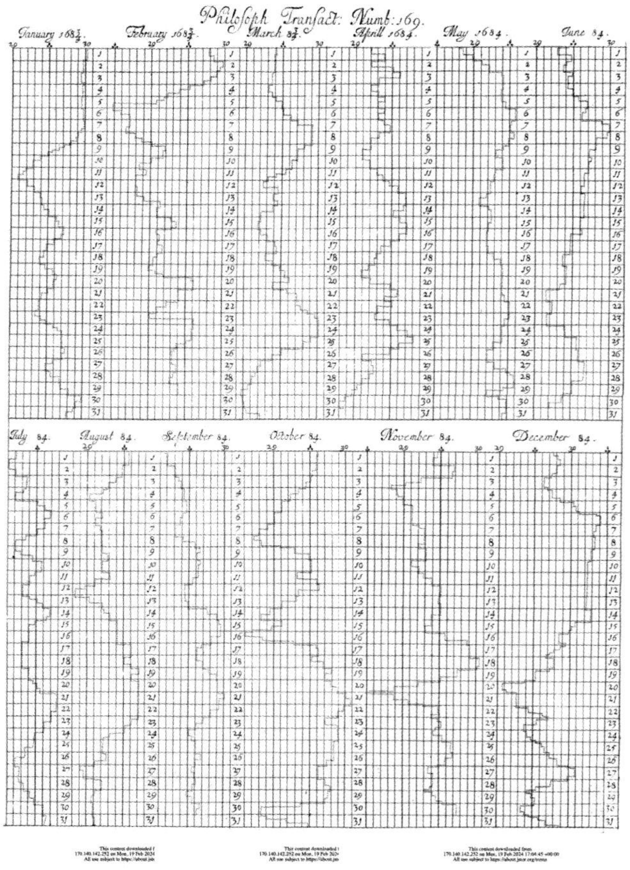

FIGURE 0.16
An early graph of daily barometric pressure in Oxford, England, that appears in Robert Plot's "History of the Weather." Published in *Philosophical Transactions* (1683–1775). © 1685 Royal Society.

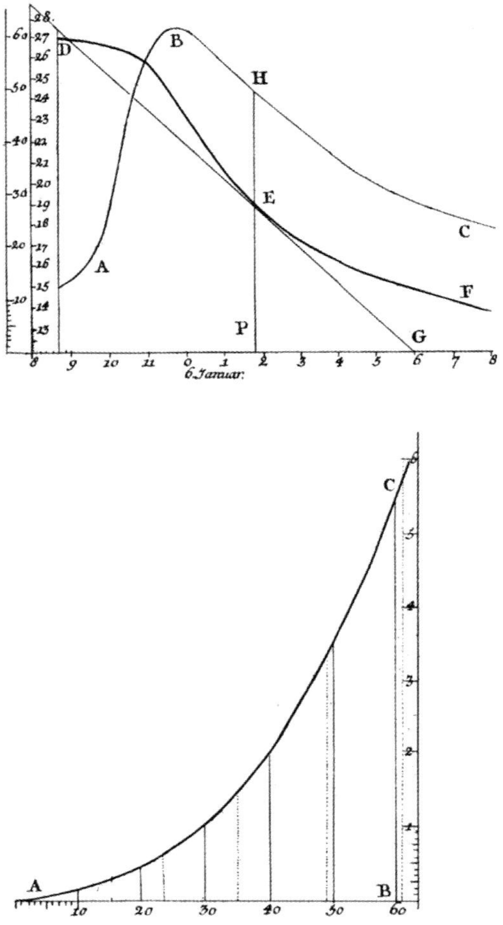

FIGURE 0.17
Two graphs from Johann Lambert's hygrometric studies, which show the relationship between air, water, heat, and evaporation. Published in "Essai d'hygrométrie ou sur la mesure de l'humidité," *Nouveaux Mémoires de l'Académie Royale des Sciences et Belles-Lettres de Berlin* (1771).

experiments, combined with the empiricist philosophies that would soon be articulated by Enlightenment thinkers such as Francis Bacon, John Locke, and David Hume, provided both the evidence and the epistemological framework for more and more people to understand vision as a lens through which to create knowledge.[41]

While there had been earlier examples of chartlike forms—an anonymous tenth-century diagram of the changing locations of constellations in the night sky **(FIGURE 0.14)**, and a set of fourteenth-century charts of the relationships between quantities created by French philosopher Nicole Oresme **(FIGURE 0.15)** are often cited—it was none other than René Descartes who first had the idea of "representing points, lines, and geometric figures by equations that could be visualized in diagrams and used to solve problems," as Michael Friendly and Howard Wainer explain in *A History of Data Visualization and Graphic Communication*.[42] Descartes was a vocal critic of empiricism. But even as his mathematical diagrams were intended to support deduction—that is, the process of moving from the abstract to the particular, as opposed to the inductive approach favored by Bacon and Locke—a crucial conceptual leap had been made. With Descartes's examples (if not his epistemology) as a guide, a growing number of people—including the perfectly named Robert Plot, who was also, incidentally, credited with discovering (but not identifying) the first dinosaur bone—began to plot other forms of numerical data, such as barometric pressure (a favorite subject of Plot; see **FIGURE 0.16**), mortality statistics (Christiaan Huygens), and soil temperature (Johann Lambert; see **FIGURE 0.17**).[43] And so, the story goes, with this expanding array of visual reference points, a growing abundance of data, and an increasingly pervasive empiricist worldview—to many, the pinnacle of Enlightenment thought—the flowering of data visualization was set to begin.

Indeed, nearly all extant histories of data visualization converge from this point on. William Playfair, the Scottish political economist whose work opened my own eyes to the rich history of data visualization, takes center stage in an account of heroic innovation and singular achievement. Hailed as the "father" of modern data visualization, Playfair's graphs and charts constitute the earliest of a small set of examples that are consistently upheld as evidence of visualization's first "golden age" (**FIGURE 0.18**).[44] Other well-known examples include the British physician John Snow, whose "most worthy" epidemiological maps, created in the mid-1850s, are often credited with identifying the mechanism of cholera transmission (erroneously so, as both the mechanism and the mapping technique were previously known) (**FIGURE 0.19**);[45] the British nurse and statistician Florence Nightingale, whose coxcomb charts of the casualties of the Crimean War (1858) provided visual evidence of the preventable nature of the majority of battlefield deaths (**FIGURE 0.20**);[46] and the French civil engineer Charles Minard, whose famed chart of Napoleon's failed Russian campaign (1869) prompted Edward Tufte, himself described as the "da Vinci of data" by the *New York Times*, to proclaim it perhaps "the best statistical graphic ever drawn" (**FIGURE 0.21**).[47]

With this triumphal account, reinforced in the visualization literature for decades, it is easy to leave the Ñudzahui mapmaker far behind. But this is where our alternate story converges quite directly with the standard account. For there are several significant connections between the ideas and examples that prompted the emergence of modern data visualization and the project of European colonial expansion that prompted the map made for the king. For one, colonialism was functionally dependent on data visualization, in the form of mapping, both in order to provide navigational aids to would-be explorers and to assert claims over the land and resources they sought to possess. In fact, some of the earliest known examples of "data maps" were enlisted for these purposes, such as the 1686 diagram of trade winds created by Edmond Halley, of Halley's comet fame, which was based on observations from his own time spent in the South Atlantic (**FIGURE 0.22**). (Halley dropped out of college for two years in order to live on the island of Saint Helena, a little over a thousand miles off the coast of Angola.)

Halley's trade wind diagram has been heralded by Edward Tufte, among others, as among the earliest examples of "placing a measured quantity on the map's surface at the intersection of two threads [of latitude and longitude] instead of the name of a city."[48] (There is that attempted distinction between charts and maps again.) But with the colonial context of its creation layered in, we can now not only recognize the *what* and *how* of the map's innovation, but also begin to see the *why*. For Halley himself, most likely, the reason was the expansion of science; but for King Charles II of England who sponsored the mission—and later personally appealed to the University of Oxford to grant Halley his degree on the basis of these contributions in lieu of returning for a third year—it was the expansion of his empire.[49] Always seeking new opportunities to acquire land and resources so as to extract additional profit for the English Crown, the king saw very clearly how this goal would first require detailed and accurate knowledge about where, and how, those resources could be found.[50]

This practical dependency on data and mapping, and the tools and methods of measurement that were required to produce them, had long been a hallmark of the European colonial project. It is not an exaggeration to say that this process of quantifying local knowledge as data and collating it into tables and charts was often an early step in a longer process of dispossession and, ultimately, death. The devastation that the Ñudzahui mapmaker was required to reckon with over his lifetime was endemic: of the estimated fifty to one hundred million people who inhabited the lands we now call the Americas before the time of first contact, only one-fifth to one-tenth remained by the end of the seventeenth century, when Halley created his trade wind map.[51] It would take until the early twentieth century for the Indigenous population of the Americas to begin to rise.[52]

But there is also a conceptual convergence between this furthering of European empire and the rise of empiricism; in other words, between our expanded origin story and the standard account of visualization's rise. For the request that resulted in the "Mapa de Teozacoalco" was, as it turns out, one of many. Since the king was unable to directly observe the geography of the colonies he sought to govern, his cosmographer had not only requested this particular map; he had issued a large-scale survey to every single locality claimed by the Spanish Crown—over 200 of them, stretching as far south as modern-day Peru.[53] He asked questions about terrain, natural resources, and demographics, and he elicited quantitative as well as qualitative responses. As a result, the *Relaciones geográficas*, as the survey would come to be known, has earned the distinction of being the first (recognized) statistical study of the Americas.[54]

The public had yet to experience the "avalanche of printed numbers" that, as historian of science Ian Hacking has famously described, heralded the start of the modern statistical age.[55] But the efforts of the Spanish king were of a piece with those of other European colonial rulers who sought to gain knowledge about their territorial holdings, and the people who lived there, by collecting data. The story of how these early data collection efforts were consolidated into national statistics bureaus, and from there prompted the development of new methods of statistical analysis hand in hand with the most reprehensible of eugenicist worldviews, is another dark path that leads from this same welter of ideas and influences. We will also explore this story in the chapters to come.[56] But for our present purposes, the key point is how King

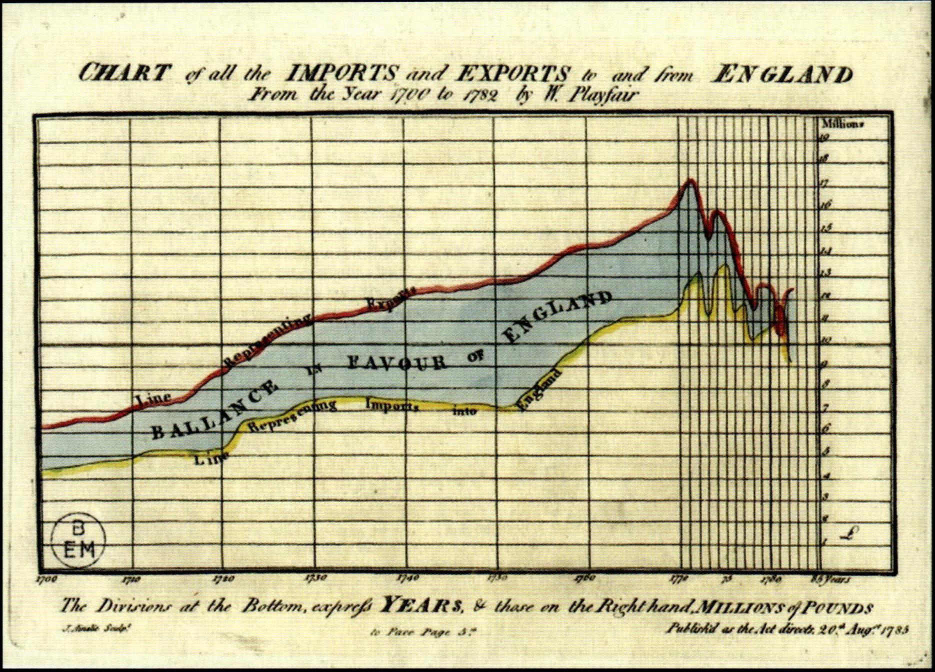

FIGURE 0.18
William Playfair's "Chart of All the Imports and Exports to and from England from the Year 1700 to 1782," from the third edition of the *Commercial and Political Atlas* (1801).Image courtesy of Wikimedia Commons.

FIGURE 0.19
A map from John Snow's "Report on the Cholera Outbreak in the Parish of St. James, Westminster, during the Autumn of 1854," which visualizes a central water pump as the epidemic's source. Image courtesy of Wikimedia Commons.

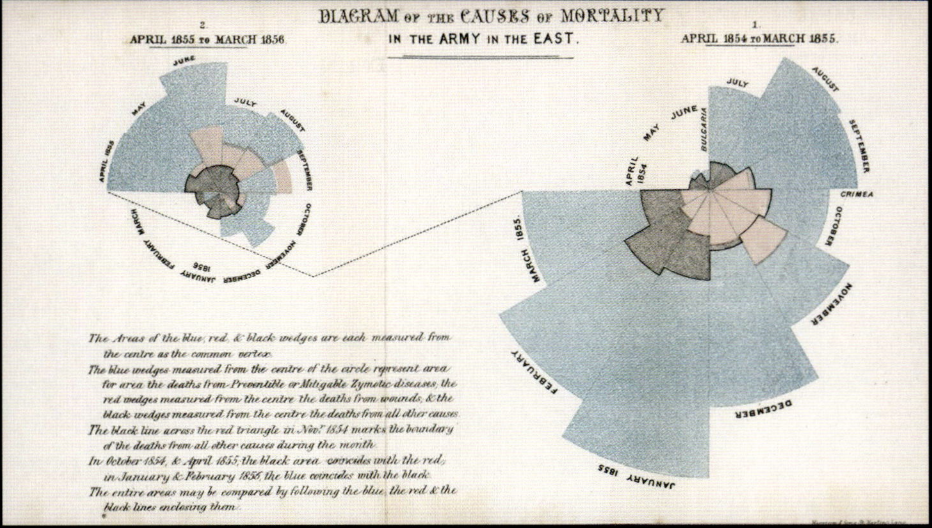

FIGURE 0.20
Florence Nightingale's "Diagram of the Causes of Mortality in the Army in the East," created in 1858, which emphasizes the number of otherwise preventable British military deaths due to poor sanitation. Image courtesy of Wikimedia Commons.

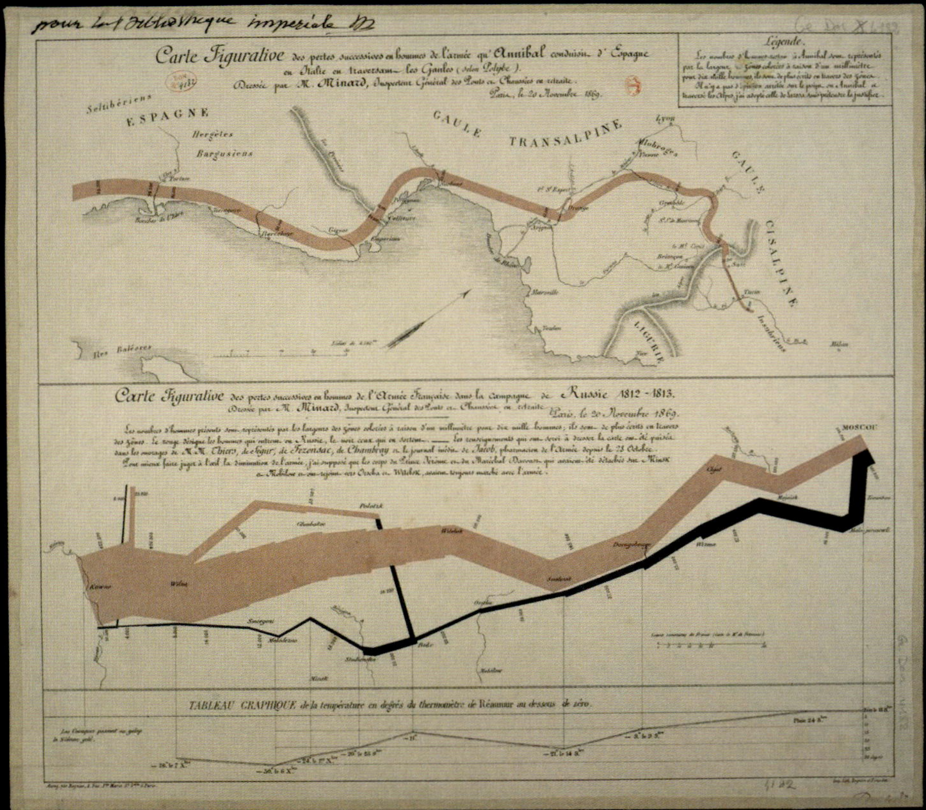

FIGURE 0.21
Charles Minard's "Carte Figurative des pertes successives en hommes de l'armée française dans la campagne de Russie 1812–1813 comparées à celle d'Hannibal durant la 2ème Guerre Punique," from 1869, which Edward Tufte proclaims perhaps "the best statistical graphic ever drawn." Image courtesy of the Bibliothèque nationale de France.

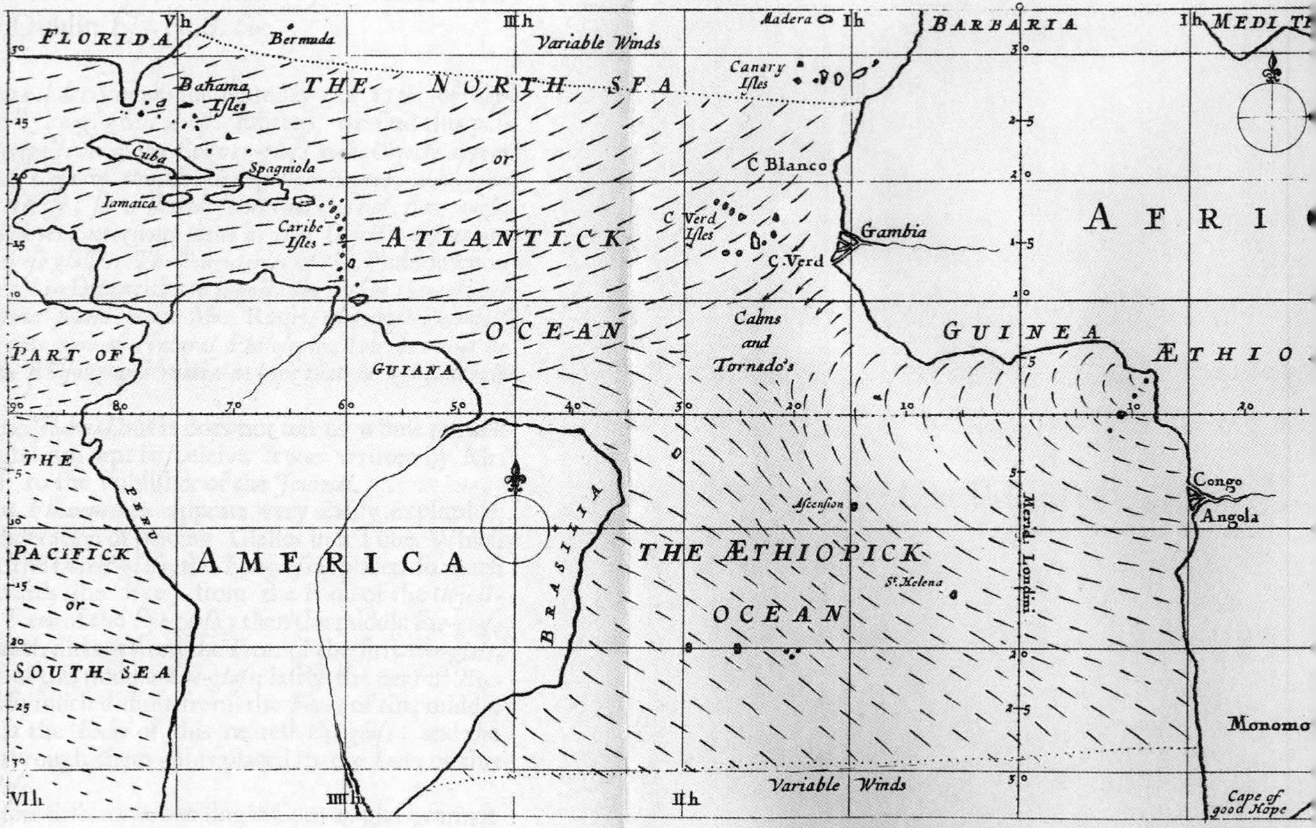
FLORIDA
Bermuda
Variable Winds
Madera
BARBARIA
MEDI TE
Canary Isles
Bahama Isles
THE NORTH SEA
Cuba
or
Spagniola
C Blanco
Jamaica
Caribe Isles
ATLANTICK
C Verd Isles
Gambia
AFRI
C. Verd
Calms and Tornado's
OCEAN
GUINEA
ÆTHIO
PART OF
GUIANA
THE
PERU
Congo
Ascension
Angola
PACIFICK
AMERICA
BRASILIA
THE ÆTHIOPICK
Merid. Londini
St. Helena
or
OCEAN
SOUTH SEA
Monomo
Variable Winds
Cape of good Hope
Vh
IIIh
Ih
VIh
IIh

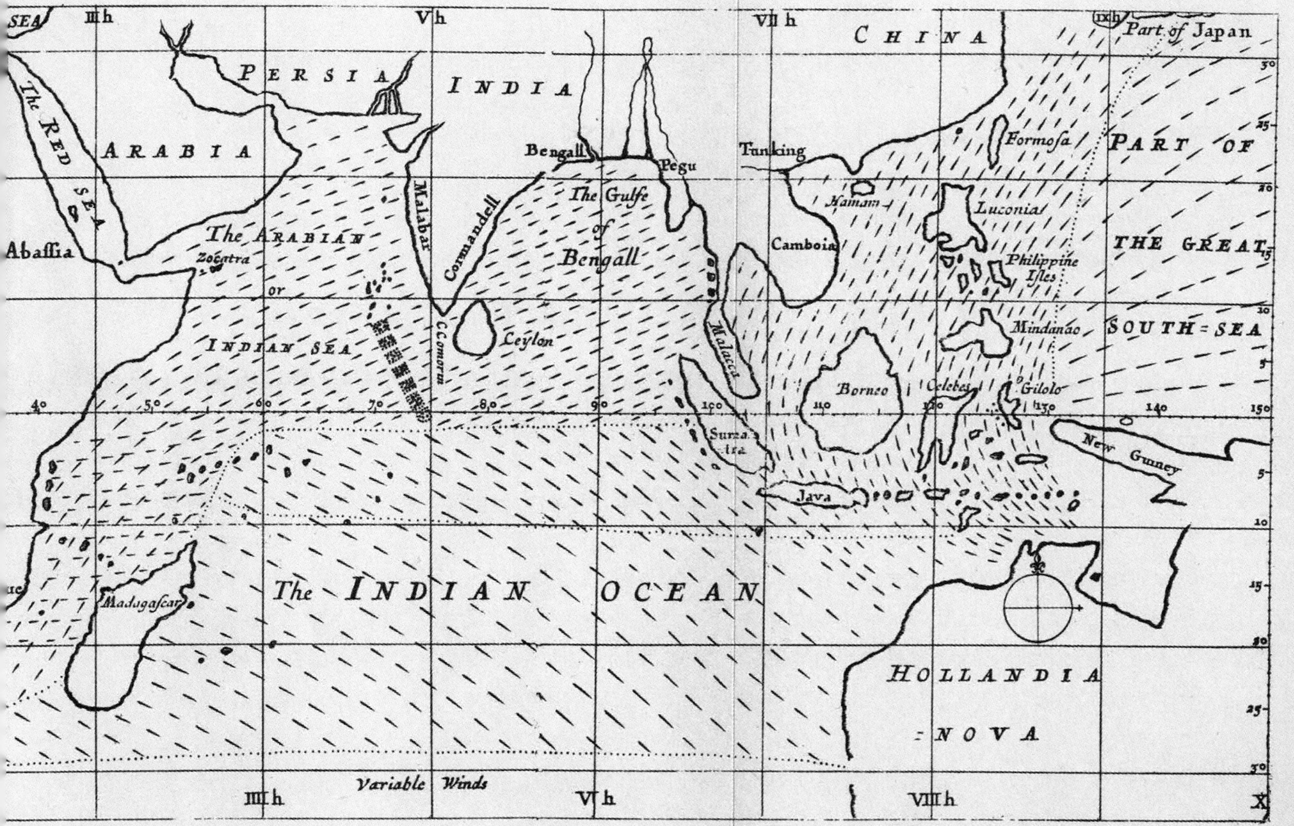

SEA
IIIh
Vh
VIIh
IXh
Part of Japan
CHINA
PERSIA
INDIA
The RED SEA
ARABIA
Bengall
Pegu
Tunking
Formosa
PART OF
THE GREAT
SOUTH=SEA
Hainam
Luconia
The Gulfe of Bengall
Malabar
Cormandell
Camboia
Philippine Isles
Abassia
The ARABIAN or INDIAN SEA
Zocatra
C.Comorin
Ceylon
Malacca
Mindanao
Borneo
Celebes
Gilolo
Sumatra
New Guney
Java
The INDIAN OCEAN
Madagascar
HOLLANDIA NOVA
Variable Winds
Vh
VIIIh
X

Philip II approached the task of gaining knowledge about his colonies by a method very similar to how Galileo approached the sky—that is to say, through empirical evidence. His motivation for the *Relaciones* reflected his belief that data derived from direct observation, enhanced by visual representations of the same, would lead to more complete knowledge of his domain. And for the king, more complete knowledge, in turn, led to more effective rule over his domain.

At this juncture, I can imagine some objections. Certainly, not all visualizations are colonial. And what of all the valuable scientific knowledge that empiricism helped to usher in? Should we throw away our knowledge of the seas and stars and planets and plants in one fell swoop? The answers to these questions are easy: correct and no. The takeaway from this expanded origin story is, once again, a lesson about the importance of considering context. More specifically, it is a lesson about the historical moment that gave rise to today's default palette of data visualization techniques. This context is *both* epistemological *and* political—because, at the risk of getting too heady, the epistemological *is* political.[57] And what this history offers us, in the present, is, in short, incontrovertible evidence of how the practice of data visualization has always involved significant ethical stakes.

The methodological similarity between how King Philip II approached the task of gaining knowledge about his colonies and how Galileo approached the sky (or Boyle the air pump) should heighten our awareness of how much it matters to ask—and, crucially, to answer—questions about the purposes of any particular visualization project we might view or undertake: What is the nature of the knowledge that it has been designed to surface? To whom does that knowledge belong? For what purposes will it be put to use? Guided by whose values? And for whose goals? While we cannot change the fact that the roots of modern data visualization are enmeshed in the European colonial project, we can move forward with a different charge: to view any future visualization that we encounter, and to approach any we may design, with intention and care—to ask what knowledge it is intended to lead toward, and what additional context is required to bring that knowledge to light.

What Is "Modern" about Modern Data Visualization?

The "Mapa de Teozacoalco" is not, of course, the earliest instance of data visualization that we know. We have already discussed a contemporary example of khipu—a system that derives from the Inkan empire, or Tawantinsuyu, which employs knots on colored strings to track data about people, land, and resources; archaeologists have found examples of khipu that date to as early as 2600 BCE (FIGURE 0.23).[58] Earlier still, in Mesopotamia, archaeologists have uncovered clay tablets dating from between 3100 and 2900 BCE that employed proto-cuneiform combined with numerical symbols to record information about the distribution of grain (FIGURE 0.25). And a recent project out of Peking University employed state-of-the-art image detection techniques to identify around 180,000 examples of historical visualizations from China, the earliest dating to 550 BCE—a project explicitly intended as a corrective to the dominant "Eurocentric view."[59]

These images join previously celebrated examples from that region, such as the set of ivory counting rods (ca. 475–221 BCE) found in Chang'an, Shaanxi province, that represented groups of numbers so as to assist in complex math (FIGURE 0.24). Or, contemporaneous with the focus of this book, we might consider the official

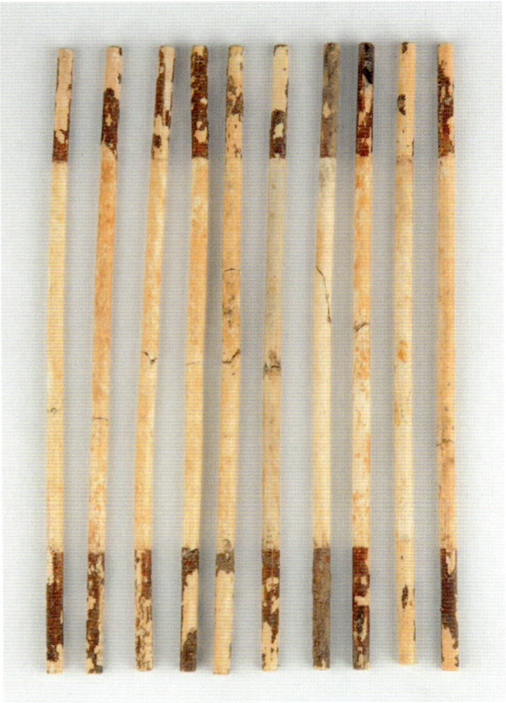

FIGURES 0.23–0.25
Examples of data visualization that predate our story. **CLOCKWISE FROM TOP LEFT:** An example of Inkan khipu, image courtesy of the National Museum of the American Indian, Smithsonian Institution (14/3866), photo by NMAI Photo Services. Chinese counting rods recently discovered in Chang'an, Shaanxi province, photograph by Tianen Zhang (张天恩). A proto-cuneiform tablet documenting the distribution of grain, image courtesy of the Metropolitan Museum of Art.

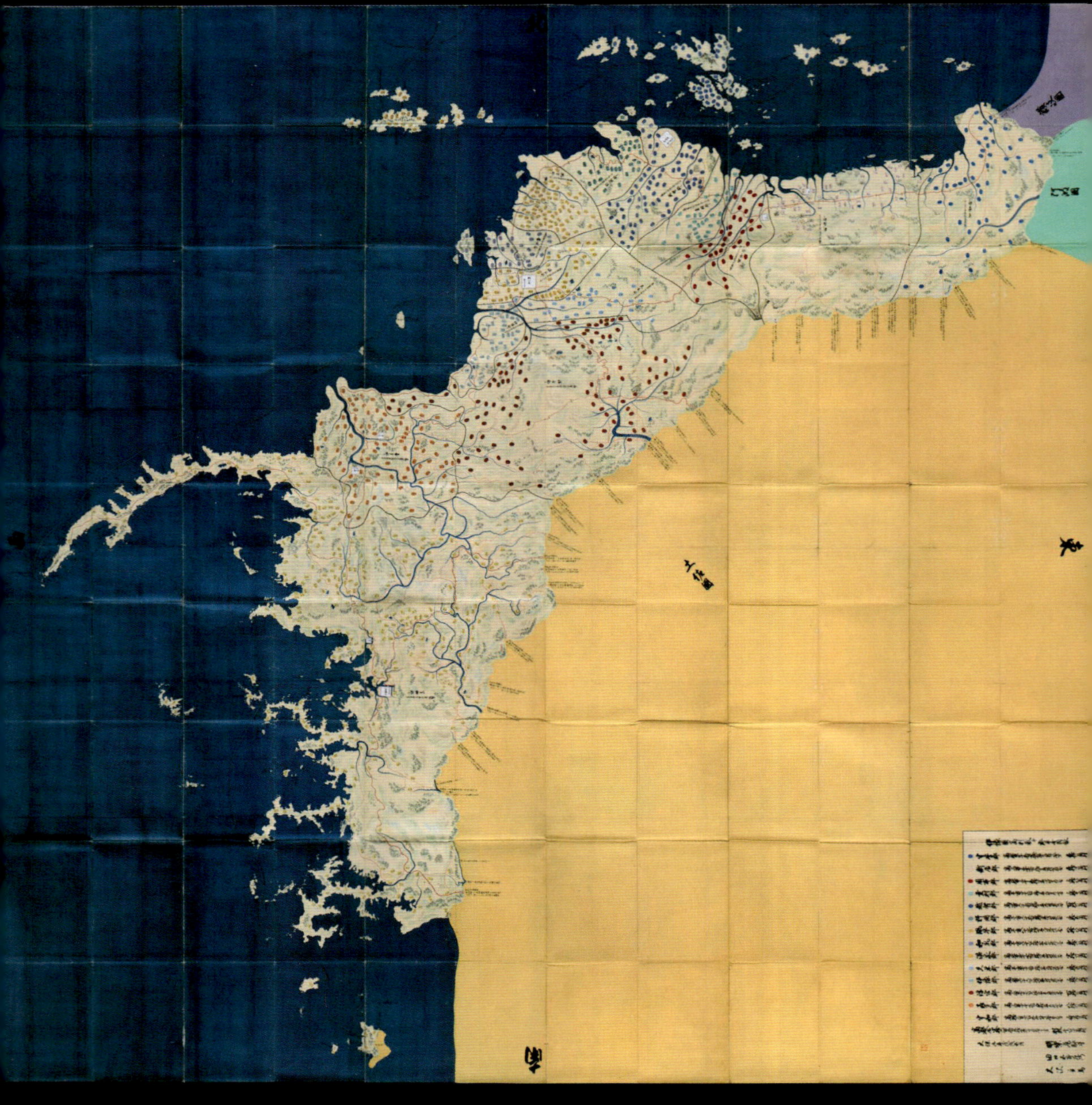

FIGURE 0.26
An 1838 shogunal map from Japan.
Image courtesy of the National Archives of Japan Digital Archive.

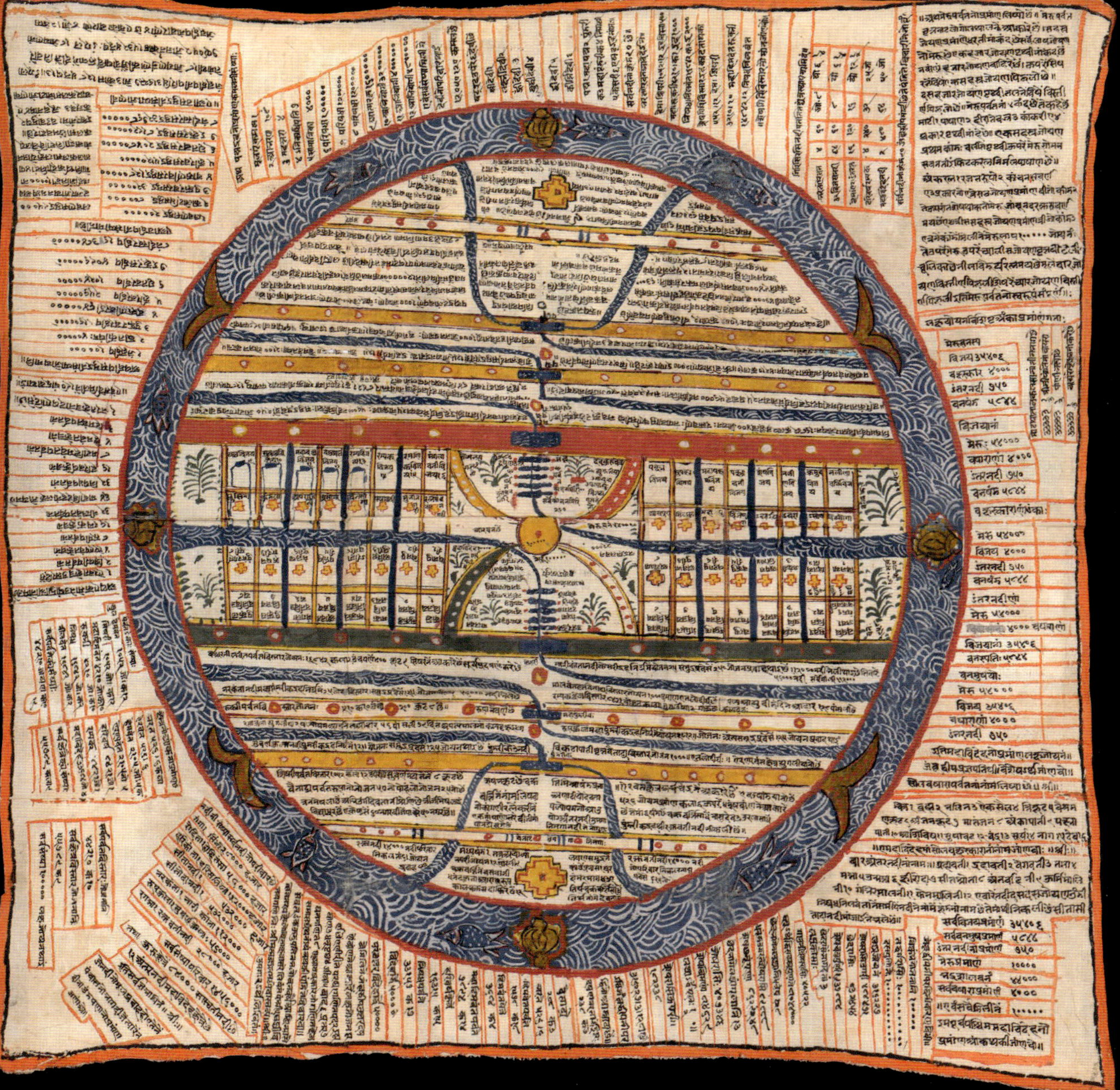

1838 shogunal map from Japan, which employed both points and polygons to map the locations of villages and checkpoints (FIGURE 0.26); or the ca. 1850 Jain diagram of Jambudvīpa, which condenses philosophical, spiritual, and cosmological information into a poster-sized fabric sheet (FIGURE 0.27).[60] As these examples collectively attest, societies from all over the world have long recognized the generative potential unleashed by translating quantitative information into visual form.

This global history, even in highly abridged form, prompts a key question: what do I mean by the phrase "modern data visualization"? The term "modern," here, seems to be doing a lot of work. Indeed, it is a term that is both deeply problematic and, as we'll see, essential to understanding the world we inhabit today. In its most general sense, "modernity" describes the period of time ushered in during the sixteenth century, corresponding with the major transformations in science, philosophy, and governance, among others, that have just been discussed. Modernity is also implicitly (and at times explicitly) European.[61] This geographical constraint is, of course, exclusionary. But what and who are excluded also tell us a lot.

More specifically, this "modern" way of thinking—which included the celebration of human reason, the articulation of individual rights and political liberties, and the development of the contemporary nation-state, among other things—was, paradoxically, accompanied by a narrowing of which types of humans were capable of exercising reason, or enjoying rights and liberties, or other protections of the state. As decades of scholarship have shown—and, for that matter, many philosophers' actual writing on these subjects—it is not a coincidence that today's racial and gender hierarchies also solidified during that time.[62] This moment was fundamentally destabilizing. It must have been scary to shift from a worldview in which both knowledge and governance were divinely received to one in which humans made their own decisions about what to believe and how to rule (or be ruled). It followed, then, psychologically as much as philosophically, that many of the same thinkers who first worked to valorize human reason and the virtues of self-governance would move on to consider questions about which humans they thought, more precisely, exemplified these ideals. It is depressing, but not surprising, that the people they identified as best exemplifying these ideals were people just like themselves: European, white, and male.

These modern thinkers, with their philosophies and their flaws, have profoundly shaped the world we live in today. As is also true for each and every visualization that we will explore in this book, what is key to understanding European modernity is to acknowledge its transformative power, and to seek out the story beyond the (in this case philosophical) frame. There is *always* more we can learn by returning to any particular primary source. One of the major methodological contributions of scholarship on the Enlightenment and its reverberations throughout the world has been to show how there are always *counterhistories* that can be surfaced if we look closely enough. These are histories that run alongside the dominant account, with each actor, and each moment, called in to challenge or otherwise destabilize the story that is generally told.[63] The history of modern data visualization (always, as you can now understand, with scare quotes implied) has not yet received this counterhistorical treatment. *Data by Design* is, in no small part, an attempt to remedy that.

Indeed, we have already begun to recognize the tremendous power in choosing how to visualize data; there is similarly tremendous power in choosing how to tell a story. For there are many paths that take us from any starting point to any finish. These paths are rarely linear, and—as we have already begun to see with the history narrated here—they are very often traveled at the same time. Some of these paths, like the dominant story of visualization's rise, are well-worn and familiar, while others, like the Ñudzahui mapmaker's, are far less trafficked, and still others have not yet begun to be explored. These less-trodden paths are powerful because they lead us to worlds not yet or not yet fully realized. These worlds may very well be preferable to the present one, but for them to be achieved will require us, in the words of sociologist Ruha Benjamin, to push "beyond the constraints of what we think, and are told, is politically possible."[64] These are precisely the worlds I seek to open up with this project: worlds of possibility for the practice of data visualization, and for no less than the way we understand knowledge—worlds of possibility that encourage a more liberatory future for us all.

The Charts in Our Story

Our story follows a chronological arc from the final years of the eighteenth century, where—as we have learned—the starting point of modern data visualization is generally placed, up to the turn of the twentieth century, when this particular age of data visualization begins to crest.[65] Along the way, we will pass through many more of the major developments that defined this era: the emergence and entrenchment of the slave trade, itself an outgrowth of the economic ambition of the European colonial class; the resistance movements in both England and the Americas that led to its abolition; the age of revolutions that saw the establishment of democratic governments in the United States, France, and Haiti; the continued dispossession of Native lands across the Americas, brought about by the territorial expansion of the United States; the nation's subsequent fracturing over the issues of slavery and sovereignty alike, which culminated in the US Civil War; and the attempts at Reconstruction that followed until the century's close.

This chronology is paralleled by a methodological progression: as the chapters advance, each also considers an additional aspect of the practice of data visualization: data, image, people, knowledge, change (as in social or political change), and labor (as in the actual work of visualization design). In this way, each chapter layers in additional complexity, building upon the ideas and examples discussed to that point. These layers form the conceptual foundation of this book. They are accented by unique colors, intended to serve as visual indicators of your reading position in this expanding conceptual frame. It is also important to note that the additional historical context you will encounter serves a different purpose than the sweeping history that has been offered thus far: to set the stage for the single chart (or set of charts) that anchors each chapter, so as to allow its own story—and its conceptual significance—to more fully unfold.

Chapter 1 begins with an indelible image, "Plan of an African Ship's Lower Deck" (ca. 1789), which visualizes the data of the slave trade by depicting each anguished body that the numbers represent. Here we learn how racial capitalism was (and remains) what authorizes colonial powers to continue to accumulate wealth and resources, and in the process to enforce a view of objects, actions, and even people as "goods" that can be converted into data and reduced to a price.[66] By beginning with this visualization, and not a more canonical example, I seek to set the story of modern data visualization on a more ethical path. This path demands that those who design visualizations, and those who view them, consistently consider what is made visible in relation to what is not. In this particular case, what is made visible is horror at scale. What remains unseen is the full humanity of the lives that the data represents. Here we also probe a conceptual tension at visualization's core: its remarkable capacity, on the one side, to distill complex information into clear and efficient insight; and its potential for harm, on the other, in how the abstraction required to distill this insight often comes at the expense of the details of the data, and the context in which it was produced.

With this key tension of data firmly in place, the clean data lines of William Playfair's time series charts—the subject of chapter 2 ■—become far more complex. Analyzing Playfair's charts alongside the text that introduces the third (and most widely circulated) edition of his *Commercial and Political Atlas* (1801) confirms how the political instability brought about by the age of revolutions, as much as an adherence to Enlightenment empiricism, inspired his influential visualization techniques. By recreating Playfair's charts with a contemporary visualization library, D3, we further show how the data lines of his charts—so often upheld for their objectivity and precision—in fact embellish the story told by the data alone. Far from delegitimizing the value of his charts, this finding serves to advance one of this project's core claims: about how an attention to context—historical, political, and in this case material—allows us to learn more about the nature of the insights, and the knowledge, that any particular chart seeks to prompt.

Chapter 3 ■ moves from England and Scotland, where "Plan of an African Ship's Lower Deck" and Playfair's time series charts were created, to North America, where the central examples of all subsequent chapters are placed. Here we meet Shanawdithit, a Beothuk woman who created a series of maps in the final months before her death, in British captivity, in 1829. The argument I make here is both more subdued than the previous chapters' and more profound: that all visualizations that we create—and not just those explicitly enlisted in the service of empire—can be refracted through a colonial frame. As in the previous chapters, we ask the same questions of the field of data visualization as we do of our own project team. Here we consider our own visualization work, as exemplified by the charts and diagrams we've designed for this book, in the context of colonialism's present-day effects. We continue to probe the uses and limits of data visualization, but here we expand our frame to consider the people whom any particular visualization serves. In the end, the lessons of this chapter tilt more toward limits than uses, but they are no less important to behold. They teach us about our obligations to our work and to the people who enable it, as well as about what visualization cannot—or should not—help us to understand.

If the emphasis of chapter 3 is the limits of current visualization practices, chapter 4 ■ opens the door to the future possibilities. Its focus is on the chronological grids of educator and editor Elizabeth Palmer Peabody, who employed color and position in order to represent historical events in time. Peabody's charts, as described in her *Polish-American System of Cartography* (1850) and documented in her *Chronological History of the United States* (1856), were designed to be abstract rather than intuitive, to promote sustained reflection rather than immediate insight, and to provoke a unique imaginative response in each viewer. These epistemological principles diverge sharply from the ideas about knowledge acquisition that have shaped visualization best practices to this point.[67] We use the example of Peabody's alternative perspective to consider more broadly what additional strategies we might envision—and what additional insights we might prompt—if we open ourselves up to a wider range of processes by which knowledge can be produced. We introduce our own data physicalization project, as well as several digital recreations of Peabody's original charts, to show how multiple methods of visualizing data contribute to more complete knowledge overall.

Chapter 5 takes us to the century's close, considering the charts designed by the esteemed sociologist W. E. B. Du Bois and his team of students from Atlanta University, including one young woman named Lula Iola Mack. Recently reclaimed by historians of data visualization as a corrective to otherwise white and Eurocentric accounts, the charts created for the 1900 Paris Exposition in fact tell a more complex story—one that acknowledges the power of visualization and simultaneously admits its limits. By exploring the charts in the context of Du Bois's writing on race and racism and his sociological theory, and by recreating them in ways that account for the students whose knowledge and labor helped to shape them, we close with a reassertion—through both text and image—of how data visualization, the conceptual conditions of its emergence, and its political consequences are fundamentally intertwined. Drawing a discussion of race, data, and political agency into the present, we conclude with an optimistic assessment of what data visualization can accomplish once we learn to recognize, and account for, its uses and its constraints. Indeed, visualization can do more than reveal the significance of data; it can bear witness to instances of oppression, it can render visible more just possible futures; and it can—if intentionally designed—hold space for what cannot be conveyed through data alone.

A concluding chapter homes in on the issue of labor, a through-line of the project overall. It takes as its focus not the history of visualization but its practice in the present, in the form of the making of the book you are reading now. While the words on this page are written by one person, as the preface explains, the original visualizations in this book have been designed collaboratively, and implemented by an intentionally interdisciplinary research team. Our expertise spans the fields of art and design, computer science, and literary, cultural, and media studies. As this project has evolved, and as the team has undergone multiple iterations, we have come to see ourselves and our

FIGURE 0.28
A guide to understanding the reading features of *Data by Design*. Designed by Tanvi Sharma.

A)
An eye-and-arrow combination indicates that the text continues on the next (colored) page.

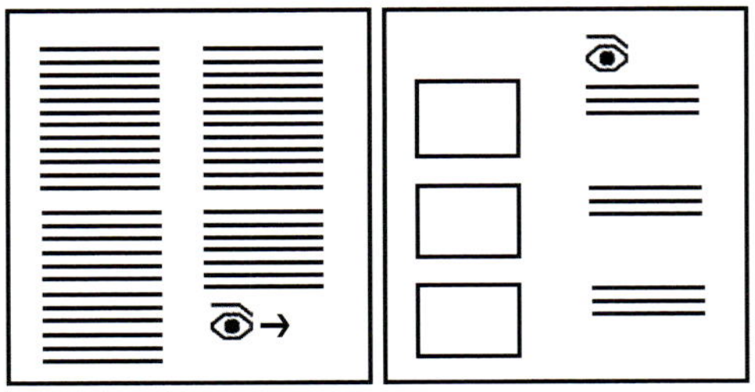

B)
A dashed border indicates a detail view of a larger image.

C)
A chart icon indicates an annotated image.

processes as embodying the ideas that emerge from our historical examples. The final chapter thus represents our attempt to document our work and credit all contributors, while providing the first extended example (in addition to the more self-contained examples throughout the book) of how the lessons of *Data by Design* can inform contemporary visualization practice.

Before moving forward, we would also like to offer a brief note about how this book has been designed (FIGURE 0.28). We have already alerted you to the small colored squares ■ ■ ■ ■ ■ ■ that anchor each of the chapters. We have designed additional icons to guide you through some of the other visual and reading features that are unique to this book. An eye icon 👁 followed by an arrow → indicates that you should continue reading on the next page, even if it seems like you should be able to skip ahead. These special pages, adapted from their interactive form, have been designed to slow down your reading experience, guiding you through a detailed analysis or design process step by step. Dashed lines around images that you may encounter indicate that they are detail views of a larger image that has been previously introduced.

A final feature, indicated by an icon of a chart, indicates a different form of visual engagement. These charts have been annotated with details drawn from the chapter text. They can be consulted as you read, before, or after; the same information is contained within the chapter text. Together, we hope these features will make for an engaged and illuminating reading experience, one that further enhances the insights of the charts themselves.

Toward a Liberatory Practice of Data Visualization

We are a team of visualization designers, software engineers, and humanities researchers. As such, we do many different kinds of work. But our efforts are linked by a common aspiration: that the ideas of this project find utility in the world. We envision *Data by Design* both as a model for how to analyze visualizations with more intention, and as a guide for how we might design more intentional visualizations ourselves. In this way, we see our project as liberatory: as advancing a practice of visualization design and interpretation that is honest about its origins, intentional in its purpose, and attentive to a range of contexts, historical and contemporary. The goal of this project is to contribute to the transformation not merely of "best practice" but—as audacious as it may sound—of the baseline conditions of our unequal world.

From the very first story we tell in this book, of the "Mapa de Teozacoalco" and its intrinsic power as a data visualization—as well as the power wielded by its maker and the power it accorded to the Spanish king—we examine the role that visualization can play in this transformative project. Even more, we seek to show how this project has already begun. Put more directly: it is not a coincidence that, as we move through the major examples that structure this book, we also move through some of the major struggles for social justice that define this era: the British abolitionist movement, the US Civil War, and Black Reconstruction, among several others. For just as data, its collection, and its visual display have long been wielded as weapons by powerful actors, they have just as often been wielded to push back. What we will both say and show is how a more liberatory approach to data visualization is possible when intention and ethics come together with contextual awareness and responsible design.

More concretely, we believe we can work toward a liberatory practice of data visualization when we insist upon historical contextualization, conceptual theorization, and detailed analysis—that is to say, when we embrace a humanistic approach to visualization design. This approach requires us to identify our existing assumptions and recognize how these assumptions can serve to narrow rather than expand what is possible for the future. As difficult as it may be to accept that data is not always beautiful, and visualization is not always—or ever—a neutral art, it does us nothing to dismiss this basic truth. Without a more contextualized understanding of the emergence of modern data visualization, we run the risk of believing that our work can only ever contribute to new knowledge—and not also, at times, narrow what is possible to know. Indeed, when we presume that our charts and interfaces will clearly point our viewers to what is most important, we run the risk of reinforcing the empiricist view that what can be seen represents the full extent of what can be known. As we will demonstrate throughout this book, there is no single dataset nor any single visual or interactive form that can give us the full picture of the phenomenon we seek to explore. This realization, however, is not a reason to reject visualization. On the contrary, it is through the continued pursuit of learning more—a pursuit that involves the identification of *both* new sources of knowledge *and* additional and perhaps unfillable gaps—that we arrive at the most complete understanding of the topic at hand.

What you will encounter in the chapters to come are, at their core, stories—stories told through visualization and told through words. Taken together, these stories seek to expand our understanding of the history of data visualization so that we can shape its future with more intention and care. When we understand the history of the field, both its range of innovations and its limits, we will gain a more complete—and therefore more accurate—picture of what visualization can do.

My strongest hope for this project, shared by the entire team, is that you will come away from the project with new ways of viewing visualization, with all that the term entails. If you are a designer, I also hope you will come away with new methods for structuring your design processes, and new contexts for putting both visualization and process to use. For while we will learn that visualization *on its own* is rarely enough to achieve the goal of a more just and equal world, we will also learn how additional knowledge-making practices and processes can amplify each other, working together to achieve this goal we share. The practice of data visualization that is modeled here begins not with data, or even with design, but with the recognition that data visualization is only as powerful as the purposes for which it is put to use.

Indeed, some of the stories we tell in this book are dispiriting. We have already begun to see how the history of data visualization directly intersects with the histories of slavery, capitalism, and colonialism. But in surfacing this history, our goal is not to suppress the use of visualization or to limit its future growth. On the contrary, we seek a more informed and more intentional—and, as a result, a more precise and effective—practice of data visualization, both in our own work and in yours. This practice is one that prioritizes the particularities of each dataset, that considers its social, political, and historical context, and that attends to the people who made it possible. It is one that, moreover, continually recalls the power—and therefore the responsibility—that we hold as designers and viewers of visualizations ourselves.

CHAPTER 1

DATA

The Society for Effecting the Abolition of the Slave Trade and Responsible Design

Before there is data, there are people.
How can the lives behind the data guide our visualization work?

FIGURE 1.1
A portrait of Thomas Clarkson, painted by Carl Frederik von Breda in 1788. © National Portrait Gallery, London.

FIGURE 1.2
A portrait of Olaudah Equiano that appears as the frontispiece of his 1789 autobiography, *The Interesting Narrative of the Life of Olaudah Equiano, or Gustavus Vassa, the African, Written by Himself.* © National Portrait Gallery, London.

It was a long and circuitous—and often painful—path that, in January of 1777, led Olaudah Equiano to London. The famed author and abolitionist was born in the Igbo area of the kingdom of Benin, in what is now southeastern Nigeria.[1] Kidnapped from his birthplace at the age of 11 and carried to the Atlantic coast, he was then forced aboard a slave ship. There Equiano encountered "a multitude of black people of every description chained together," as he recalled in his autobiography, the captives packed in quarters "so crowded that each had scarcely room to turn himself."[2] He was taken to Barbados and then to Virginia, where he was sold. Equiano would remain enslaved for almost 20 years.

By the time that Equiano settled in London, he had been free for over a decade; in 1766, Equiano had purchased himself, and therefore his freedom, with earnings from his personal trading business. But his experience of enslavement was never far from his mind. And so in the late 1780s, as the British antislavery movement began to coalesce, Equiano grew increasingly involved. Through this activism Equiano became acquainted with Thomas Clarkson (FIGURE 1.1), a leading white abolitionist and key member of the London Committee of the Society for Effecting the Abolition of the Slave Trade (SEAST).[3] Thus when, in early 1789, Clarkson received a copy of a diagram created by the Plymouth Committee of SEAST entitled "Plan of an African Ship's Lower Deck with Negroes in the Proportion of Only One to a Ton," he knew exactly whom he should ask to confirm the truth of what he saw.[4]

The diagram that Clarkson showed to Equiano depicted the configuration of captive bodies in the hold of a slave ship—a "scene of horror almost inconceivable," as Equiano, in his autobiography, described his own first view into the hold (FIGURE 1.2). Indeed, Equiano's friend and fellow antislavery activist Quobna Ottobah Cugoano had internalized his similar experience as such a trauma that, in his own 150-page treatise on the subject, he chose not to describe it at all.[5] But in the diagram that Clarkson received from the Plymouth Committee, and that he shared with Equiano, Clarkson believed he'd found the key to conveying the depth of this inhumanity to his white compatriots. The diagram, which Clarkson would go on to revise and extend, publishing it in March of that year as "Description of a Slave Ship," would create an "instantaneous impression of horror upon all who saw it," he later recalled, and would compel them to join the abolitionist cause.[6]

The "instantaneous impression of horror" that Clarkson hoped his diagram would prompt was, like all carefully engineered viewing experiences, the result of countless hours of research and design. Given Clarkson's choice of words—in particular, the "instantaneous impression"—it is safe to say that the diagram represented the culmination of over two centuries of thinking about the value of empirical evidence and the impact of giving it visual form.[7] As we learned in the introduction, the idea that quantitative information could be visualized as data, perceived by the eyes, and then processed into knowledge had been steadily coalescing from the earliest expressions of empiricist philosophy, and from an increasingly wide range of examples of how information could be abstracted into images that the eye could perceive.

This intellectual genealogy in fact easily aligns with the standard story we tell about the origins of modern data visualization. But with the "horror" that Clarkson engineered his chart to produce, the expanded origin story of visualization—the one that includes its connections to colonialism and, as we will learn in this chapter, slavery—also enters the frame. SEAST's powerful slave ship diagrams require that we account for this additional context, and in so doing demonstrate the importance of a braided narrative for our understanding of the field. This expanded account of visualization's origins underscores its tremendous power to distill complex information such that insight can easily and efficiently emerge. At the same time, it reminds us—both those who design visualizations and those who perceive them—how the abstraction that is required to efficiently produce this insight always, and *necessarily*, comes at the expense of additional detail.

Setting SEAST's diagrams at the center of the story we tell about the emergence of modern data visualization also reminds us that the context of any particular visualization always matters—and that context carries with it both social and political force. More basically, and more profoundly, these diagrams remind us that before there is data there are people—people who contribute to the production of data, and people who themselves become data, not always with consent. As we will see as this chapter unfolds, the act of transforming people into data, and of putting that data on display, are both acts of power. But that power is not guaranteed to be a benevolent force. Centering SEAST's visualization work in the story we tell about data visualization is therefore long overdue, for it requires that we recognize the responsibility that comes with the power of visualizing data—the power to frame what it is possible to see, and thus to know.

Insight and Abstraction

The original engraving that Clarkson showed to Equiano, and that provided him with the visual model for his own, is attributed to the Plymouth Committee as a collective. But it is generally believed that the image was designed by a single man: William Elford, the Committee's chairman and a veritable polymath. A banker by trade, Elford was also an acclaimed painter whose landscapes had shown at the Royal Academy of Arts, as well as an amateur scientist whose experiments had earned him membership in the Royal Society, the highest scientific honor of the time. (Locke, Boyle, Plot, and others mentioned in the introduction were also members.)

Supplementing this range of skills were both domain expertise and class privilege: Elford had family ties to the Royal Navy, and as a result had early access to a report being prepared by one of its naval captains. Earlier in the year, a group within the British Parliament had tasked the captain, Parrey, with investigating the ships docked in Liverpool that were involved in the transatlantic slave trade.[8] This investigation took both quantitative and qualitative form; in addition to taking precise measurements of the ships and scrutinizing muster rolls—lists of crew members aboard any particular ship—Parrey also interviewed the captains and sailors of the ships themselves, learning otherwise undisclosed information about the perils faced by both captives and crew.

The art historian Cheryl Finley, whose book *Committed to Memory: The Art of the Slave Ship Icon* explores the origins and evolving significance of these charts in extensive detail, speculates that among the "most useful" artifacts from Parrey's report were the hand-drawn diagrams of the ships created by the captains themselves.[9] Intended as "a type of visual shorthand apparently used to increase the efficiency of packing ships," these sketches, for their abolitionist viewers, likely became a graphic call to arms.[10] Elford, for one, saw a direct visual depiction of the dehumanization that was required in order to reduce human lives into commodity goods on the part of slavery's profiteers.

In the online version of this book, we made the decision to hide the visual details of Elford's "Plan," and the other sensitive images discussed in this chapter, until our viewers could be more prepared for what they might see. We made this choice because of the range of harms that can be brought about by engaging with the archive of slavery without warning, context, or consent.[11] (FIGURE 1.3A AND B).

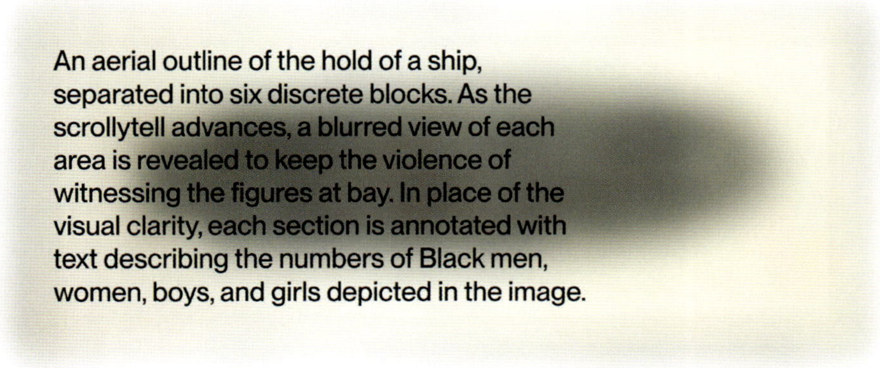

FIGURE 1.3A
The blurred image of "Plan of an African Ship's Lower Deck" as it appears in the web version of *Data by Design*. Visitors to the site must opt in before viewing the sensitive images in the chapter or reading the more descriptive alt text. Design by Tanvi Sharma. Implementation by Jay Varner. Alt text by Margy Adams.

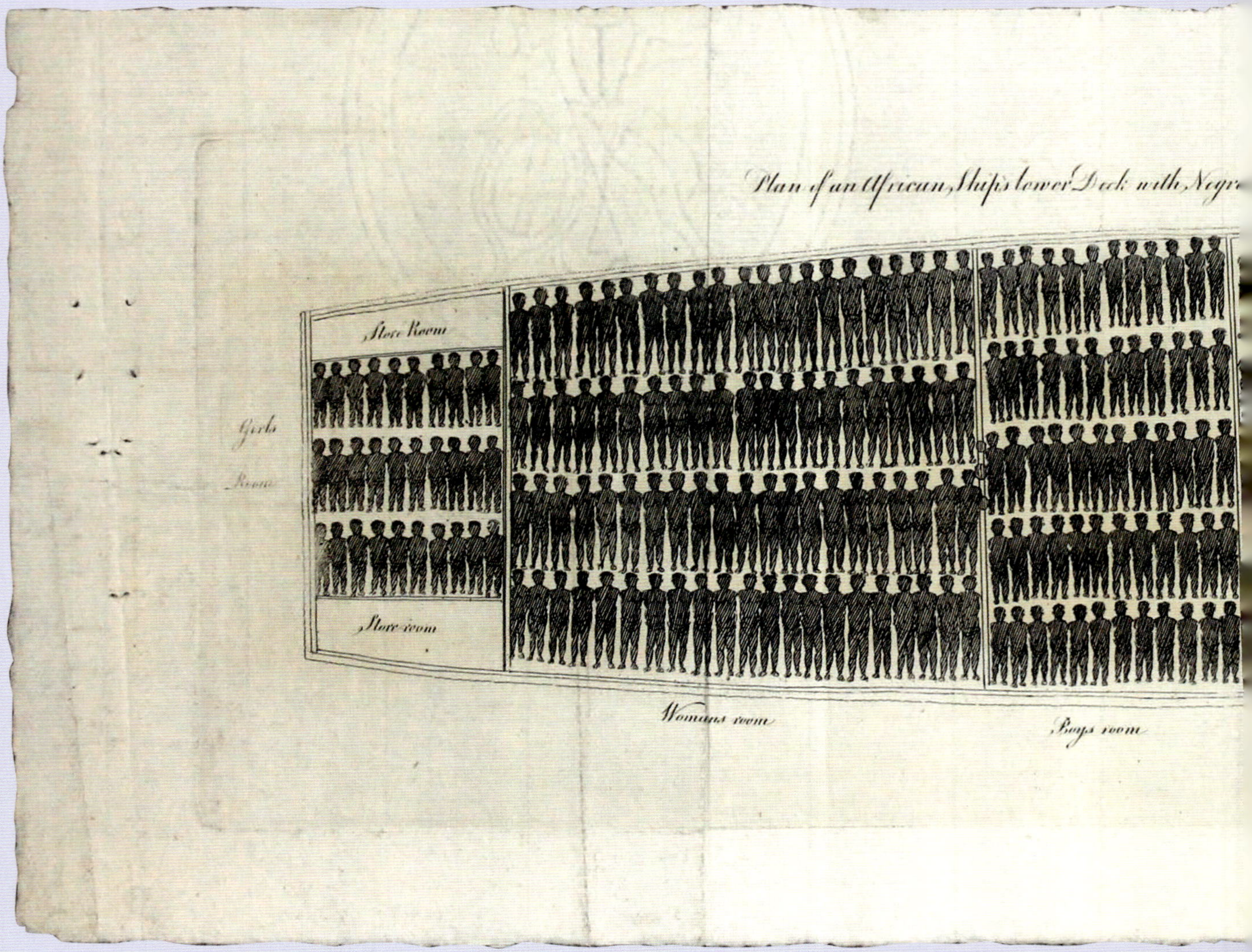

The hold is divided into four distinct areas:

↑
At the stern is the "Girls Room," the figures shorter and squatter than the boys, arranged in three rows of 10.

↑
To its right is the larger "Womens room," depicting figures the same size as the adult men, but with breasts. They are depicted in four rows of 21, representing 84 women in total.

↑
In the middle is a narrow area labeled "Boys room" which depicts shorter figures in five rows of 12.

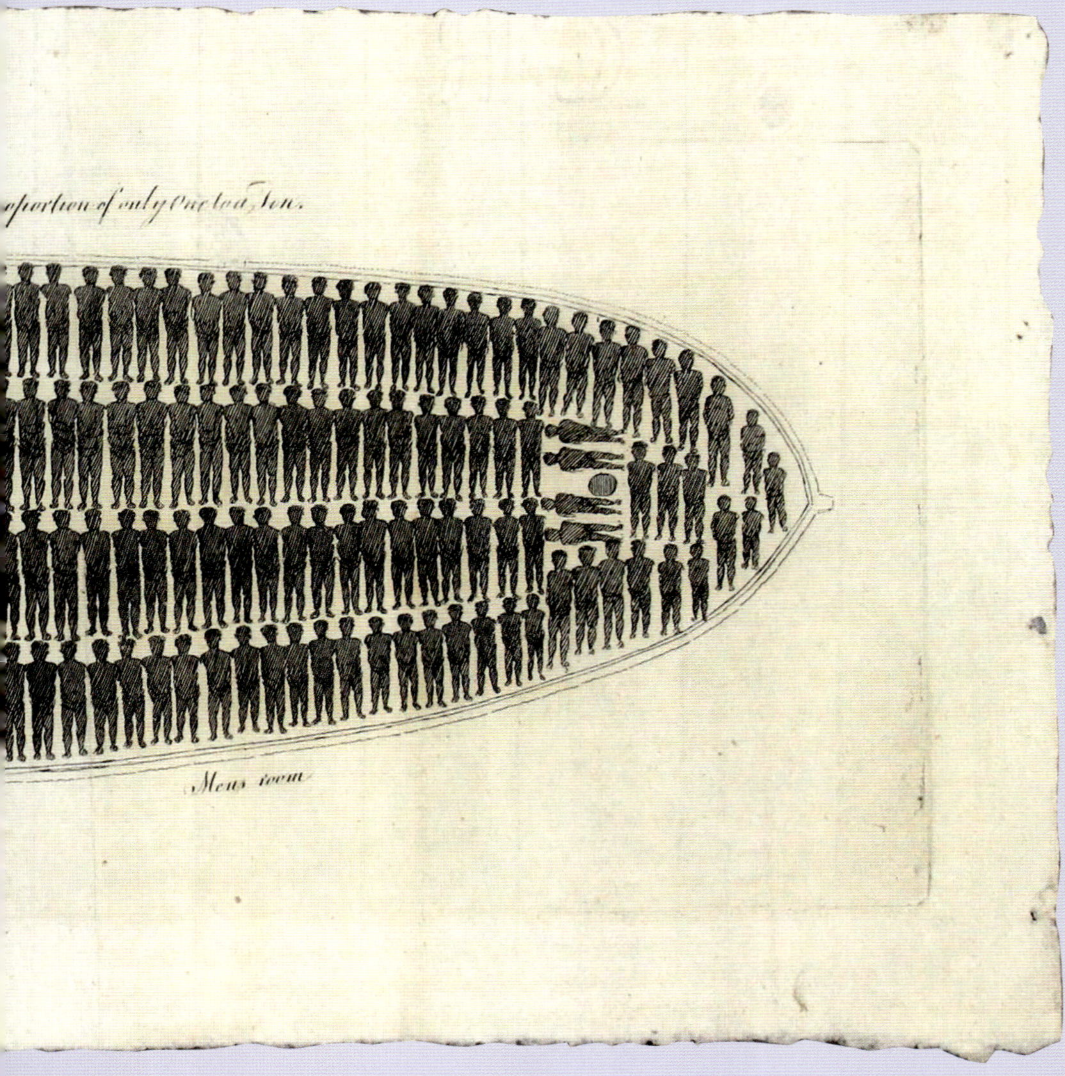

FIGURE 1.3B
The first published version of the slave ship diagram, "Plan of an African Ship's Lower Deck with Negroes in the Proportion of Only One to a Ton," created by the Plymouth Committee of the Society for Effecting the Abolition of the Slave Trade in England in 1788. © Britain Yearly Meeting.

↑
The largest section, in the bow of the ship and occupying the entire right half of the diagram, is labeled the "Mens room." It depicts 120 men's bodies in four rows of 24, with another 24 figures fit into the curved space closest to the front of the bow.

These "scaled inequalities," as Black feminist theorist Hortense Spillers describes them, literalize the process of "dehumanizing, ungendering, and defacing" at the heart of the Middle Passage.

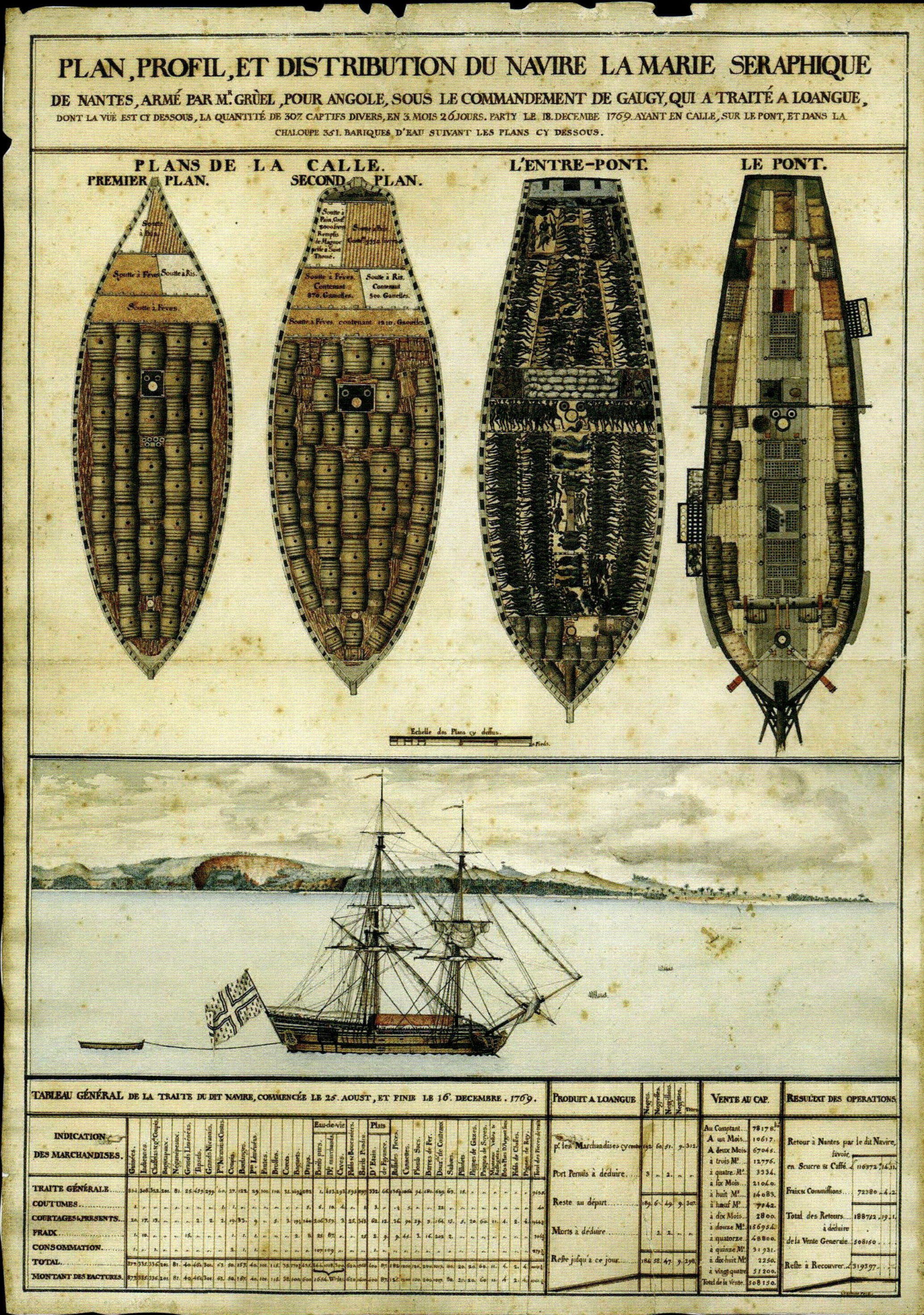

Nantes (ca. 1770), that makes use of an aerial view of the ship's hold, displaying its human captives alongside its goods for sale. Image courtesy of Wikimedia Commons. Digitized by the Musée d'histoire de Nantes.

But this interaction is not possible to replicate in print. Instead, we have enacted a form of *progressive disclosure*, a concept formulated by Silas Munro, a member of our project team. By this he means a mode of presentation that gradually introduces individual elements of the most sensitive images included this chapter, so that the reader is not subjected to the same experience of visceral and immediate horror that SEAST intended with their charts. While this decision runs counter to the Society's original goal, our encounter with these images over two hundred years after they were first shown serves a different purpose: to understand how the charts' designers were able to leverage a combination of abstraction and detail in order to achieve their catalyzing effects.

This is the reason that we delayed reproducing any of the SEAST diagrams until after this chapter's introduction. We wanted to give you, the reader, time to prepare for what you would see. But the diagrams remain as viscerally affecting as they are visually impossible. In the original Elford diagram—the one on the previous set of pages, and that Equiano saw—viewers see the ship from above, as if they are gods in the heavens. The top deck of the ship has been removed, so that the viewer can see directly into the hold. (Later we will return to this strategy of what is, in fact, a "god trick," as feminist philosopher Donna Haraway describes.)

The hold is divided into four distinct areas: the largest, in the bow of the ship, and which occupies the entire right half of the diagram, is labeled the "Mens room," and depicts 120 men's bodies in four rows of 24, with another 24 figures fit into the curved space closest to the front of the bow. In the middle is a narrow area labeled "Boys room" which depicts shorter figures in five rows of 12. To its left is the larger "Womens room," depicting figures the same size as the adult men, but with breasts. They are depicted in four rows of 21, representing 84 women in total. At the stern is the "Girls Room," the figures shorter and squatter than the boys, arranged in three rows of 10. Together, these "scaled inequalities," as Black feminist theorist Hortense Spillers describes them, literalize the process of "dehumanizing, ungendering, and defacing" at the heart of the Middle Passage.[12]

We do not know with certainty whether Elford was familiar with an earlier, more literal depiction of a slave ship, the *Marie Séraphique*, which dates to around 1770 (FIGURE 1.4). This image, commissioned by the ship's owner to commemorate the "successful" return of its first slaving voyage, presents a view of the hold—and the captives within it—as one of four plans of the ship's levels. The image also includes a watercolor of the ship arriving in Loango, in what is now the Republic of Congo, where the 312 captives were first purchased by the French captain, as well as data about the voyage presented in what was by then standard tabular form.[13] The largest table, on the bottom left, provides a breakdown of the counts of various "items of trade," indicating which were intended for sale, as gifts to the king, consumed along the way, and so on. But next to that section, in near-identical form, is another table labeled "Products of Loango." The numbers here refer to people, sorted by gender and age and whether or not they survived the voyage.

This visual presentation of people as data—a literal headcount, as project team member Margy Adams has observed—reinforces the claims made by historians of slavery, such as Jessica Marie Johnson, that data itself "is the evidence of terror, and the idea of data as fundamental and objective information ... obscures rather than reveals the scene of the crime."[14] We will return to this observation as we confirm the link between these data tables and contemporary digital databases when we discuss our own design process. Here, however, we focus on the visual elements of this repellent "Plan." Unlike the "cartoon figures" in Elford's diagram, as Spillers describes them, the captives in this image are each individually drawn.[15] They are shown lying on their sides. Most are naked, but several are clothed. The men are shackled—some with their arms and legs shackled to each other; others shackled two by two. One woman nurses a child.

Whether or not Elford saw this particular "Plan," he clearly considered what he would give visual form—and also what he would not. As evidence, consider how, in contrast to the diagram of the *Marie Séraphique*, the two areas of Elford's diagram labeled "store room" are left blank, even as they were presumably packed with barrels and other dry goods similar to those depicted in the plan of the *Marie Séraphique*. The white space of the storerooms instead emphasizes the rhetorical point of Elford's diagram: that nothing was more important than the packing of *people* as cargo on these ships.[16]

At the same time, Elford could not but participate in this process of dehumanization, however inadvertently. A large part of how the diagram achieves its visual impact is the way its 294 human souls are drawn as nearly identical figures—what literary scholar Marcus Wood describes as a "mass of black human flesh."[17] The figures' collective rather than individual significance is further accentuated by their being set against the clean outlines of the ship.[18] The labels associated with each area, engraved in neat script, accentuate this reduction in complexity—which is, of course, a reduction in humanity as well.

It is here that the significance of the diagram for the larger practice of data visualization begins to cohere: Elford's design achieves its success because of its strategic use of abstraction. It represents the captives as almost proto-Isotypes, rather than individual people with unique bodily features and distinct inner lives.[19] In so doing, he "induce[s] the viewer to think about the substance" of the data of the slave trade, to employ Edward Tufte's language of graphical excellence, rather than specific people on specific ships.[20] Yet Elford achieves his visual tour de force by stripping away the individual lives behind each datapoint—the lives that, paradoxically, he designed his charts in order to support.

Not all data visualizations take on this most odious episode in human history, of course. But the lesson about the irresolvable tension between abstraction and detail is one that we can apply to visualization as a whole. Put another way: the abstraction that is required to efficiently prompt insight *always*—and just as important, *necessarily*—comes at the expense of the full detail of what it seeks to represent. What should we do about this tension, either as viewers of visualizations or as designers of them? The answer is not that we should reject visualization out of hand. It is, rather, that we must always consider what is lost in the process of visualizing data at the same time that we consider what is gained.

"Plan of an African Ship's Lower Deck" also reminds us, with its subject most profound, that there are always aspects of human experience that visualization cannot convey. No diagram can ever express the full extent of the brutality and degradation that was required to enforce the enslavement of otherwise equals. No diagram can ever fully communicate the "horror almost inconceivable"—to return to Equiano's chilling words—to those who did not personally experience it; nor can any diagram appropriately convey the additional trauma that others, like Cugoano, chose to keep suppressed. As stated above, the lesson is not of the futility of visualizing data. It is, rather, one of hope: that with the image of the slave ship indelibly etched in our minds, we will keep both the power and limits of data visualization together in view.

Indeed, the power to produce an "instantaneous impression" in the eyes of viewers remains among the greatest strengths of data visualization. In *Data Feminism*, Catherine D'Ignazio and I quote visualization designer Fernanda Viégas—who created the "Wind Map" that appears in this book's introduction—in describing this strength as a "superpower."[21] Her words could not be more revealing. Because just as we consider this tremendous power, we must also consider our responsibility for the insights that our visualizations produce, as we do for the detail and context—and in this case, the lives—that visualizations on their own cannot convey.

Graphical Authority and the God Trick

As William Elford was scrutinizing the Parrey report, Thomas Clarkson was pursuing his own research into the pernicious nature of the slave trade. Like Parrey, Clarkson began by transforming the available records into data. He visited merchant halls in order to examine the muster rolls stored there, and used them to compute mortality rates among the sailors aboard the ships. (The dangers of the slave trade to the sailors, who were predominantly white, would become a highly persuasive piece of evidence in the argument for its abolition.) While examining the muster rolls, Clarkson also covertly transcribed 20,000 of the sailors' names. He then sought out individual sailors—primarily those who had been mistreated or maimed—whom he thought might be willing to speak about the conditions aboard the ships (both those that they personally experienced and those of the captives that they observed). These efforts at an early form of counterdata collection, complemented by what was, in effect, mixed-methods research, underscore how Clarkson understood the value of empirical evidence—qualitative as well as quantitative—in advancing his abolitionist claims.[22]

Clarkson's awareness of the value of evidence of multiple forms, and of the value of multiple forms of display, strongly influenced his revisions to the original diagram. While Elford and the Plymouth Committee had first printed the diagram as a companion to a four-page abolitionist pamphlet, and later as a broadside version with the image at the top, it was Clarkson who insisted that the London Committee's version also include data tables. The tables included measurements of the ship, the *Brooks*, that had been used as a model for the diagram, and a conversion scale that indicated precisely how much square footage had been intended to be allocated to each captive on the chart. A second set of tables enabled a comparison between the number of captives who had actually been held on the original ship and the smaller number depicted in the diagram. This additional information was intended to "give a representation of the trade against which no complaint of exaggeration could be brought."[23]

Clarkson's attention to the strategic deployment of data also underscores the fact that "Description of a Slave Ship" was, like its predecessor, a data visualization (FIGURE 1.5). The SEAST diagrams have at times been dismissed from the visualization pantheon because of their explicit political objectives, which go against the presumption of visualization's neutrality—a myth we will continue to dismantle in chapter 2.[24] They are also sometimes left unconsidered because they are presumed to picture the data of one specific ship—that is to say, a form of direct rather than abstract representation. But Clarkson's extensive commentary

DESCRIPTION OF A SLAVE SHIP.

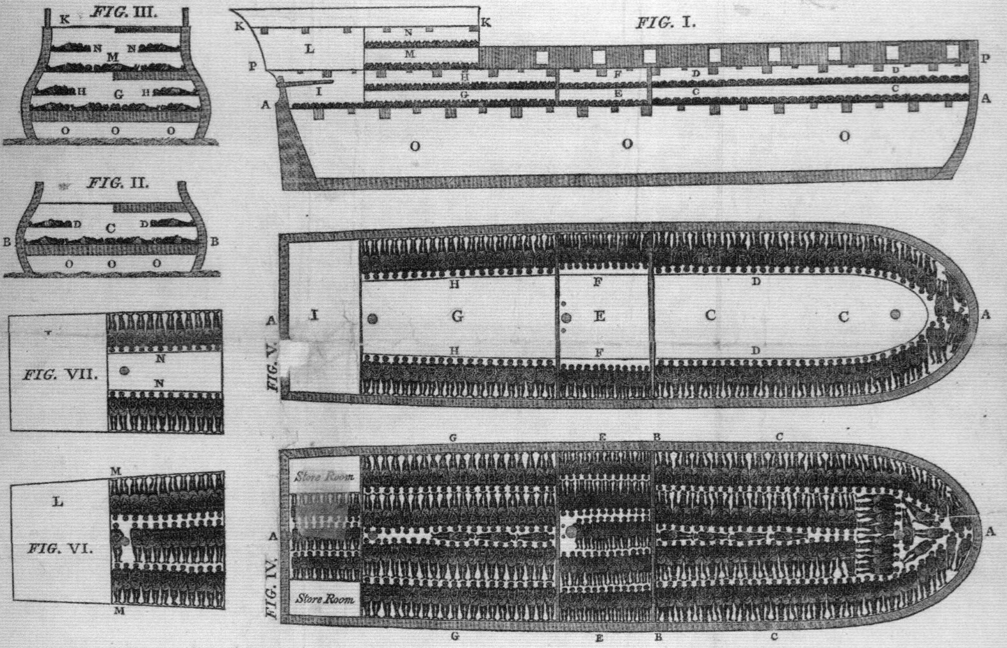

The PLAN and SECTIONS annexed exhibit a slave ship with the slaves stowed.* In order to give a representation of the trade against which no complaint of exaggeration could be brought by those concerned in it, the *Brooks* is here described, a ship well known in the trade, and the first mentioned in the report delivered to the House of Commons last year by Captain Parrey, who was sent to Liverpool by Government to take the dimensions of the ships employed in the African slave trade from that port. These plans and sections are on a scale of the 8th of an inch to a foot.

DIMENSIONS OF THE SHIP.

	Feet	Inches
Length of the *Lower Deck*, gratings, bulk-heads, included at AA	100	0
Breadth of *Beam* on the *Lower Deck* inside, BB	25	4
Depth of *Hold*, OOO from cieling to cieling	10	0
Height between decks from deck to deck	5	8
Length of the *Mens Room*, CC on the lower deck	46	0
Breadth of the *Mens Room*, CC on the lower deck	25	4
Length of the *Platform*, DD in the mens room	46	0
Breadth of the *Platforms* in mens rooms on each side	6	0
Length of the *Boys Room*, EE	13	9
Breadth of the *Boys Room*	25	0
Breadth of *Platform*, FF in boys room	6	0
Length of *Womens Room*, GG	28	6
Breadth of *Womens Room*	23	6
Length of *Platform*, HH in womens room	28	6
Breadth of *Platform* in womens room	6	0
Length of the *Gun Room*, II on the lower deck	10	6
Breadth of the *Gun Room* on the lower deck	12	0
Length of the *Quarter Deck*, KK	33	6
Breadth of *Quarter Deck*	19	6
Length of the *Cabin*, LL	14	0
Height of the *Cabin*	6	2
Length of the *Half Deck*, MM	16	6
Height of the *Half Deck*	6	2
Length of the *Platforms*, NN on the half deck	16	6
Breadth of the *Platforms* on the half deck	6	0
Upper deck, PP		

Nominal tonnage — —	297
Supposed tonnage by measurement	320
Number of seamen —	45

The number of slaves which this vessel actually carried appears from the accounts given to Capt. Parrey by the slave-merchants themselves as follows:

Men — —	351	Total 609
Women — —	127	
Boys — —	90	
Girls — —	41	

The room allowed to each description of slaves in this plan is:

To the Men 6 feet by 1 foot 4 inches.
Women 5 feet 10 in. by 1 foot 4 in.
Boys 5 feet by 1 foot 2 in.
Girls 4 feet 6 in. by 1 foot.

* This is the usual manner of stowing the slaves, but it varies according to the position of the ship, and the practice of different commanders.

With this allowance of room the utmost number that can be stowed in a vessel of the dimension of the *Brooks*, is as follows, (being the number exhibited in the plan) and is less than 1½ to a ton, viz. †

		On the Plan.	Actually carried.
Men—on the lower deck, at CC	124	190	351
Ditto on the platform of ditto, CC DD	66		
Boys—lower deck EE —	46	70	90
Ditto—platform FF —	24		
Women—lower deck, GG —	83	183	127
Ditto—platform, HH — —	40		
Women Half deck, MM —	36		
Ditto Platform ditto, NN —	24		
Girls Gun room, II — —		27	41
General total		470	609

The principal difference is in the *men*. It must be observed, that the *men*, from whom only insurrections are to be feared, are kept continually in irons, and must be stowed in the room allotted for them, which is of a more secure construction than the rest.

In this ship the number of men actually carried was —	351
The number of men stated in the plan at 1 foot 4 inches each	190
Difference —	161

As the ship on this plan would stow 42 women boys and girls in the places here allotted them more than she did carry, supposing that number of men taken from the mens room, and placed in their stead, this will reduce the number of men to 309 in the mens room; of course the room allowed them, instead of being 16 inches as in the plan, was in reality only 10 inches each; but if the whole number 351 were stowed in the mens room, they had only 9 inches each to lay in.

The men therefore, instead of lying on their backs, were placed, as is usual, in full ships, on their sides, or on each other. In which last situation they are not unfrequently found dead in the morning.

The longitudinal section, fig. I. shews the manner in which the slaves were placed on all the decks and platforms, which is also further illustrated by the transverse sections, fig. II. & III. By which it appears, that the height between the decks is 5 feet 8 inches, which, allowing 2 inches for the platform and its bearers, makes the height between the decks and the platform 2 feet 9 inches; but the beams and their knees, with the carlings, taking 4 inches on an average, this space is unequally divided, and above or under the platforms cannot be estimated at more than 2 feet 7 inches; so that the slaves cannot, when placed either on or under the platforms, relieve themselves by sitting up; the very short ones excepted, nor can *they*, except on board the larger vessels. The average of 9 vessels measured by Captain Parrey, being mostly large ships, was only 5 feet 2 inches. The height of the Venus between decks was 4 feet 2 inches; of the Kitty, 4 feet 4 inches, both of which had platforms. In these smaller vessels therefore they they have not 2 feet under or upon the platforms.

In fig. 1. under the upper deck PP, and the lower deck AA, the beams and the intervening carlings are represented by shaded squares. The beams are also introduced on one side of the transverse sections II and III, in order to shew the space which a slave placed under a beam has to lie and breathe in.

† It must be noted, that every possible advantage of stowing is allowed in the plan. There are or ought to be in each apartment one or more necessary tubs; there are also stanchions to support the platforms and decks; for which no deduction is made; but the deck is supposed clear of every incumbrance whatever.

It may be expected, from this mode of packing a number of our fellow-creatures, used in their own country to a life of ease, and from the anguish of mind their situation must necessarily create, that many of them fall sick and die. Instances sometimes occur of horrible mortality. The average is not less than 1-5th, or 20 per cent. The half deck is sometimes appropriated for a sick birth; but the *men slaves* are seldom indulged the privilege of being placed there, till there is little hope of recovery. The slaves are never allowed the least bedding, either sick or well; but are stowed on the bare boards, from the friction of which, occasioned by the motion of the ship, and their chains, they are frequently much bruised; and in some cases the flesh is rubbed off their shoulders, elbows, and hips.

It may not be improper to add a short account of the mode of securing, airing, and exercising the slaves.

The women and children are not chained, but the men are constantly chained two and two; the right leg of one to the left leg of the other, and their hands are secured in the same manner.

They are brought up on the main deck every day, about eight o'clock, and as each pair ascend, a strong chain, fastened by ring-bolts to the decks, is passed through their shackles; a precaution absolutely necessary to prevent insurrections.—In this state, if the weather is favourable, they are permitted to remain about one-third part of the twenty-four hours, and during this interval they are fed, and their apartment below is cleaned; but when the weather is bad, even those indulgencies cannot be granted them, and they are only permitted to come up in small companies, of about ten at a time, to be fed, where after remaining a quarter of an hour, each mess is obliged to give place to the next in rotation.

In *very* bad weather, some are unavoidably brought on deck: there being no other method of getting water, provisions, &c. out of the hold, but by removing those slaves who lie on the hatch-ways. The consequence of this violent change from their rooms, which are inconceivably hot, to the wind and rain, is their being attacked with coughs, swellings of the glands of the neck, fevers, and dysenteries; which are communicated by infection to the other slaves, and also to the sailors.

The only exercise of the men-slaves is their being made to jump in their chains; and this, by the friends of the trade, is called *dancing*.

To persons unacquainted with the mode of carrying on this system of trading in human flesh, these Plans and Sections will appear rather a fiction, than a real representation of a slave-ship. They will probably object, that there is no room for stowing cables, and such other utensils and stores as are usually placed between decks. In a slave ship (i. e. a full one) these articles are either deposited in the hold, or piled upon the upper deck; and from thence, in case of bad weather, or accidents, no small confusion is occasioned.—It may be also said, the slaves are placed so very close, that there is not room for the surgeon to visit and assist them: The fact is, that when the surgeon goes amongst them, he picks out his way as well as he can, by stepping between their legs. He frequently finds it to be impossible to afford them that relief which an humane man (and such there are even in this trade) would willingly give them. When attacked with fluxes, their situation is scarcely to be described. To give an instance, (as related by an eye-witness) as it serves to convey some idea, though a very faint one, of the sufferings of those unhappy beings whom we wantonly drag from their native country, and doom to perpetual labour and captivity: "Some wet and blowing weather having occasioned the port holes to "be shut, and the grating to be covered, fluxes and fevers among the "negroes ensued. While they were in this situation, my profession "requiring it, I frequently went down among them, till at length their "apartments became so extremely hot, as to be only sufferable for a very "short time. But the excessive heat was not the only thing that rendered "their situation intolerable. The deck, that is, the floor of their rooms, "was so covered with the blood and mucus which had proceeded from "them in consequence of the flux, that it resembled a slaughter-house. "It is not in the power of the human imagination to picture to itself "a situation more dreadful or disgusting. Numbers of the slaves had "fainted, they were carried upon deck, where several of them died, and "the rest were, with difficulty, restored. It had nearly proved fatal to "me also."*

Another objection which may be stated, is, that here no room is allowed for the sailors hammocks. In slave ships, while the slaves are on board, the sailors have no other lodging than the bare decks, or (in large ships) the tops. From this exposure, they often are wet for a long time together, the rains in those climates being frequent and extremely heavy. There is in wet weather a tarpawling placed over the gratings: if the sailors to shelter themselves creep under this, they are exposed to the noisome and infectious effluvia which continually exhale from the slaves below.

It appeared from the evidence given by the slave merchants last year before the House of Commons, that the employment of the seamen, viz. boating up the rivers after the negroes, guarding them on board, cleansing the vessel, &c. is of a nature offensive and dangerous beyond that of seamen in other services, and that the small-pox, measles, flux, and other contagious disorders, are frequent on board these ships.

It is therefore falsely said by the well-wishers to this trade, that the suppression of it will destroy a great nursery for seamen, and annihilate a very considerable source of commercial profit.—The Rev. Mr. Clarkson, in his admirable treatise on the Impolicy of the Trade, has proved from the most incontestable authority, that so far from being a nursery, it has been constantly and regularly a grave for our seamen; *for that in this traffick only, a greater proportion of men perish in* ONE *year, than in all the other trades of Great Britain in* TWO *years.*

Besides the time spent on the coast to complete their cargoes, which sometimes lasts several months, the slaves are from six to eight weeks on their passage from thence to the West-Indies.

Now let any person reflect on the situation of a number of these devoted people, thus managed and thus crammed together, and he must think it dreadful, even under every favourable circumstance of an humane captain, an able surgeon, fine weather, and a short passage. But when to a long passage are added, inhuman treatment, scanty and bad provisions, and rough weather, their condition is miserable beyond description. So destructive is this traffick in some circumstances, particularly in bad weather, when the slaves are kept below, and the gratings covered with tarpawlings, that a schooner, which carried only 140 slaves, meeting with a gale of wind which lasted eighteen hours, no less than 50 slaves perished in that small space of time.

As then the inhumanity of this trade must be universally admitted and lamented, people would do well to consider that it does not often fall to the lot of individuals, to have an opportunity of performing so important a moral and religious duty, as that of endeavouring to put an end to a practice, which may, without exaggeration, be stiled one of the *greatest evils at this day existing upon the earth.*

* *Falconbridge's Account of the Slave Trade*, page 31.

LONDON: PRINTED BY JAMES PHILLIPS, GEORGE-YARD, LOMBARD-STREET. M,DCC,LXXXIX.

"Description of a Slave Ship," this version is generally attributed to Thomas Clarkson. It expands upon the original diagram by including additional cross sections, as well as statistics and explanatory text. Image courtesy of the Beinecke Rare Book and Manuscript Library, Yale University.

on his methodological choices makes clear that his "Description," like Elford's "Plan," involved a synthesis of data from multiple sources. More than that, it satisfies both of the core criteria of that era's definition of data visualization, as Michael Friendly and Howard Wainer propose: making "previously invisible phenomena subject to direct inspection," and making those phenomena "palpable and concrete."[25]

To wit: just as Clarkson's reintroduction of data to the chart underscored its basis in empirical evidence and buttressed its claims to the truth, so too did his visual modifications. More specifically, he shifted the view of the hold lower down the page, and added a series of cross sections that showed additional views of the ship from the side. The inclusion of the cross sections drew from the conventions of naval architectural plans, which were by that time strongly established. This had the rhetorical effect of securing the chart's "graphic authority," as Marcus Wood describes it, in ways similar to how drawing new national borders on an existing map leverages the power of its seemingly documentary form.[26] Unlike a typical naval diagram, however, Clarkson's cross sections also included people. These represented the captives themselves, whom Clarkson included as a way to show from multiple perspectives precisely how they were confined.

Here we might recall how Elford's diagram depicted the captives' bodies with minimal differentiation, which scholars have largely interpreted as reflecting "an abolitionist cultural agenda which dictated that slaves were to be visualized in a manner which emphasized their total passivity and prioritized their status as helpless victims."[27] In Clarkson's version, however, the captives are shown wearing loincloths, among other additional elements of visual detail. The men are shackled together in pairs, by both their hands and their feet. The women, on the other hand, remain unbound, their breasts exposed.[28]

The combined effect of these changes was a diagram that engaged the viewer through two very different epistemological registers, both at the same time: first, as a representation of the data that, through its use of naval convention, conveyed its graphical authority; and second, as an intentionally evocative graphic that was intended to elicit a combination of sympathy and shock. The desired result of the diagram, driven by both empirical evidence and emotion, was that the viewer would perceive the "inhumanity of the trade" through the eyes and the heart at the same time and, prompted by the "instantaneous impression" that it made on the senses, be compelled to act.[29]

At this point, one previously unstated detail must now be made explicit: these viewers, like the captives, had a race. They were white, predominantly British and predominantly men, with lives far removed from the experience of enslavement. In fact, Clarkson had an even more specific audience in mind with his design: the members of the British Parliament, who were scheduled to vote on a motion to abolish the slave trade in several weeks' time. One of the London Committee's own members, William Wilberforce, was among them, and he believed, and said as much, that if the MPs "could actually *see* one thousandth part of the evils of that practice which they have, for so many years, under one pretense or another, been prevailed on to suffer to be continued," they would quickly commit themselves to the abolitionist cause.[30]

Here is where we return to the idea of the "god trick," mentioned at the outset of this chapter. The "god trick" is an idea developed by the feminist philosopher Donna Haraway to describe the false sense of neutrality conveyed through the default perspective of data visualization, the "view from above."[31] The view from above can seem godlike, but it is a "trick" because it preys upon our general tendency not to notice, let alone question, any perspective that adheres to the default. In this particular case, the seemingly godlike perspective is, in fact, the perspective of "predominantly white and male abolitionists and lawmakers," as sociologist Simone Browne observes.[32] In her field-defining work on race and surveillance, Browne analyzes Clarkson's "Description" in terms of the god trick, and employs it to underscore Haraway's primary point: while the "view from above" may seem like a view "from nowhere," it is in fact a view from somewhere; all viewpoints are.[33] For Browne—and, we hope, for you now as well—the "Description" provides unassailable evidence of this fact.

Given Clarkson's own writing on the subject, he would likely not disagree with Browne's assessment of his intended audience. Clarkson celebrated how his chart "brought forth tears of sympathy in behalf of the sufferers, and it fixed their sufferings in [the viewer's] heart."[34] But Browne's analysis of the power relations embedded in the "Description," and more specifically of "the primacy given in these abolitionist texts to white gazes and vantage points to the trauma of slavery," pushes us as twenty-first-century viewers—with our more nuanced understanding of how power is refracted along racial lines—to see even more.[35] Looking closely:

> One can see that each of the tiny black figures are not replicas of each other; rather, some have variously crossed arms, different gestures, or seem to turn to face one another, while some stare and look back at the gaze from nowhere, and in so being the *Description of a Slave Ship* can also be understood as depicting black looks and the trauma of Middle Passage as multiply experienced and survived.[36]

Did Clarkson intend to depict the captives looking back at the viewer, challenging their gaze? Or was it just that, in his commitment to accurate representation, he could not help depicting the captives' resistance, in ways both large and small? These questions lack definitive answers, but their possibilities are what matter more. What would it mean to visualize the experience of the Middle Passage from the perspective of the captives themselves? Is this a task that should be undertaken, and if so, by whom? What data would be required? Or, in the absence of data, what information—or what context—might be marshalled instead? What new knowledge, if any, might this visualization bring to light? And for whom might this be new knowledge? For whom might this, instead, further perpetuate harm?

Visualizing Resistance

These were the very questions that I brought to the *Data by Design* project team. I also brought a dataset: a full download of the transatlantic portion of the Voyages Database, which (as of spring 2026) contains records of 36,150 unique slaving voyages that took place between the years 1514 and 1866, resulting in the captivity and forced migration of an estimated 10.6 million souls (FIGURE 1.6).[37] Ever since the first release of the database, in 1999 via CD-ROM, scholars and designers have attempted to give this powerful data visual form.[38] Perhaps most prominently, in 2016, Andrew Kahn and Jamelle Bouie produced an animated visualization of the data for *Slate*, the online magazine, which depicted each of the 20,528 voyages that were then in the database as small black dots that were seemingly pulled from the west coast of Africa to the Americas as if by magnetic force.[39] While intended to "give a sense of the scale of the trans-Atlantic slave trade across time, as well as the flow of transport and eventual destinations,"

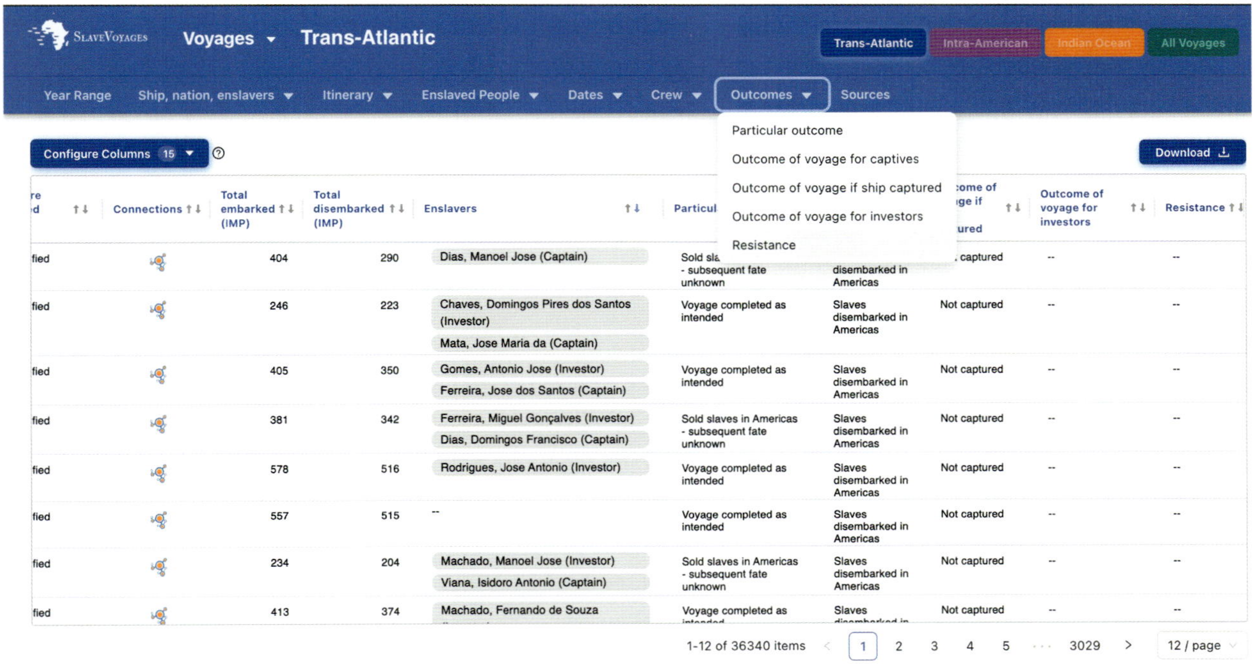

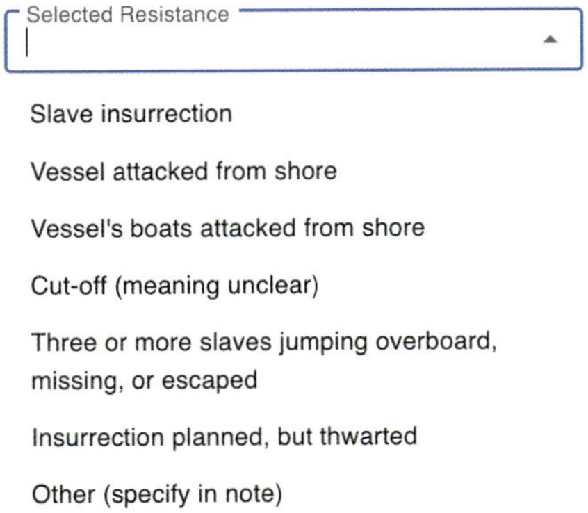

FIGURES 1.6 AND 1.7
The fields included in the Voyages Database as viewed through its web interface. **ABOVE:** The "outcome" field, which includes the category of "resistance." **LEFT:** The seven subcategories of "resistance," reflecting details noted by the historians who contributed to the database.

the visualization was roundly critiqued by scholars of slavery for the god's-eye perspective that it adopts. As literary scholar Britt Rusert asserts, "It's as if a series of 'invisible hands' operate the trade," rather than specific people who should be condemned for their acts.[40] Although she does not formulate her critique in these exact terms, what Rusert identifies is another version of the "god trick" at work.[41]

Reflecting on this visualization, as we considered whether a different depiction of the Middle Passage might be possible, enabled us to clarify our task. We first considered the provenance of the dataset, and contemplated how the shipping logs and other data tables that served as its primary sources—like the table associated with the plan of the Marie Séraphique—were created by the enslavers and not the enslaved. This reminded us of the power relations embedded

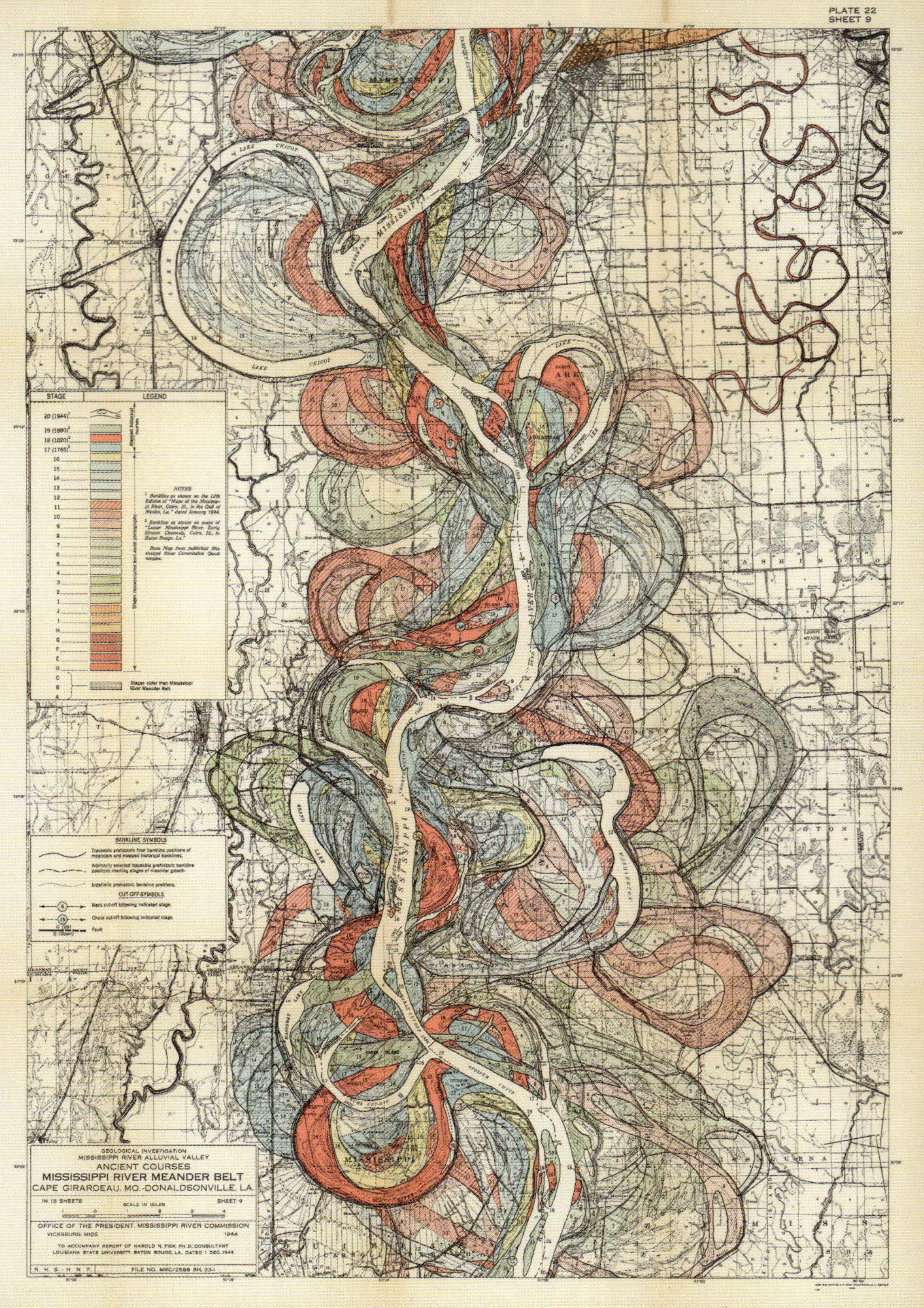

FIGURE 1.8
One of the 15 maps included in Harold Fisk's *Geological Investigation of the Alluvial Valley of the Lower Mississippi River* (1944). Image courtesy of the US Army Corps of Engineers, Engineer Research and Development Center.

in the data, those that could never be removed. It also reminded us of how that power contributed to a range of omissions and gaps. In addition, we considered our own subject positions, and the fact that, among the five of us collaborating on this particular visualization, only one of us was connected to ancestors who had themselves been enslaved. This prompted a recognition on behalf of the group that there were certain stories about the data that were not ours to tell, and, following scholar of slavery Saidiya Hartman, should be left for others to convey.

Finally, we returned to Jessica Marie Johnson's powerful statement, in her "Markup Bodies" essay, that "there is nothing neutral, even in a digital environment, about doing histories of slavery."[42] This called us back to the range of harms that can be brought about by engaging with this history, as discussed earlier in this chapter, and to how data visualization is not immune. We saw the potential for "second-order violence," as Hartman terms it, both in the act of reanimating a dataset that, in its original form, conscribed the people it represented to living death; and in the act of controlling the layout and motion of this dataset, which itself documents the forced migration of so many. From this process, our goal became more refined. In order to better represent the enslaved as they lived, and not further reinforce how they were reduced to data, we would need a visual strategy that could draw out meaningful facets of the dataset while simultaneously reinforcing just how much about each life the data could not show.[43]

We began our prototyping process, as many do, by examining the summary statistics of our dataset. But having articulated our ethical framework and our design goals, we found that few facets of the data—counts of ships and people, and, more chillingly, prices—seemed to hold any possibility of being reframed in ways that would not further reinscribe the violence of the original numbers.[44] However, one variable, labeled "resistance," seemed to provide a small opening that we thought we could further expand. It contained seven different subcategories of resistance that might have taken place on any particular journey. These included attempts at insurrection while at sea, both those that were successful and those that were thwarted while under way; attacks on the ship (or larger fleet) while on shore; and unspecified acts of resistance undertaken by "three or more" captives at a single time (FIGURE 1.7). Could we use this "resistance" variable to create a visualization that, to recall Browne's phrase, "looked back at the gaze from nowhere"?

We began by pulling out the voyages that had any form of resistance associated with them, as well as six of the more basic variables that were associated with each journey. The additional variables we selected—the voyage's start date, its end date, the total number of individuals who "embarked" on the voyage, and the total number of the individuals who "disembarked"—would allow us to provide a basic picture of each of the 572 voyages that contained a documented act of resistance.

Because our initial motivation was to visualize the dataset from the perspective of the enslaved, our design process began by focusing on a frequent observation made about the experience of the Middle Passage: that the captives did not experience time as linear while in the hold of the ship.[45] Drawing visual inspiration from Harold Fisk's alluvial diagram of the Mississippi River (FIGURE 1.8), which, as artist and scholar Romi Morrison explains, "deemphasizes the linearity of the river" in favor of showing a comparative view of its various paths over time, we decided to use bends and turns to represent the nonlinearity of the Middle Passage.[46]

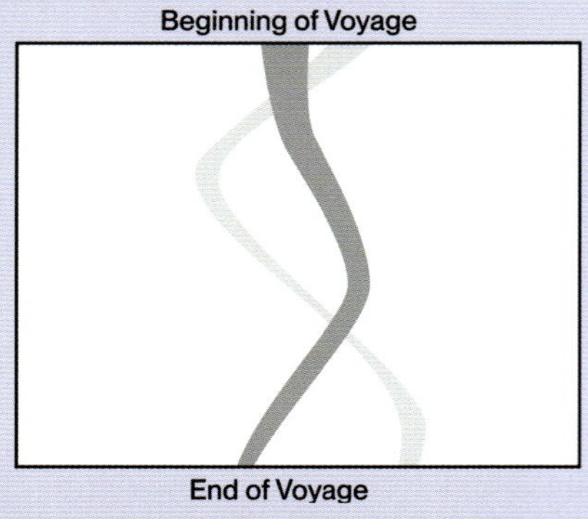

FIGURE 1.9
(A) The vertical orientation of our visualization, with the beginning of each voyage anchored to the top of the visualization and the end to the bottom.

We retained the vertical orientation of the Fisk diagram so as to ensure that the viewer could not interpret the paths of the voyages as corresponding to any actual location on a map (FIGURE 1.9A).

(B) Each voyage is plotted as a snaking line, or "bind," from top to bottom. The bind's color is selected from a Fisk-inspired color palette.

We also borrowed the color palette of the Fisk diagram but muted the colors, in the hope that this visualization of trauma would not inadvertently become too beautiful. We represented each voyage as a snaking line, its color randomly selected from our Fisk-inspired palette (FIGURE 1.9B). Shiyao Li, the primary computer science researcher on our team, had the idea to call these rivulets "binds," since what they represented was not the freedom of movement that comes with flowing water, but its opposite: a form of bondage often described as "living death."[47]

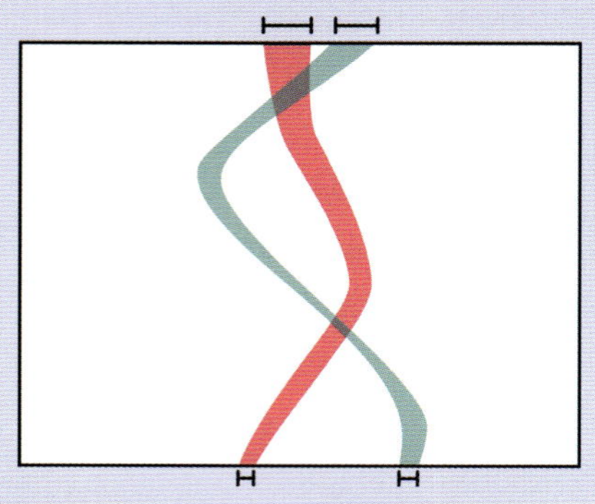

(C) The width at the top of each bind indicates how many captives were taken from Africa on each voyage. The width at the bottom indicates how many survived the voyage.

In the final visualization, the width at the top of each bind corresponds to the number of captives who were taken from any location in Africa. The width at the bottom corresponds to the number who arrived in the Americas having survived (FIGURE 1.9C).

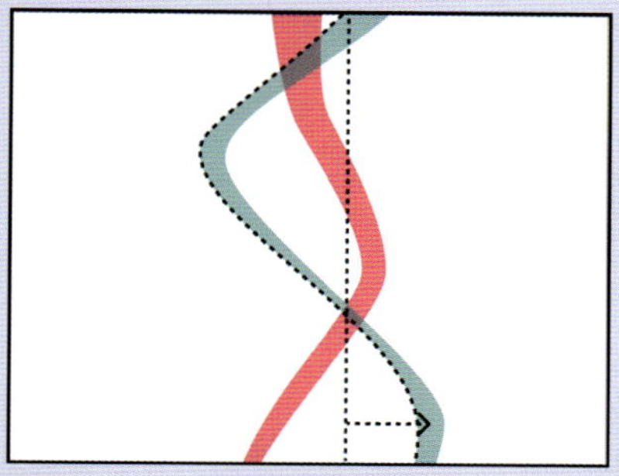

(D) The duration of each voyage is conveyed through the amplitude of each bind, plotted side to side.

The duration of each voyage is conveyed through the amplitude of each bind, but plotted from side to side. Encoded in this way, the longer voyages have wider sideways curves, and the shorter voyages narrower and more direct paths (FIGURE 1.9D).

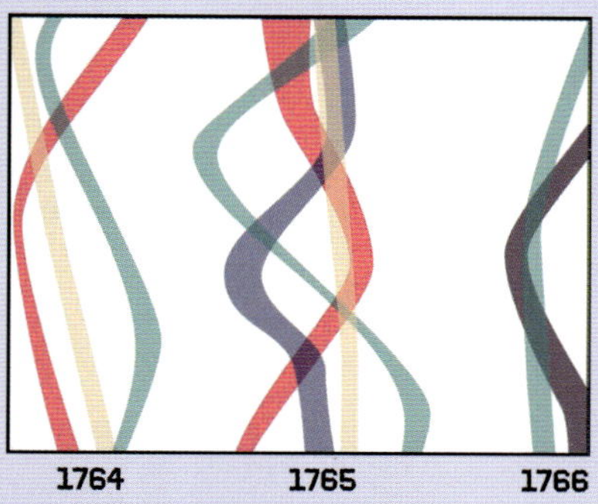

(E) The voyages are arranged chronologically from left to right, starting with the first recorded act of resistance aboard a slave ship, in 1565, and ending with the last in 1865.

Diagrams by Tanvi Sharma.

While Fisk's original design superimposes the floodplains of the Mississippi from all points in time on a single image, we chose to retain the start and end date of each voyage, since the rise and fall of the slave trade—and the resistance that met it throughout—seemed like a crucial feature of the data to convey. In our visualization, the binds are arranged chronologically, from the first recorded act of resistance aboard a slave ship in 1565, on the far left, through the last in 1865, on the far right (FIGURE 1.9E).

→

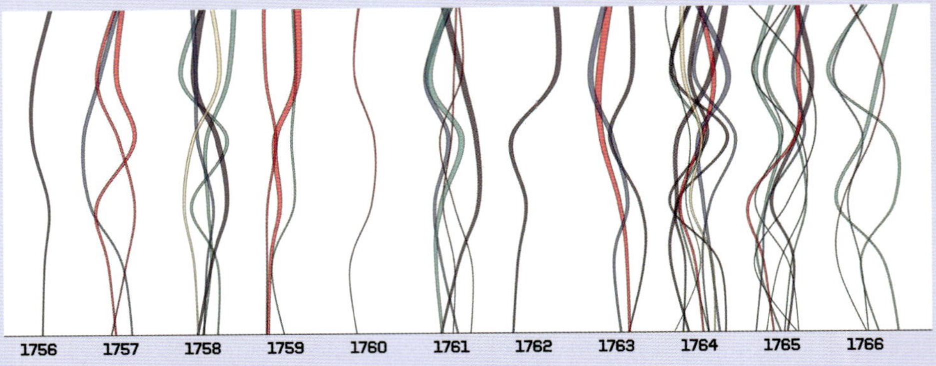

FIGURE 1.10
The interactive visualization as it appears in the web version of *Data by Design*, showing the decade between 1756 and 1766, when Equiano was enslaved.

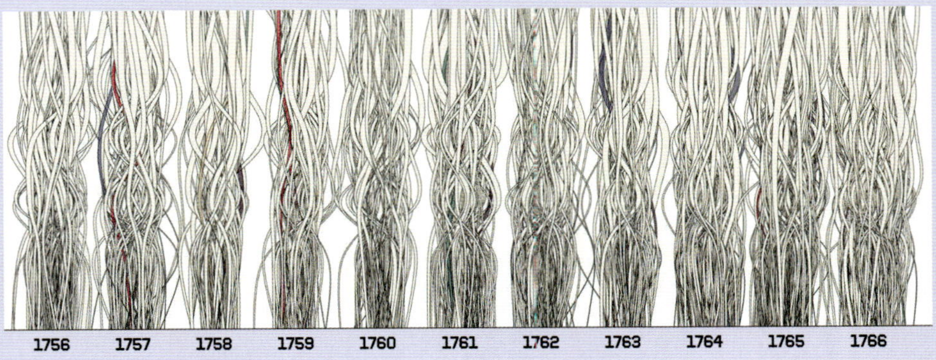

FIGURE 1.11
The same visualization showing all of the slaving voyages that took place between 1756 and 1766. The “resistance” voyages are shaded with a color fill, while the voyages without documented acts of resistance are displayed only in outline. In this state, the “resistance” voyages appear as life-affirming arteries within the sinews of human flesh.

FIGURE 1.12
The same visualization of the slaving voyages. Here, all of the voyages are shaded with a color fill, underscoring how resistance took many forms, some of which were recorded as data and some of which were not.

Visualization design by Tanvi Sharma. Implementation by Shiyao Li and Jay Varner.

Consider the time span between 1756 and 1766, the decade during which Olaudah Equiano was enslaved (FIGURE 1.10). Within each year, we can see that the binds overlap—what is called "occlusion" in visualization design. While occlusion is generally viewed as a problem and something to avoid, we made the decision to not further separate the voyages, since viewing them together communicates their collective force.

The voyage that took Equiano from Benin to Barbados and on to Virginia is not pictured in this chart, however, because it did not include a form of resistance that was documented in the dataset. But it might have been included among the 35,578 additional voyages that the dataset contains. This was our inspiration to plot these voyages alongside the "resistance" voyages, further troubling the distinction between the voyages that included documented acts of resistance and those that did not (FIGURE 1.11). With these additional voyages plotted, distinguished from the others by their lack of color fill, we observed a visual transformation that we had not intended, but that aligned with our conceptual goals: the binds were transformed into life-affirming arteries, enclosed within the sinews of human flesh.[48]

As evocative as this visualization may be, there are many other forms of resistance that it still does not represent. In his autobiography, for example, Equiano recalls observing acts of resistance that were set in motion but ultimately "prevented by the ship's crew." These acts of resistance likely had concrete effects, both for the crew which—in their need to suppress them—perhaps also attuned them to their own complicity; and for the captives, who perhaps might have taken heart in—or inspiration from—the possibility of pushing back against their likely fate. But as unfinished acts of resistance, they would have gone unrecorded in the dataset, as they did not meet its criteria for inclusion. We might also consider the myriad smaller acts of resistance, including those in which Equiano himself engaged. Upon first being captured, for example, Equiano refused to eat; and when his captor attempted to rename him, depriving him of his identity and his Ibo roots, he explains, "I refused to answer to my new name."[49] These more "quiet" forms of resistance, as literary scholar Kevin Quashie might term them, were also real, and also meaningful, even as they remain difficult to represent as data at all.[50]

In our final view, we decided to leverage the power of visualization to amplify these quiet acts (FIGURE 1.12). We display all 36,150 voyages with a color fill, selected from the same color palette as the visualization of only the resistance voyages. The implication of this choice, we hope, is clear: that every single journey documented in the Voyages Database involved acts of resistance—some that were recorded as "resistance" voyages, some that took place but went unrecorded as "resistance," and some that defied recording at all. What we were visualizing in the end, our process allowed us to see, was not actually the Middle Passage, but the data it had left in its wake.

In order to underscore this point, we drew from an array of common visualization techniques. But we did not use them in common ways. For example, instead of providing an overview first, and then "details on demand," our visualization inverts this process. We begin with the details of the "resistance" voyages, but as the additional voyages are layered in, the viewer can no longer see clearly enough to deduce anything about a specific voyage. This is intentional, because the insight that we seek to prompt exceeds the data on display. Our insight is this: that there are certain phenomena, such as the trauma of the Middle Passage, that we—in the present—can never fully understand (FIGURE 1.13).

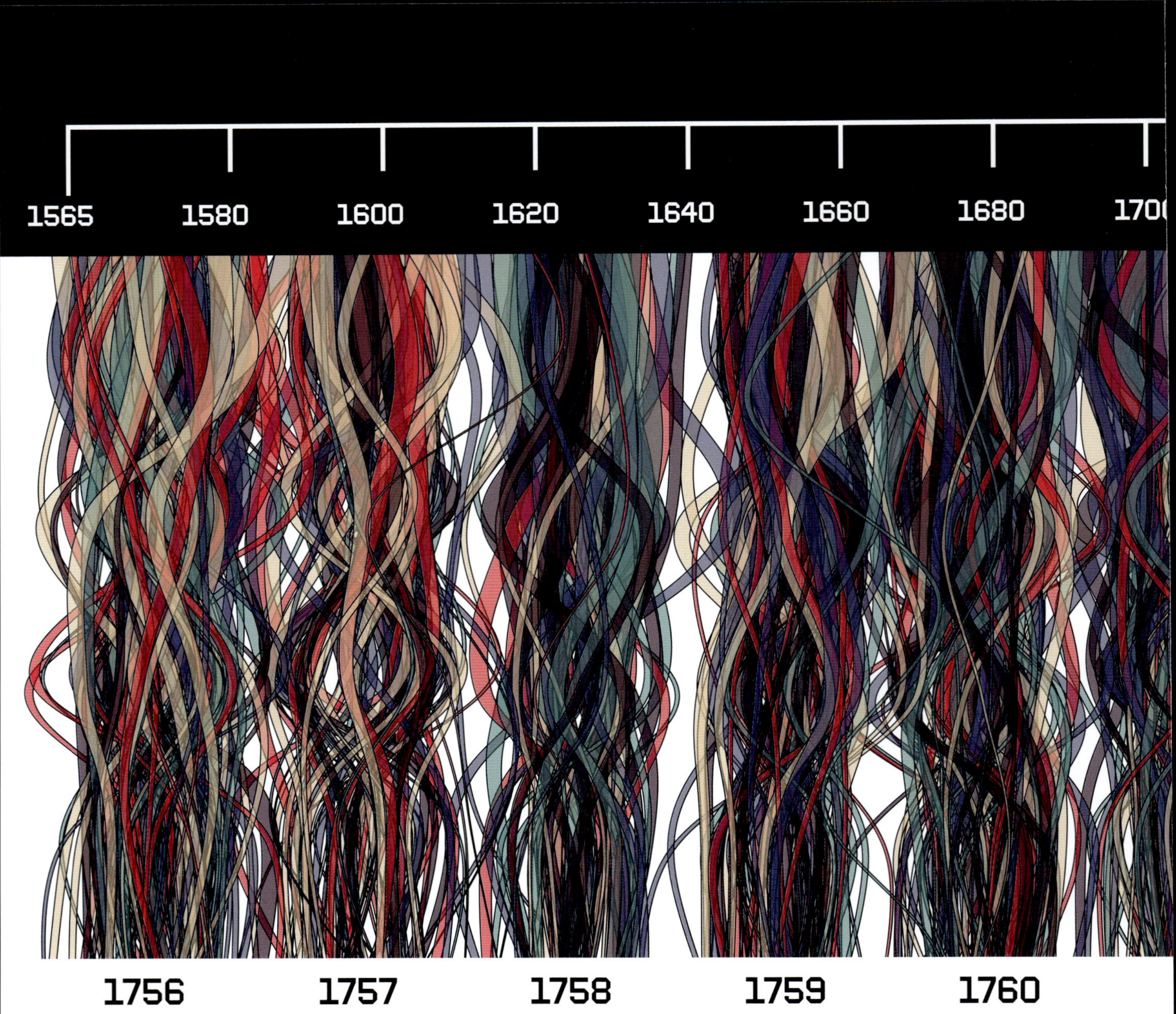

FIGURE 1.13
The final interactive phase of the visualization that appears in the web version of *Data by Design*. The user can adjust a slider to see the slaving voyages that took place during any decade between 1565 and 1865. Design by Tanvi Sharma. Implementation by Shiyao Li and Jay Varner.

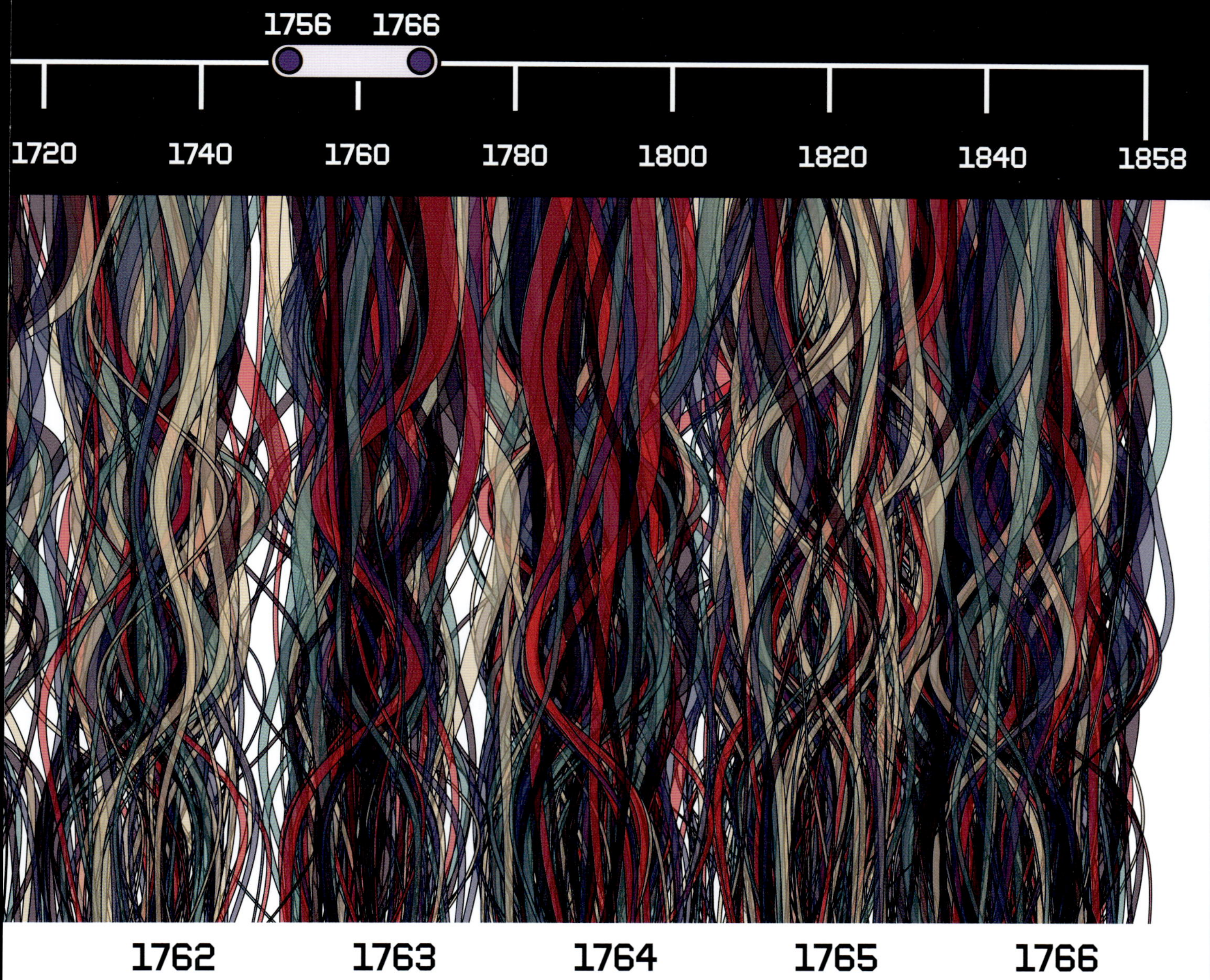
1756
1766
1720
1740
1760
1780
1800
1820
1840
1858
1762
1763
1764
1765
1766

FIGURE 1.14
A diagram of the British and American abolitionist movement, depicted as a network of rivers, created by Thomas Clarkson for his book *The History of the Rise, Progress, and Accomplishment of the Abolition of the African Slave-Trade by the British Parliament* (1808). Image courtesy of Wikimedia Commons.

The Alternate Streams of Data Visualization

While it would take another nine years for the British Parliament to formally abolish the slave trade—and, according to the Voyages Database, an additional 1.5 million people or more forcibly separated from their homelands—historians generally credit "Description of a Slave Ship," and Thomas Clarkson in particular, for playing a significant part in the campaign's success.[51] Clarkson himself likely contributed to this narrative; in 1808, one year after the formal Act of Parliament, he published a celebratory two-volume tome, *The History of the Rise, Progress, and Accomplishment of the Abolition of the Slave Trade by the British Parliament*, which provided his own firsthand account of British abolition.[52]

Interspersed among the *History*'s nearly 1,200 pages were four images. Among them were a revised version of "Description of a Slave Ship" and a new visualization that Clarkson himself designed (FIGURE 1.14). It depicts the history of abolition in the form of a watery network. Each of the "springs and rivulets" is labeled with the name of a significant abolitionist. The streams are arranged from top to bottom, roughly according to time, with horizontal lines separating efforts that took place before 1650, 1700, 1740, and 1787, respectively. As the viewer follows the streams down the page, they converge into "two great political rivers, representing the abolitionist movement in England and America," which, while outside of the bounds of the page, presumably meet somewhere in the Atlantic.[53]

On the surface, the metaphor of abolition as a network of streams and rivers flowing into a common sea may seem like an apt representation of a social movement which drew strength from countless individuals and collectives over an extended period of time. This resonance must have been intentional, given how Clarkson elsewhere analogized the British antislavery movement to the distributed functioning of a human body, with "every limb ... essentially necessary for the completion of a perfect work."[54] And yet, Clarkson's sense of "completion" is irrevocably flawed: certain key contributors are missing from his chart. Most notably, the name of Olaudah Equiano, or of any other Black abolitionist, is nowhere to be found.

As with "Description of a Slave Ship," the question of audience returns to the fore. This chart was also not designed for those who were enslaved, but rather for a white British viewership—those who sought to valorize their own role in the fight to end the slave trade. In his analysis of the image, Marcus Wood speculates that this goal—and its disconnection from the actual facts—is the source of the confusion that the visualization ultimately imparts. Clarkson's "attempt to create a self-sufficient cultural history for the cause requires a coherent descriptive model which is capable of overriding the heterogeneity of the names and writings he has furnished" in the text of his volume, Wood asserts.[55] Indeed, Clarkson's account overrides the same heterogeneity of the figures involved in the abolitionist movement overall. It follows, then, that in his diagram, the success of the movement "belong[s] to no single abolitionist but to a mysterious sea, into which, in a strange reversal of the generative metaphor, all the tributaries pour."[56] The sea metaphor is further clouded by the role that actual water—namely, the Atlantic Ocean—played in the slave trade. Why would Clarkson use water—the very site of the dehumanization that was required of the condition of enslavement—as the anchoring metaphor of his chart?

As it turns out, Clarkson was anchoring his diagram in some stable ground: namely, the then-prevalent

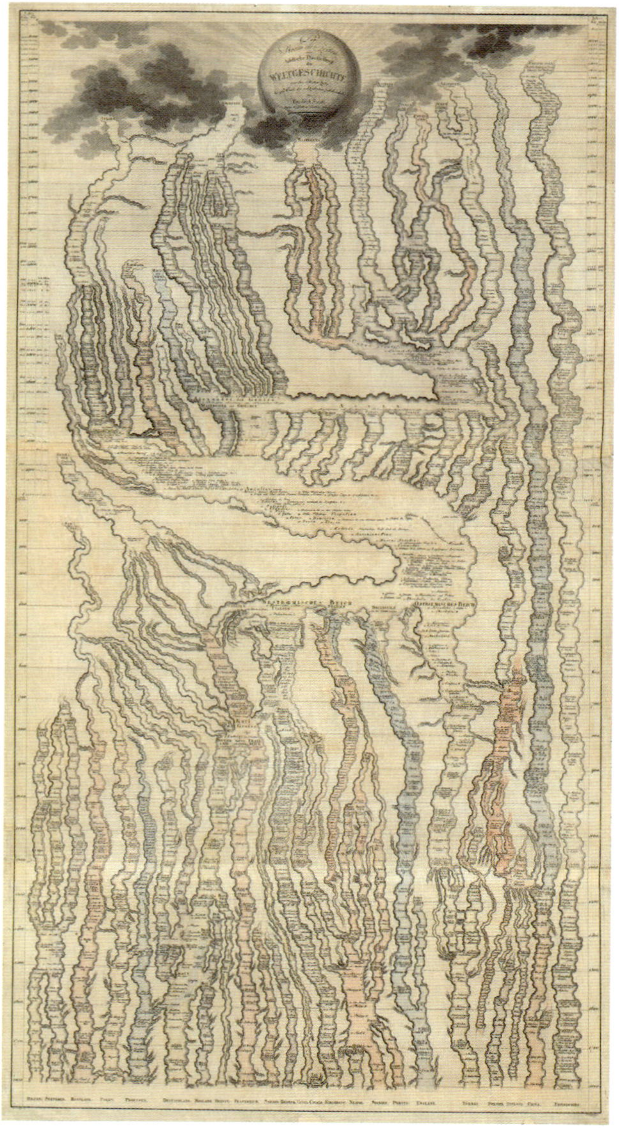

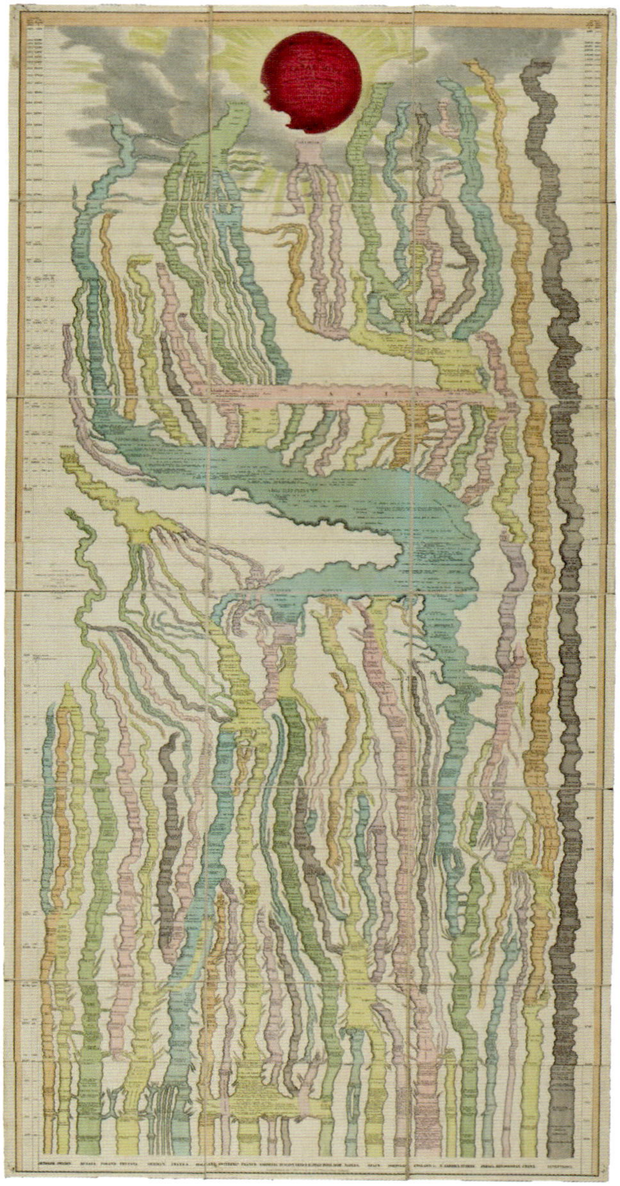

FIGURE 1.15
An early and influential "stream diagram" created by Friedrich Strass in 1803. The diagram was translated into multiple languages and became a model, in both Europe and the United States, for how chronology might be depicted. Image courtesy of the David Rumsey Map Collection, Cartography Associates.

FIGURE 1.16
An English translation of Strass's "Strom der Zeiten," published in 1849 by William Bell. Image courtesy of Cotsen Children's Collection, Department of Rare Books and Special Collections, Princeton University.

use of water and streams to visually represent the passage of time. In *Cartographies of Time*, Daniel Rosenberg and Anthony Grafton position Clarkson's diagram as a direct adaptation of the ideas expressed by the German chronologer Friedrich Strass, who had published his own widely circulated "Strom der Zeiten" (Stream of Time) only a few years earlier, in 1803 (FIGURE 1.15).[57] Translated into multiple languages, including English, and widely circulated across Europe and the United States, "Strom der Zeiten" was almost certainly a chart that Clarkson saw. Even if he did not, Clarkson clearly intuited how the water metaphor "gives greater liveliness to the ideas, and impresses events more forcibly on the mind, than the stiff regularity of the straight line," as William Bell, the English translator of Strass's chart, explains (FIGURE 1.16).[58]

Indeed, there is a version of the history of data visualization that, without too much reconfiguring, provides Clarkson with a more prominent place. Many scholars, including Rosenberg and Grafton, have made the case that chronological charts such as those created by Strass and Bell, as well as their more linear antecedents, "cleared the way for statistical graphics" by introducing the idea of consistent scale.[59] This argument is confirmed by the purported "pioneer" of statistical graphics himself, William Playfair, the subject of the next chapter, who praised the genre for "making space represent time," and for using "a line of proportional length and in a suitable position."[60] As we will learn, Playfair described the purpose of his own visualizations in terms quite similar to Clarkson's: to unite "a number of separate ideas ... under one simple impression of vision, and consequently, one act of memory."[61]

But to simply slot Clarkson into the standard history of the field would miss this chapter's main point. For Clarkson's visualizations, both "Description of a Slave Ship" and his map of abolition, matter for the questions as much as the insights they prompt. When we ask about the perspectives of the people for whom these charts have been designed, the people who will benefit from looking at them, and the people who are merely looked at, we begin to see whose data is truly on view in any particular chart. And when we consider that additional context, and the potential for violence or harm, that a visualization might, however inadvertently, bring about, we come to see the responsibility that we hold as viewers of data visualizations—and as designers of them as well. If these considerations seem weighty, it is because they are. Visualizations of data are indeed powerful, as both the political efficacy and emotional charge of "Description of a Slave Ship" affirm. But it is precisely because of this power that we must continue to probe the nature of the insights that visualization can prompt, and each of our roles in bringing them to light.

We will never know if Equiano posed questions like these to Clarkson upon seeing the original "Plan"; there is no record of the contents of the conversation that transpired. Several weeks later, however, Equiano published a letter in *The Public Advertiser*, a prominent London newspaper, in which he acknowledged "having seen" the Plymouth Committee's chart.[62] Interestingly, Equiano does not comment directly on the image, choosing instead to affirm the work of the abolitionist movement overall. But I believe that we can still learn from Equiano's statement about what he saw. In the diagram, Equiano saw clear evidence that its designers sought to "contribute to so important a moral and religious duty as that of" ending the slave trade. But he also saw that the chart alone would not be enough. As a person who had himself been enslaved, Equiano understood firsthand that ending "one of the greatest evils now existing on earth" would require more than the chart alone could convey.

With this statement we return, for a final time, to this chapter's epigraph—about how before there is data, there are people—people with lives that data alone cannot convey. With the story of Equiano lending context to this claim, we can now also see how, if the goal of any particular visualization is to bring about change, then it must necessarily be accompanied by action. We will return to this topic in the chapters to come, as we consider the ways that design processes (chapter 3) and design outputs (chapter 5) can better serve the people whose struggles for justice and dignity we seek to support. Here, however, we—the authors of this chapter—seek to make a broader point. Not all visualizations are designed with a goal of liberation in mind, and that is perfectly fine. In fact, most visualizations are designed with a more focused goal of eliciting insight about a specific dataset, or through a specific visual or interactive form. Placing the SEAST diagrams at the center of the story we tell about the emergence of modern data visualization matters in these contexts as well—and perhaps even more. For the SEAST diagrams do not ask but *demand* that we recognize the responsibility that comes with designing and viewing data visualizations. By attempting to picture the most monumental of humanity's stakes, the SEAST diagrams serve as a constant reminder to always ask about the forms of knowledge that a visualization can illuminate—and what it cannot show.

Takeaways

FOR VIEWERS

Reckon with the power of visualization

Ask for and by whom the visualization was designed

Consider the context of the data being visualized

Remember that there is always more than data alone can show

FOR DESIGNERS

Balance abstraction and detail

Ask who may benefit and who may be harmed

Probe any missing data and ask why it is missing

Own your responsibility for framing what we can know

CHAPTER 2

IMAGE

William Playfair and the Politics of the Simple View

Visualization has never been neutral or objective.
How can we learn to recognize the decisions that shape each design?

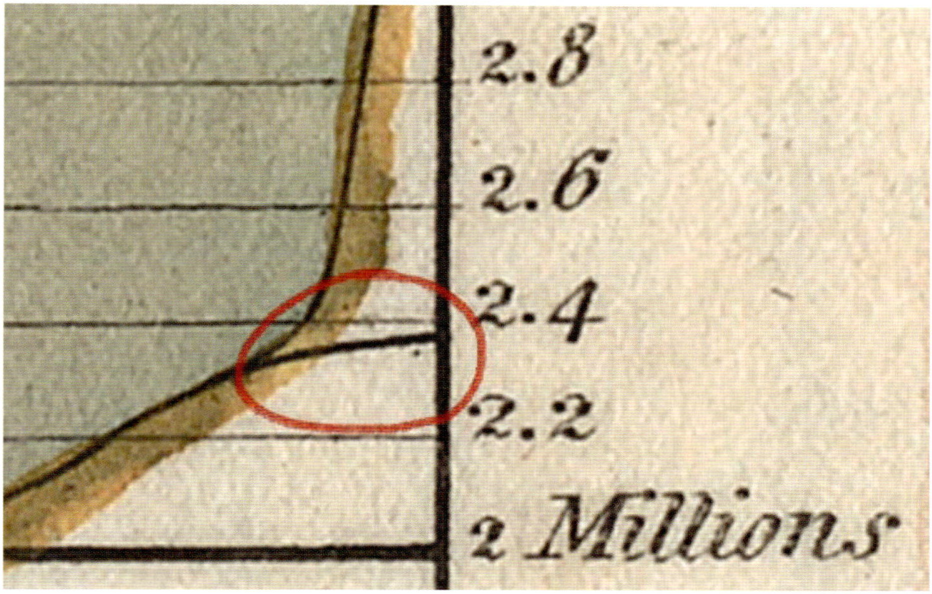

FIGURE 2.1, DETAIL
William Playfair's chart of "Exports & Imports to and from all North-America," published in the third edition of the *Commercial and Political Atlas* (1801), showing an engraving error, likely the result of a tired or sweaty hand. Image courtesy of the Library Company of Philadelphia.

We can only imagine the choice words William Playfair exclaimed when he realized the error that he had engraved into his most recent chart, "Exports & Imports to and from all North-America" (FIGURE 2.1). It was 1801, 12 years since SEAST first published "Plan of an African Ship's Lower Deck" and 15 since Playfair had published the first edition of *The Commercial and Political Atlas*, in which an early version of his own chart first appeared. Engraving was then, as now, an incredibly time-consuming process. Albrecht Dürer, the Renaissance printmaker credited with elevating engraving to an art form, took over three months to complete his famed *Knight, Death, and Devil* (1513), a print not much larger than an iPad. In the case of Playfair, however, it was not merely the time he had invested in producing the 28 plates for the third edition of his *Atlas*; it was also the expense.

Today, Playfair, who lived from 1759 to 1823, is widely celebrated for his leading role in the development of modern data visualization. His bar charts (FIGURE 2.3; compare FIGURE 2.2), pie charts (FIGURE 2.4), and time series charts (FIGURE 2.1) are often heralded as among the earliest of their kind.[1] Ever since they were "discovered" in the 1860s by the British economist William Stanley Jevons, Playfair's charts have consistently been prominently featured—and consistently fulsomely praised—in accounts of the field's emergence.[2] But in his own time, Playfair remained "largely unacknowledged" for his innovations.[3] More to the point, he was almost always nearly broke.[4] So while Playfair chose to commission one of the most skilled engravers in all of London, Samuel John Neele, to produce the plates for his book, he also likely requested that Neele work at speed so as to minimize the costly detailing and other flourishes for which he was known. It is hypothesized that Neele engraved the charts' decoration, framing, titles, and other lettering, leaving Playfair—who had previously trained as an engineer—to engrave the lines of imports and exports by himself.[5]

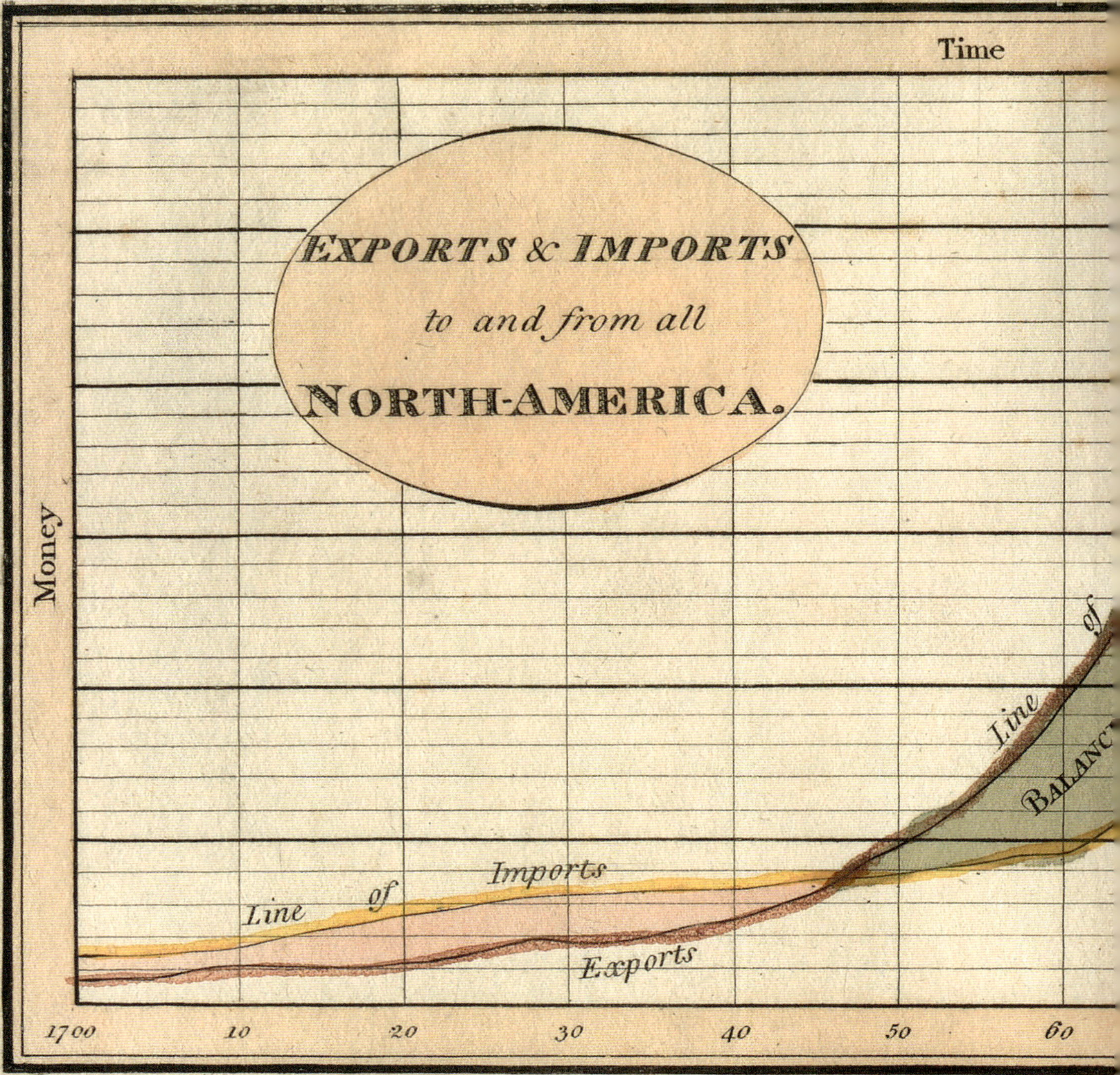
Time
EXPORTS & IMPORTS
to and from all
NORTH-AMERICA.
Money
Line of Imports
Exports
Line of
BALANC
1700
10
20
30
40
50
60

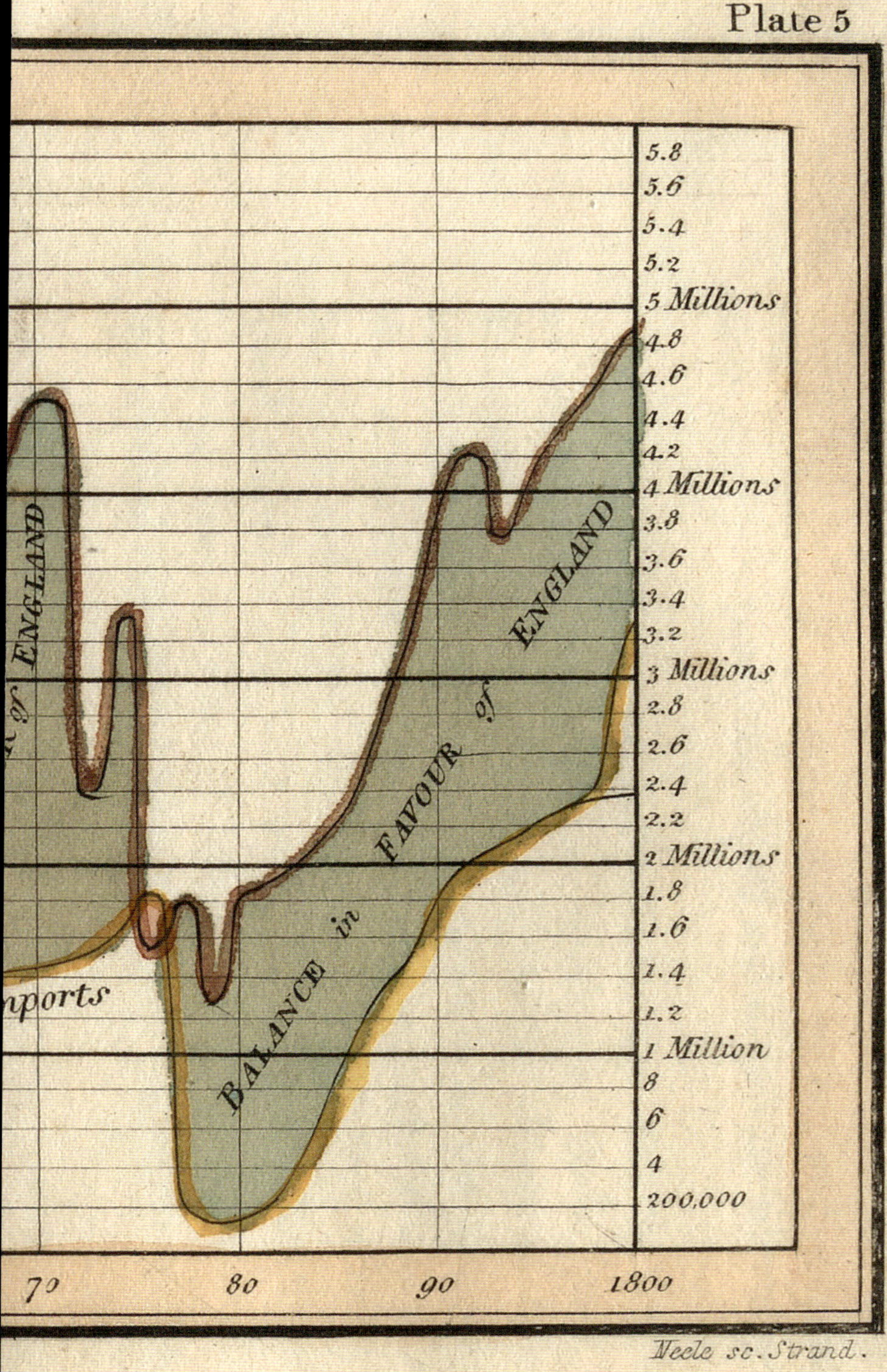

FIGURE 2.1
William Playfair's chart of "Exports & Imports to and from all North-America," published in the third edition of the *Commercial and Political Atlas* (1801). Image courtesy of the Library Company of Philadelphia.

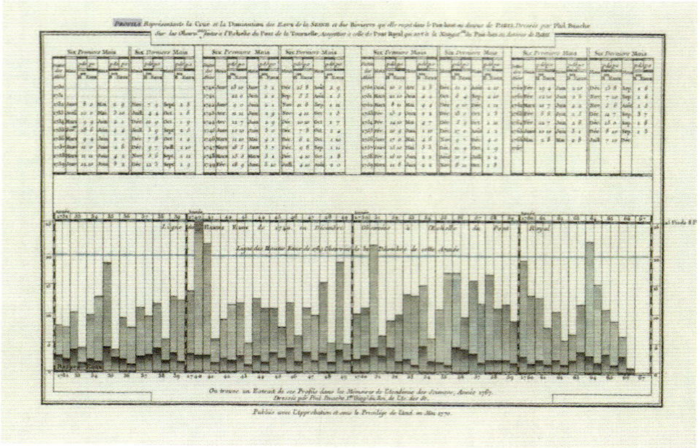

FIGURE 2.2
A bar chart from 1770 created by the French mapmaking team of Philippe Buache and Guillaume de L'Isle. Image from the David Rumsey Map Collection, courtesy Stanford University Libraries.

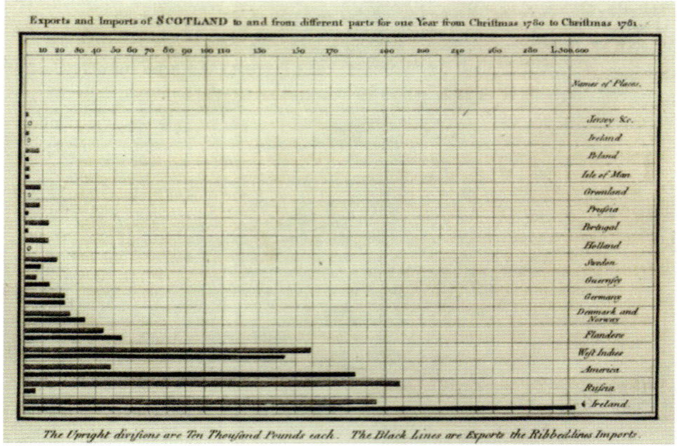

FIGURE 2.3
The bar chart included in the first edition of Playfair's *Commercial and Political Atlas* (1786), also considered one of the first examples of the bar chart form. Image courtesy of Wikimedia Commons.

To produce engravings like Playfair's, a thin copper plate is first coated with a "ground." Typically, this is a layer of wax or varnish, chalk, or soot. Using a stylus, the engraver traces an outline of the design in mirror image into the ground. The wax (or equivalent) layer is then removed, but a faint impression remains, which the engraver then uses as a guide to carve the image into the copper plate. The image is engraved with a metal tool called a burin, which, somewhat counterintuitively, is held still while the engraver rotates the underlying plate.[6] Playfair's error was thus a common one, a slip of a tired hand. It wouldn't even be very noticeable once the colored paint was overlayed. But neither of those excuses would have made it more tolerable to the man who was already, by his own account, "long anxious" to be acknowledged for the "invention" of what we now call data visualization.[7] Unlike the array of software platforms and programming libraries used to create visualizations today, each of which allow for infinite revision—at least until the ship date—the engraving process employed by Playfair resulted in an image that was irreversibly inscribed into copper. Considering the time and money also invested in the work, it might as well have been set in proverbial stone.

This chapter takes up the processes involved in transforming data into image, both material and conceptual, in order to continue our exploration of the relationship between data and its visual display. For everyday viewers, it can be tempting to assume that any particular visualization—or, at least, any good one—offers a direct representation of the data underneath; that its visual form is neutral; and that what we're looking at reflects an objective view of the data at hand. In the interest of full disclosure, I will admit that I, a person who should know better, still sometimes find myself seduced by the "aura of objectivity" that visualizations so often convey.[8] But that feeling of objectivity is often carefully engineered, as the SEAST ship diagrams of chapter 1 have begun to show. And those assumptions about transparency and neutrality are only just that—assumptions. Turning now to Playfair's time-series charts, so often upheld as the epitome of neutral and objective form, we will together come to see how human hands—and human decisions—shape the

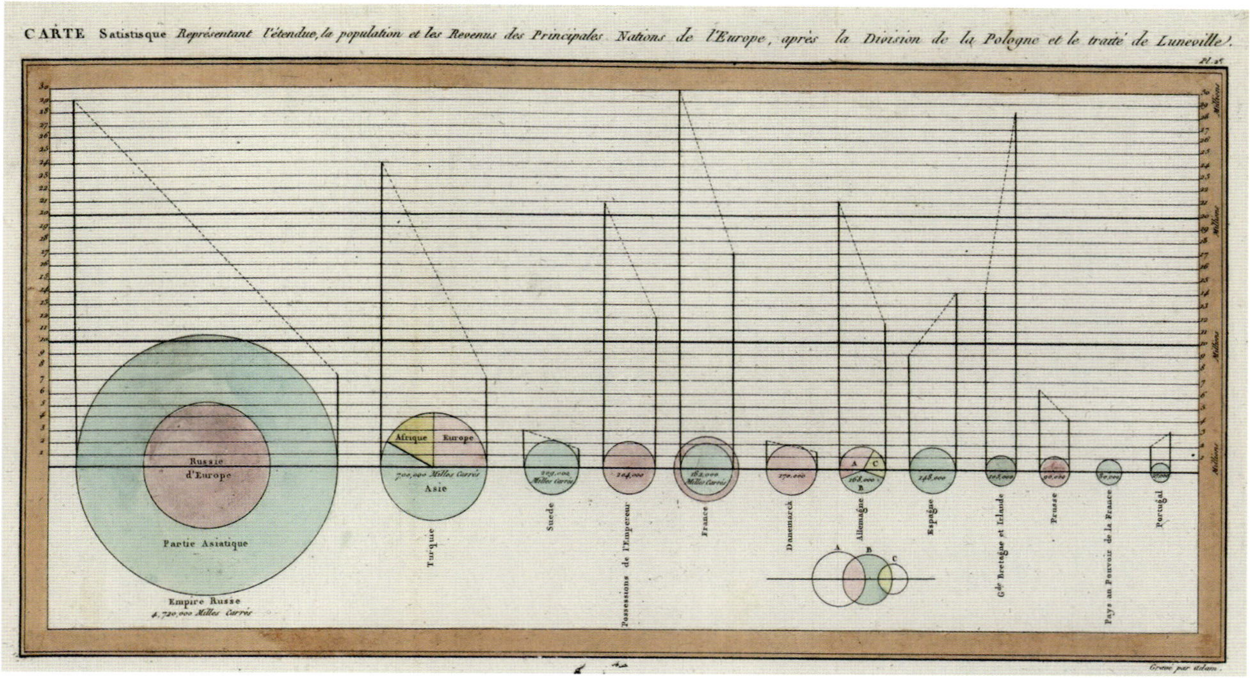

FIGURE 2.4
"Chart Representing the Extent, Population & Revenues, of the Principal Nations in Europe, after the Division of Poland & Treaty of Luneville," from the *Statistical Breviary*, French edition (1802). Image from the David Rumsey Map Collection, courtesy Stanford University Libraries.

design of each and every chart. We will further confirm how the act of visualizing data, in Playfair's time as today, carries with it the power and peril of presenting these human decisions as unassailable, objective fact.

By examining the material processes by which Playfair created his charts, together with his writing, this chapter will surface the range of decisions that contributed to the charts in their final form. More than that, this chapter will show how these decisions reflected Playfair's personal view about the value of visualization and the people who, he believed, could best benefit from it. Very crucially, this perspective on the uses of visualization does not invalidate the insights that Playfair's charts so clearly and efficiently convey. On the contrary, it opens up to a lesson more profound: about how the practice of visualization is always shaped by specific people, their tools and processes, and the ideas that underlie them, just as much as it is by the specific dataset being given visual form.[9] With our increased knowledge about Playfair's ideas and processes, his charts become all the more meaningful for the history of the field. For they not only exemplify the power of visualization to elicit fresh insight from data, but also affirm its power to shape the production of knowledge itself.

The Value of Visualization

The charts that William Playfair created for his *Commercial and Political Atlas* have long appeared to speak for themselves. Their bold data lines, enhanced

Imports and Exports to and from NORTH AMERICA from 1700 to 1780.

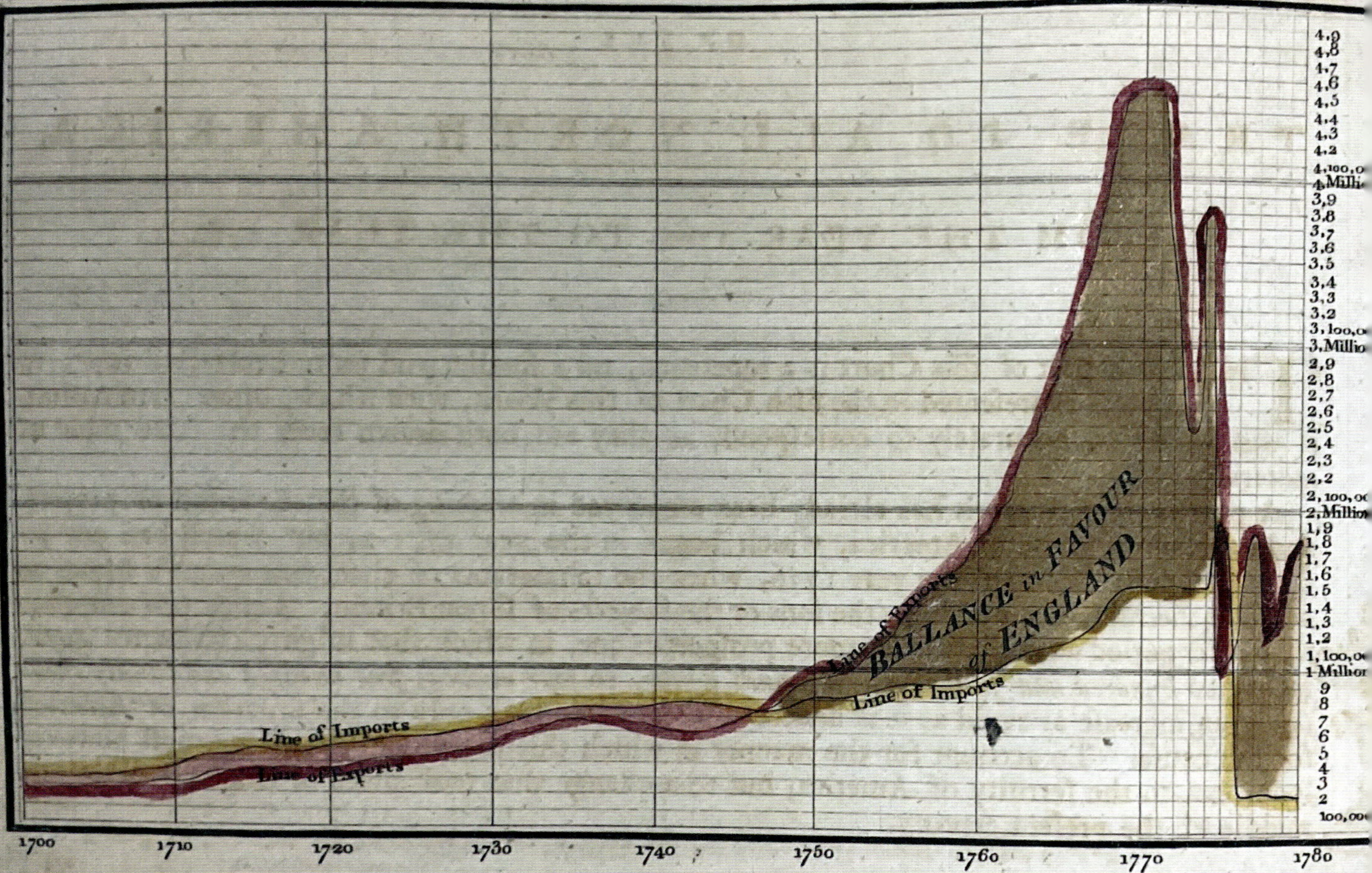

The Bottom Line is divided into Years that on the Right hand into £100,000 each.

Published as the Act directs, 1st May 1786, by Wm Playfair.

Neele sculpt 352 Strand, London.

by hand-tinted color fills and set against sharp black gridlines, suggest nothing less than statistical certainty. But Playfair did not intend to include his charts' underlying data in the *Atlas*. In his mind, his charts stood on their own. It was only after soliciting feedback from James Watt, inventor of the steam engine—and a man for whom Playfair worked in his youth—that he decided to render the data associated with his charts in tabular form. "It might be proper," Watt advised, "to give in letter press the Tables from which the Charts have been constructed ... for the charts now seem to rest on your own authority, and it will naturally be enquired from whence you have derived your intelligence."[10]

Watt's concern reveals what Playfair's charts conceal: they are not transparent windows into economic or political truths. Rather, they are visual interpretations of overall trends—trends that require statistical validation in order to be confirmed. In response to Watt's comments, Playfair thus dutifully compiled data tables to accompany each of his charts, which documented the figures he'd "derived from the papers and books" of the London custom house, and he included these as appendices in the first and second editions of the book.[11]

But by the book's third edition, Playfair had gained enough confidence in the form and function of his charts that he no longer felt obligated to include the associated data tables, as Watt had initially advised. Having reflected on the value of his contributions in the 15 years since the book's first edition, Playfair had come to see the function of his charts as quite distinct from that of the associated data, or "figures," as he termed them.[12] In the introduction to the third edition, he explains:

> The advantage proposed by this method, is not that of giving a more accurate statement than by figures, but it is to give a more simple and permanent idea of the gradual progress and comparative amounts, at different periods, by presenting to the eye a figure, the proportions of which correspond with the amount of the sums intended to be expressed.[13]

In his explanation—which was required because so few of his readers had ever seen a data visualization before—Playfair is explicit about the fact that his charts are "not more accurate" than his data tables. Rather, he understands the value of his charts as offering a different contribution: the ability to impart a "more simple and permanent idea." As in Thomas Clarkson's writing, discussed in chapter 1, here we also see the strong influence of Enlightenment empiricism.[14] Playfair asserts that his charts will present an image "to the eye," in keeping with the Lockean view of how sensory experience—and vision in particular—serves as the substrate from which knowledge is formed. This image, as he further explains, is one that can then be processed by the mind into a "simple, accurate, and permanent idea."[15]

Throughout the introduction, Playfair continues to employ language that echoes Clarkson's and the empiricist influences they shared. Here we might recall Clarkson's stated aim of creating an "immediate impression." Similarly, Playfair expresses his hope that "a sufficiently distinct impression will be made, to remain unimpaired for a time, and the idea which does remain will be simple and complete."[16] This statement encapsulates Playfair's ideas about how visualization leads to knowledge—in short, his epistemological theory. For Playfair, visualization prompts insights that are "distinct" from the insights that derive from scanning data tables, or from the statistical analysis of same. The knowledge that visualization leads toward is "more simple," but it is also easier to understand. As a result, it is easier to remember—and valuable in its own right.[17]

	PLATE V. AMERICA.		
Years.	Imports.	Exports.	Balance.
1770,	1,480,000	4,550,000	3,070,000
1771,	1,430,000	4,630,000	3,200,000
1772,	1,445,000	3,600,000	2,155,000
1773,	1,465,000	2,465,000	1,000,000
1774,	1,435,000	3,840,000	2,300,000
1775,	2,065,000	985,000	1,080,000
[illegible]	245,000	1,190,000	945,000
1777,	230,000	1,880,000	1,650,000
1778,	265,000	1,150,000	885,000
1779,	295,000	1,370,000	1,075,000
1780,	300,000	1,805,000	1,505,000
1781,	385,000	1,545,000	1,160,000
1782,	295,000	905,000	610,00[illegible]

FIGURES 2.6
The data table included in the second edition of *The Commercial and Political Atlas* (1787) that shows annual data for the years 1770 to 1782. Image courtesy of the Library Company of Philadelphia.

Playfair's goal of presenting a "simple and complete" idea, over and above any particular data point, is visible in the revisions that he made to "Exports & Imports to and from all North-America" for the third edition of the *Atlas*.[18] The tables in the first and second editions of the *Atlas* include annual data for the years between 1770 and 1782 **(FIGURE 2.6)**.

PLATE XI. ALL NORTH AMERICA.		
Years.	Exports.	Imports.
1700,	200,000	280,000
1710,	250,000	370,000
1720,	350,000	520,000
1730,	500,000	630,000
1740,	620,000	780,000
1750,	930,000	820,000
1760,	1,750,000	950,000
1770,	4,550,000	1,480,000
1780,	1,805,000	300,000

FIGURES 2.7
The data table that shows decennial data for the years 1700 to 1780. Image courtesy of the Library Company of Philadelphia.

For the years between 1700 and 1770, there is only data for each decade **(FIGURE 2.7)**.[19]

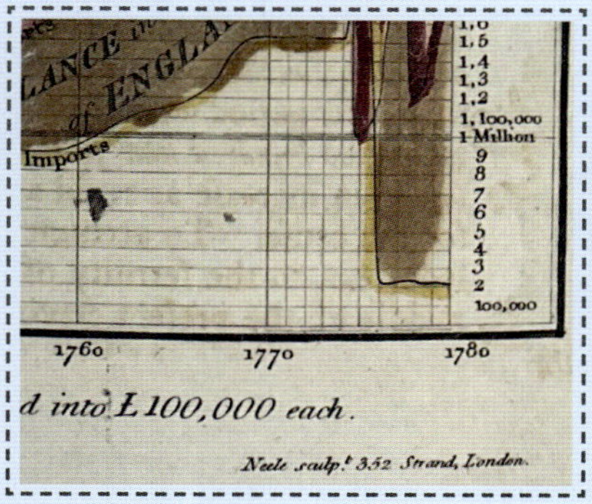

FIGURE 2.8A
Detail view of the first version of "Imports and Exports to and from North America from 1700 to 1780," showing the minor gridlines for the years between 1770 and 1782.

In the accompanying chart in the first and second editions of the *Atlas*, Playfair includes both major and minor gridlines along the *y*-axis of the chart (FIGURE 2.5). He includes minor gridlines along the *x*-axis only for the twelve years for which he possesses annual data (FIGURE 2.8A). Presumably, this indicates the greater granularity of those years' data.

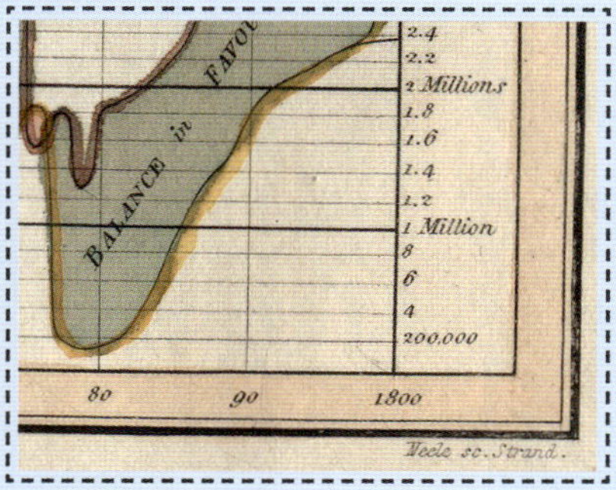

FIGURE 2.8B
A detail view of "Exports & Imports to and from All North-America" from the third edition, showing the minor gridlines removed.

In the third edition, these gridlines disappear—along with the data tables (FIGURE 2.8B). Playfair's revisions create an effect of continuity, consistent with the "more simple" view he sought to produce and in spite of the actual discontinuities that his two datasets contained.[20] He also extends the endpoint of the *x*-axis to 1800—to what was then the present. In addition, the data lines become smoother and less precise, either a reflection of his desire to convey a "more simple" idea, or of his improved engraving technique, or both.

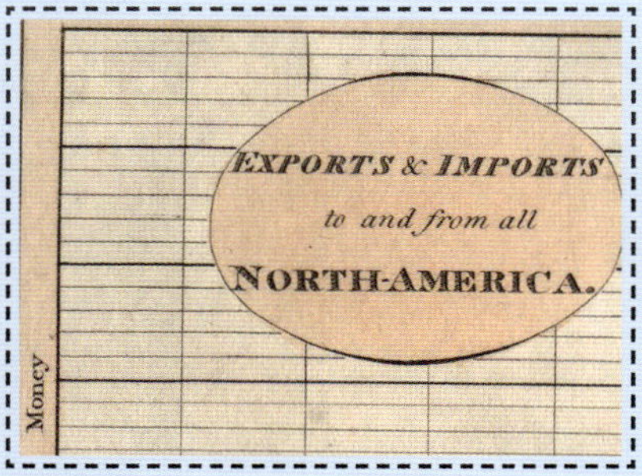

FIGURE 2.8C
A detail view of the chart from the third edition, showing the title likely engraved by Samuel John Neele.

Playfair also makes significant improvements to the chart's design. Some of these improvements can likely be credited to the master engraver, Neele, who was presumably the one to superimpose the chart's title on the otherwise blank space in the chart's top left (FIGURE 2.8C). Neele was also likely the one who engraved the labels on the axes, as the elegant cursive suggests. He or Playfair also modified the scale markers on the chart, which allowed them to remove the explanatory note below, further improving legibility.

→

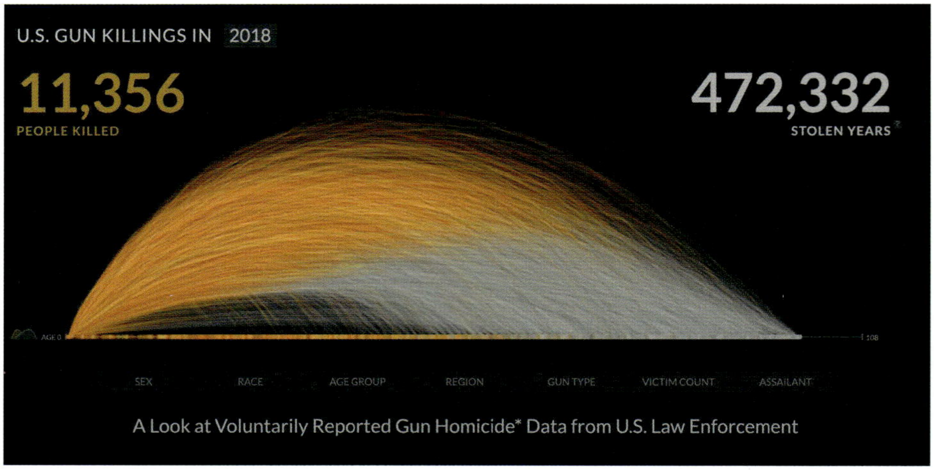

FIGURE 2.9
A screenshot of an animated visualization of the "stolen years" of people killed by guns in the United States in 2018, by the design firm Periscopic. Courtesy of Periscopic, periscopic.com.

Playfair was loath to accrue any unnecessary costs, as he had been embroiled in bankruptcy proceedings over the previous several years.[21] But he saw successful sales of the third edition of his *Atlas* as his most likely ticket back to financial solvency, so he replaced the less costly shading techniques that he had employed in the book's second edition with the hand-stained color that he had used in the first, and which had contributed to its initial acclaim. Each of these decisions, owing to a combination of personal preference, prior knowledge, and financial pressure, combined to create the effect of graphical authority that Playfair had envisioned from the start.

We can still marvel at Playfair's striking images—perhaps even more so after we recognize the range of circumstances that contributed to their final form. But we must also consider the limits of the visual epistemology that they encode. This is one that, in many ways, has been carried directly forward from Playfair to the present, in our continued belief that clarity and efficiency are what will lead us to the most objective knowledge.[22] But complete objectivity is an illusory goal. This is the reason that I find so much significance in the error engraved in Playfair's chart; it endures as a reminder of the person who created it, and the decisions that person—William Playfair—personally made. While our awareness of this error might detract from the simple view that Playfair sought to engineer, what we gain is more valuable: a more accurate understanding of what we are actually looking at—and more crucially, the reasons why.

The Politics of Playfair's Charts

Playfair created his charts at a time of intense political change, matched only by the intensity of his personal financial struggles. At the time that Playfair released the third edition of the *Atlas*, the French Revolution had only just come to an end, the result of a coup staged by Napoleon Bonaparte (who would himself go on to inspire Charles Minard's iconic flow map, discussed in this book's introduction). Playfair himself had lived in Paris for six years after the initial publication of the *Atlas*,

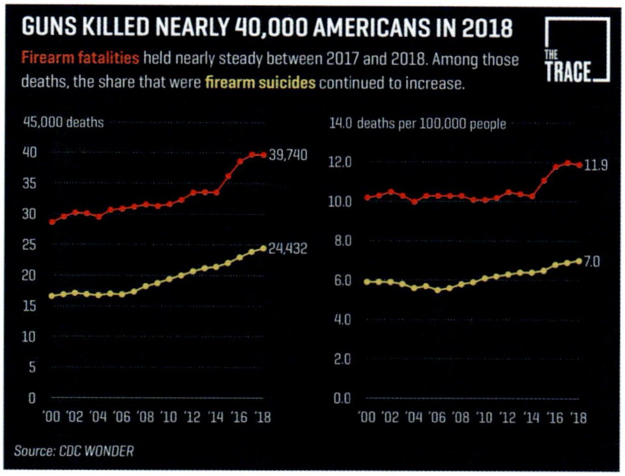

FIGURE 2.10
A pair of double line graphs showing the number of people killed by guns in the United States between 2000 and 2018, emphasizing the number of firearm suicides as a significant proportion of overall firearm fatalities. Chart by Daniel Nass for *The Trace*.

first riding high on his reputation—Louis XVI reportedly "expressed great satisfaction" in his charts—and then on the profits from a series of business schemes.[23] But in 1792, after a surge of anti-aristocratic violence, including an assault on a close friend that took place right outside his door, Playfair was forced to return to England—and to his writing career.[24]

Upon his return to writing, perhaps predictably, Playfair's finances began to decline.[25] With the exception of his 1795 *History of Jacobinism*, informed by his experience in France, Playfair's writing did not earn him the attention, or the profit, that he believed he deserved. A series of failed business ventures in the second half of the 1790s—most ignominiously, the loss of an entire bank—culminated in personal bankruptcy. Thus when Playfair explains, in the preface to the third edition, that he has "chosen the present moment" to rerelease his charts, he is referring to a moment of both political uncertainty and personal upheaval—the latter, by all accounts, more acutely felt.[26]

In fact, Playfair would go to debtor's prison shortly after the publication of the third edition of the *Atlas*. He would never see the book's proceeds, which would go to his creditors, as he explains in a court document submitted upon his release.[27] But the extent of his financial distress makes his continued faith in the power of data visualization all the more affecting. He still vehemently believed that his work could make an intervention on a global stage—a means of countering the instability that the age of revolutions had brought about. "The minds of men, the boundaries of nations, their laws and relations with one another, are all in a state of change," he writes in the third edition's preface.[28] As we have learned from this brief biography, Playfair was one to know.

Nevertheless, Playfair avoids further prognosticating in the text of the *Atlas*. He cannot (or will not) say whether "war and contention" or "art and commerce" will ultimately prevail.[29] Far from diminish the utility of his charts, however, Playfair sees this uncertainty as further underscoring their enduring value. As he explains:

> If [a future of war] turns out so, a picture of the past will be a valuable thing, if, on the contrary, commerce should still continue its progress, this will make the first part of a great whole, which, when completed on some future day, will be a most valuable work.[30]

From these lines, it would seem that Playfair thinks his charts can not only capture a clear "picture of the past" but also retain their significance in a range of possible future scenarios. Their simplicity, as it turns out, is a feature of the design and not a bug. It is by minimizing complexity—and not, per today's adage, visualizing it—that , according to Playfair, "valuable" and enduring knowledge is made.[31] Returning to the engraving error in "Exports & Imports in all of North America," it may well be that Playfair was not perturbed at all. The chart still communicated all the information that Playfair intended, which in the end was not dependent upon the accuracy of any particular line or data point but only a general trend.

Setting aside Playfair's feelings about the error, a pair of more portentous questions remain: for whom is this particular "picture of the past" truly valuable, and why must a "simple" picture be the one that remains? As Playfair continues to elaborate the impetus behind the "form and manner" of his charts, he makes clear that his intended audience is not "any person" in the world, but rather a narrower demographic: "men of high rank, or active business."[32] These men, he continues, "can only pay attention to general outlines; nor is attention to particulars of use."[33]

It is not an overstatement to say that when I first read these lines, I was stopped in my tracks. Before my own exploration of the *Atlas*, I'd been aware of many ways of visualizing data—from the most intentionally evocative (think Periscopic's animated visualization of gun deaths; FIGURE 2.9) to the most matter-of-fact (think of comparable data journalism by *The Trace*; FIGURE 2.10). Along this spectrum, I'd always assumed that unembellished line graphs were among the most neutral of possible visual forms. Of course I'd read the research.[34] I'd even contributed some of my own.[35] But I was not expecting William Playfair, of all people, to be the one to help me prove my point.

Playfair's explanation makes it incontrovertibly clear that line graphs have a politics. More than that, his writing reveals that he has made certain assumptions of his own. By designing his charts for elite men and other businesspeople, Playfair presumes that their concerns are not with complexity, or with personal impact. His own life experience had confirmed for him firsthand how, for the truly moneyed class, rank and resources consistently shielded them from any fallout from an unexpected turn of financial events. The knowledge that is recorded and visualized in Playfair's *Commercial and Political Atlas* is useful to them precisely because it is simple; because it allows people like them to ignore any details that might otherwise cloud their view.[36]

It is here that we might return to two strands of the argument made in chapter 1. First, the god trick. Whereas with Clarkson's chart, the god trick was deployed in order to enhance the persuasive force of his image, in Playfair's case it was wielded to *hide* the politics of his chart. We do not think to ask about his intended viewer because we are so seduced by the "aura of objectivity" that the chart conveys. This is, as Alberto Cairo explains, how objectivity can be weaponized—made to foreclose questions about the data provenance, design choices, intended viewership, and more.[37] But this is also how Playfair's charts, placed in their context, can open up to something wholly new: by showing how asking such questions results in a more accurate interpretation of what it is that we are looking at, as well as what forms of knowledge remain to be further probed.

Second, we might also return to our discussion of the tradeoff between abstraction and detail that was begun in chapter 1. Per Playfair's own discussion about the differences between the information conveyed through data and that conveyed through charts, he clearly considered this tradeoff as he prepared to publish the third edition of his *Atlas*, and came away on the side of abstraction. From our vantage in the present, however, we might also consider the detail—what, in the lines quoted above, Playfair describes as "the particulars"—that he abstracted away. Consulting disaggregated data sources on British trade for the same time period, from the ledger books of the London custom house, it becomes clear that the "particulars" to which Playfair chose not to give visual form are, in no small part, people: the enslaved human labor which the "balance of trade" of the British empire depended upon.[38]

Looking closely at these custom house ledgers, now digitized and made available online via the British National Archives, offers a sobering reminder of the violence brought about by reducing people to commodity form (FIGURE 2.11). Counts of people—one who was trafficked to Pocomoke, Maryland; 14 to Roanoke, Virginia; five to Charleston, South Carolina; and seven to Sunbury, Georgia—are recorded right alongside counts of mill sheets (a kind of paper) and gallons of train oil (a type of oil from whales). Put another way: it is not a coincidence that this is the second chapter, of two thus far, in which data tables such as these have appeared. The production of data is, in many ways, the logic of capitalism laid bare: the idea that people and goods can be easily exchanged when both are reduced to a price.

The intentional abstraction of Playfair's charts supports this easy exchange. And while some scholars have attempted to find room in his thinking for a more progressive view on slavery, more specifically, this view is not supported by the writing in the *Atlas*, nor by the sequence of its charts.[39] As evidence, consider the chart that directly precedes the chart of North America, the one for the West Indies (FIGURE 2.12). It is only in the final paragraph of the two-page explanation of the West Indies chart that Playfair connects the balance of trade that favors England, as depicted in the chart, to the "lives and freedom of the much injured, and wretched inhabitants of Africa."[40] In the much longer and more impassioned account of the chart of North American trade, Playfair does not mention slavery at all.[41]

For Playfair—and, it would seem, for the "men of high rank" for whom he has designed his charts—it is the failure of Great Britain to hold onto its American colonies, and the loss of profit and power that came with it, that is the more important insight to convey. The chart's design reflects this assessment, and explicitly so. "What numbers have been ruined, and how many more have been deprived of fortune, by our ill-conducted trade with America?," Playfair laments.[42] In contrast to the SEAST ship diagrams, there are no captive bodies here to illustrate this "ill-conducted trade." On the contrary, Playfair's brightly colored data lines emphasize the "numbers" ruined and the "fortune" lost. The exclusion of "particulars"—of *people*—is what makes the chart's interpretation quick and "easy." Thinking back to chapter 1, if "Plan of an African Ship's Lower Deck" represents one side of modern data visualization's initial coin, then this chart is its opposing face.

141

Exports to the British & Foreign West Indies Continued

		Grain						Iron														Oil
	Grindstones	Corn Indian	Wheat	Oats	Hay	Honey	Hides	Pig	Bar	Cast	Wrought	Indigo Carolina	Lignumvitae	Lampblack	Lime	Leather	Mahogany	D° Bedposts	Mastick	Mill Stones	Negroes	Train
	N°	bushels	bushels	bushels	C	lb	N°	tons	tons cwt qr	tons cwt qr lb	tons cwt qr lb	lb	tons	bbls	bush	pounds	feet	N°	F°	N°	N°	Tons Gals
Newfoundland;																						3.39
Quebec			200																			
Nova Scotia																						3.78
New Hampshire	10	10			10				2.1.1		...5.1				1000	2300						22.21
Falmouth																						4.87
Salem & Marblehead		179			31											580						15.27
Boston	101	520			16					1.4.2.14				203								62.72
Rhode Island	3	1383		300					2...		6.10..			180	1212							66.10
New London		3012		300					9...	..6.2..					9							16.57
New Haven		3140		4030																		2.126
New York		57469		5013	20	4110			81.15..	..5.2.26						1123	4260					32.130
Perth Amboy		2037																				
Burlington																						
Salem & Cohensy																						
Philadelphia		20571	733	1617					123...5		1.12..					5356						56.21
New Castle		1195		40		102	10		6.14...							850						
Lewis																						
Pocomoke		27232		1200					4.18.3												1	
Chester		5200																				
Patuxent		18508							22.15..													
North Potomack		1631		125				3	6....													...53
Accomack		20300		3280																		
South Potomack		679							3....													
Rappahanock		24586		178					2....													
York River		9917																				
James River L. Dist.		137921		5335		96			.13..													...126
Ditto upper dist.		10835				288			2....													
Currituck		220																				
Roanoke		20189														100					14	
Bath Town		15																				
Beaufort		10742														450				34		..94
Brunswick		177			30																	..126
Winyaw		472																				
Charles Town		29318							7....			83				6575		1			6	
Port Royal		300																				
Savannah		6312							2	.3.2.8						1010	300					31
Sunbury																					7	
St. Augustine																						
Pensacola																						
Mobile																						
Bahama													3				11524	74	1830			
Bermuda		3575													4170		1000					.220
Totals	114	432958	955	21838	107	1086	10	3	272.19.1	12.3.2.22	8.7.1	83	3	383	6391	15324	16884	75	1830	34	27	238.78

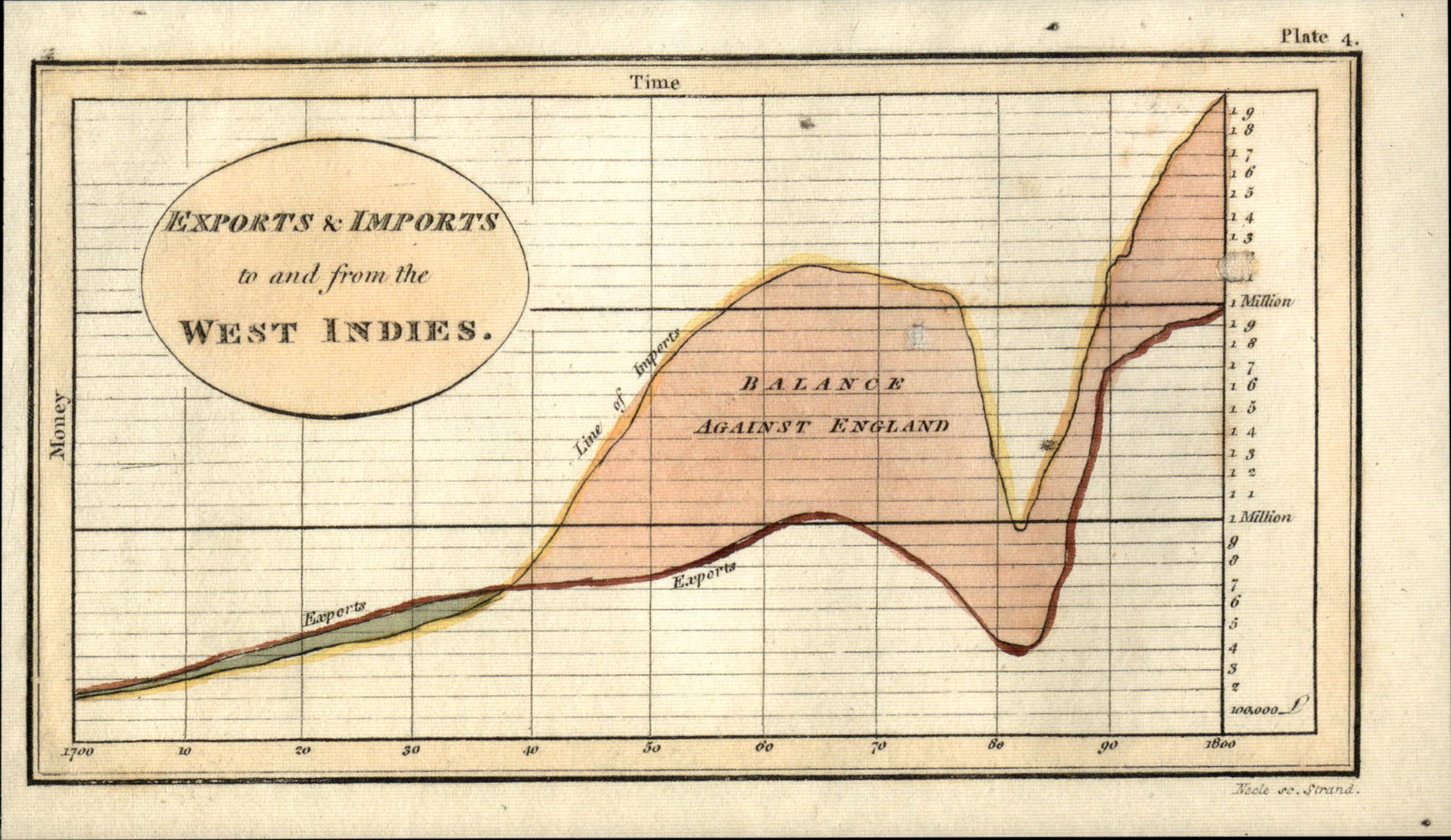

FIGURE 2.12
Playfair's chart of "Exports & Imports to and from the West Indies," from the third edition of *The Commercial and Political Atlas* (1801).

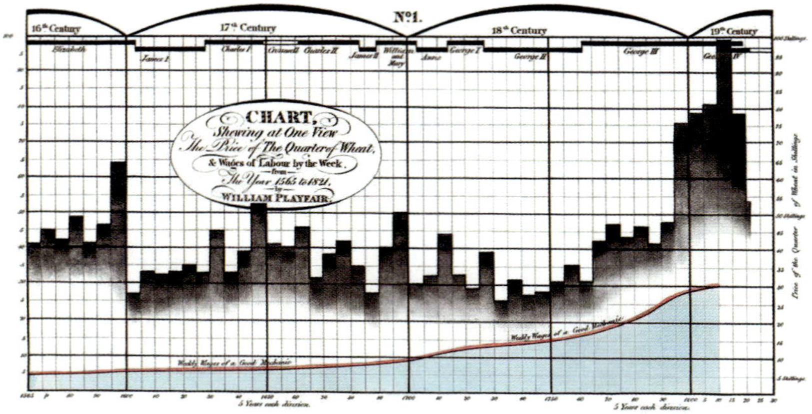

FIGURE 2.13
Playfair's original "Chart, Shewing at One View the Price of the Quarter of Wheat, & Wages of Labour by the Week, from the Year 1656 to 1821." Image courtesy of Wikimedia Commons.

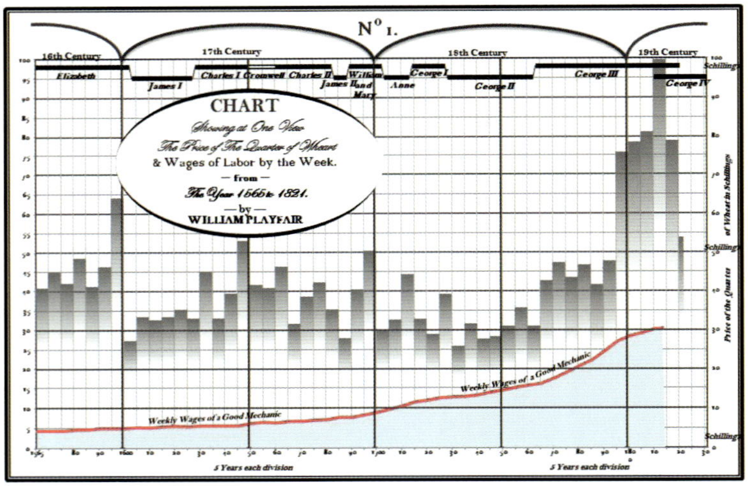

FIGURE 2.14
The same chart recreated in Microsoft Excel by Jorge Camoes. Used with permission.

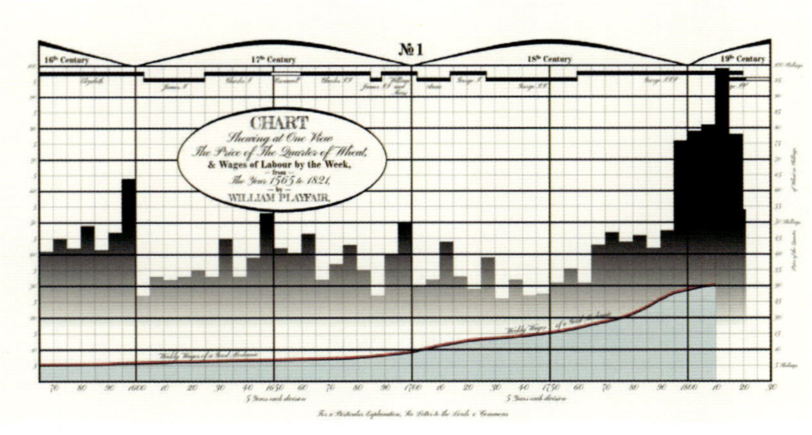

FIGURE 2.15
Playfair's chart of wheat and wages recreated by Jo Wood using Observable Plot.
Used with permission.

To be clear: very few of the myriad people who employ time series charts today do so with the same politics or aspirations as William Playfair. It is unlikely that they are even aware of the context of the charts' creation at all. But they—and we—would be well served by more fully recognizing the politics that shaped the charts' original design, as this politics—even if we do not share it—continues to shape our design choices today. An expanded knowledge of Playfair's politics can push us to consider not only the knowledge that can be gained from a "simple" view, but also the people who stand to benefit the most from it.

With this lesson, we—the authors of *Data by Design*—might layer in one of our own: to consider not only those who benefit from a simple view but those who do not. More broadly, what we gain from our newfound knowledge of the politics of Playfair's charts is a constant reminder—in the form of each and every chart—that there is never not meaning conveyed through the decision to give data visual form. It is either communicated through the inclusion of context and detail, or it is implied through the decision not to include the same.

Playfair's Charts in the Present

Perhaps because of the prominent role that Playfair now occupies in the story of visualization's emergence, his charts are frequently replicated with current libraries and tools. When I first began to notice this phenomenon, it struck me as akin to the art world model of imitating a masterwork: an exercise undertaken by a younger artist in order to learn by doing from a famed forebear, and in the process provide evidence of their own emergent skill. Nearly a decade after I first came across business consultant Jorge Camoes's recreation of Playfair's "CHART, Shewing at One View The Price of The Quarter of Wheat, *& Wages of Labour by the Week, from The Year 1565 to 1821*" in Microsoft Excel, I still find the verisimilitude that Camoes was able to achieve nothing short of astounding (FIGURES 2.13 AND 2.14).[43] With his ability to overlay the bar chart, time series, and proto-Gantt chart that the original contains, each with coloring that strongly resembles the original, he demonstrates a level of expressivity that, I suspect, few knew that Excel could achieve.

But as visualization researcher Jo Wood explains, there is another reason why recreating historical examples with contemporary tools remains a useful exercise. "It helps both to understand the expressive capabilities of the [new] package as well as focus on the details of design choices that have been made in the original," he explains.[44] In his own exploration of Observable—a state-of-the-art (circa 2026) web-based visualization authoring platform—Wood chose this same chart to work through how Observable Plot, the platform's underlying software library, could be used both to recreate the original chart and to document his own reflections (FIGURE 2.15). In so doing, Wood not only confirms the possibilities for visual expressiveness that are enabled by the software library, but also illustrates the explanatory power—not present in the original—of interspersing visualizations with inline text. Here, Wood also implicitly models an epistemology: a theory of how knowledge is produced, realized through Observable's notebook form.

We can use this same process of recreating historical charts to surface the implicit epistemological assumptions that guided their initial creation. As an example, let's return to the chart that began this chapter, Playfair's "Exports and Imports to and from all North-America," which we recreated, much like Camoes and Wood, for the web version of this book. Perhaps most significantly, we could not begin our process as Playfair had, drawing his data lines freehand. Because we were using D3, the popular JavaScript-based visualization library, we required to possess a dataset before we could produce anything that we could see.[45] We thus consulted the data tables included in the first and second editions of the *Atlas* and typed them up in a file format that was usable by D3. Only then were we able to create a prototype for our design (FIGURE 2.16).

The sequence of this process—first data, then image—is more than merely pragmatic; as we have just asserted, it also represents an epistemological claim. Whereas Playfair saw his charts as offering something different than what data could convey, the designers of D3 were guided by a more contemporary view: that any and all visualizations should be directly linked to the data they seek to represent. The three Ds of D3 stand for "data-driven documents." This claim is also reflected in the library's underlying architecture, which works by binding the dataset directly to objects in the web browser's DOM, or document object model, the conceptual configuration of a web page's elements that the browser uses to render them in visual form. For visualization designers, this low-level linkage between the dataset and the web page's basic building blocks results in greater control over the data and its visual representation, as well as greater efficiency when rendering the data in visual form. But it also represents the culmination of an epistemological evolution that has traveled far from William Playfair's original ideas: of the value of visualization as distinct from the data it represents.

The idea that visualization should directly follow from the data has taken decades if not centuries to evolve. Contemporary visualization researchers often point to Jacques Bertin's *Sémiologie graphique*, first published in French in 1967 and translated into English in 1983, as the first formalization of what's come to be known as the *encoding-decoding model*: the idea that visualization should entail a process of encoding the data with "visual variables," which the viewer would then use their mind to decode.[46] Later, Leland Wilkinson's *Grammar of Graphics* (1999) would extend Bertin's encoding-decoding model, introducing specific rules for mapping data to a range of visual representations, including colors and scales. Wilkinson also, crucially, made explicit the statistical transformation required to produce a chart, introducing "statistical transformation" as a distinct phase before visualization.[47] This more structured approach was what, in turn, enabled Hadley Wickham to operationalize a grammar of graphics in ggplot2 ("grammar" and "graphics" are the two g's in the name), another software library that remains in widespread use today.[48] While there is more nuance to each of these approaches that could be further explored, the point of linking them together here is to show the consolidation of an epistemological view—quite distinct from William Playfair's—of how visualization produces knowledge by directly translating the information documented in a particular dataset into visual form.[49]

In the case of our implementation of Playfair's chart, however, translating the original data into image was only the beginning. While it generally takes only a single line of D3 code to draw a path from one point to the next, Playfair's data lines contained more detail than were recorded in his data tables. Because our goal was to recreate Playfair's chart with fidelity to the original image, and not to the data tables that Playfair himself removed, we needed a way to convert his data lines back into more detailed data that we could then plot. To do this, Lauren first requested a high-resolution scan of the original image from the Library Company

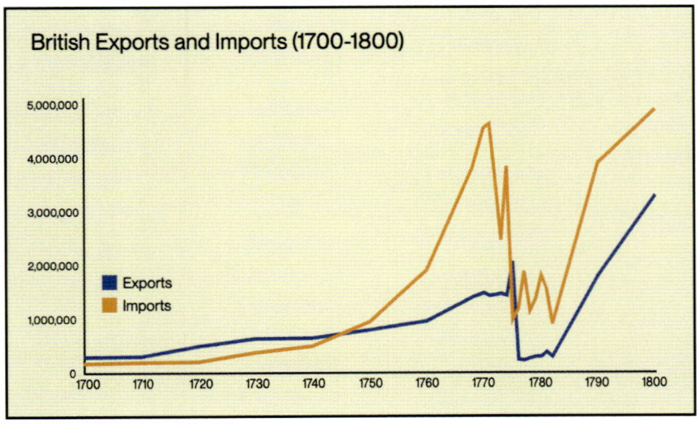

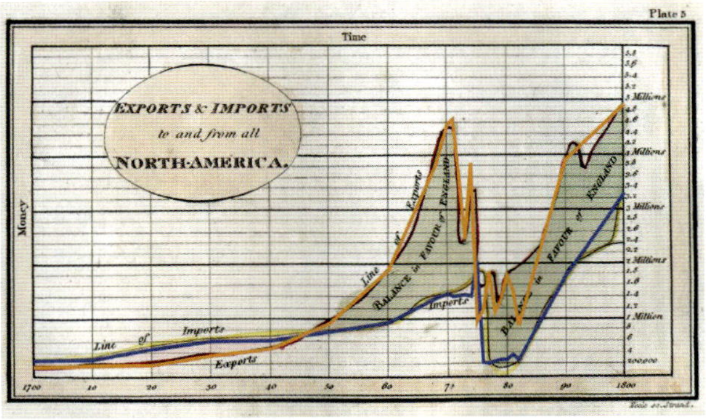

FIGURE 2.16

(A) A rendering of William Playfair's import-export chart in D3 on the basis of his recorded data alone. Design by Tanvi Sharma. Implementation by Jay Varner.

(B) The rendering superimposed on Playfair's original chart.

of Philadelphia, which Tanvi then imported into Figma, a web-based design platform (FIGURE 2.17A). Then Tanvi traced the data lines with a digital pen so that they could be saved as vector-based paths (FIGURE 2.17B). Tanvi exported each path's coordinate vector from Photoshop as a standalone file (FIGURE 2.17C), which she handed over to Jay. Jay then imported the paths as data into his D3 code. Only then could he plot Playfair's data lines on the chart as they appeared in his own published chart (FIGURE 2.17D).

If this process seems convoluted, that is a large part of the point. It underscores the degree to which D3 depends on data, as well as the degree to which, more broadly, the knowledge that we gain from visualization today is presumed to depend on the data underneath. This is for the most part a welcome development; we would not want our viewers, or ourselves, to derive false conclusions from images that are presumed to be supported by data but in reality are not. This is among the hazards of the "god trick," as discussed in chapter 1. More pointedly, it's one of the easiest ways that charts can lie.[50] But as our elaborate process of scanning and tracing and vectorizing helps to show, there are additional facets of knowledge production—those that do not derive from data—that are encoded in the act of visualization itself.

We have previously discussed how the engraving error that Playfair inscribed into "Exports and Imports of all North-America" provides insight into his personal finances as well as his epistemological claims. While the shift from physical to digital visualization tools means that, today, errors made along the way can be more easily corrected, with no visible trace of error in a chart's finished form, digital tools themselves leave data traces, which we can examine—and visualize—in order to elicit additional insight about how the process of visualization is practiced today. Chapter 5 represents our project team's attempt to do just that.

Before we get there, however, it is worth considering what more we can learn about the *original* charts—their material processes, their underlying assumptions, and the epistemologies encoded in both—when attempting to recreate them in digital form. What we learn is that, for Playfair, the "immediate

FIGURE 2.17
The process of recreating the data pictured in Playfair's original chart.

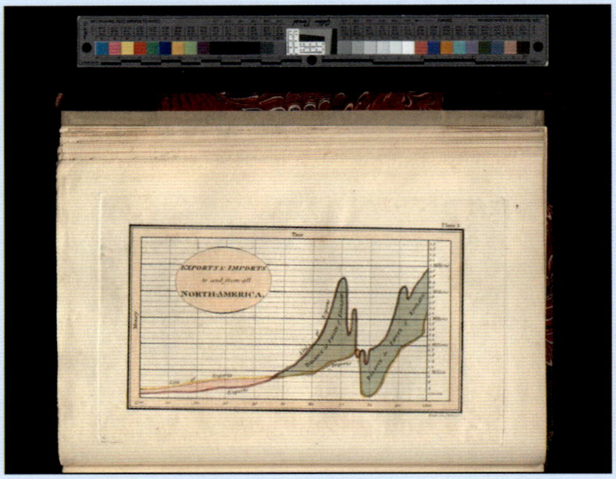

(A) Step 1: Lauren requested a high-resolution scan of the original image from the Library Company of Philadelphia, which Tanvi then imported into Figma, a design platform.

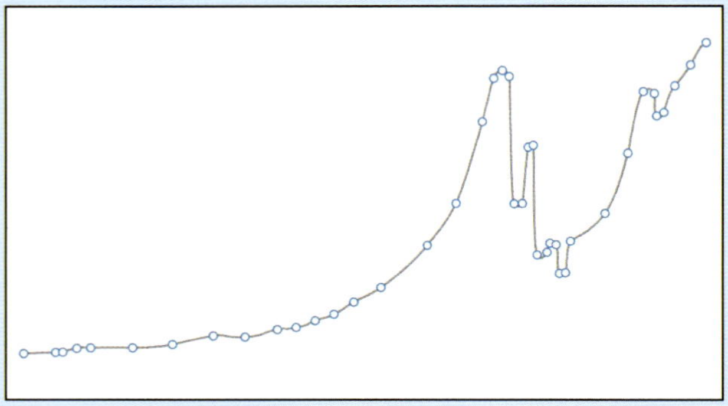

(B) Step 2: Tanvi traced the data lines with a digital pen so that they could be saved as vector-based paths.

```
<svg width="388" height="179" viewBox="0 0 388 179" fill="none" xmlns="http://ww
w.w3.org/2000/svg">
<path d="M0.0179458 178.424L18.1262 177.775C19.4707 177.803 20.8156 177.754 22.1
546 177.63C24.769 176.746 27.4336 176.02 30.134 175.456C34.2769 174.72 34.8737 1
75.139 38.0358 175.173C38.0358 175.173 55.7358 175.037 62.0708 175.132C68.4057 1
75.228 78.465 174.544 84.6032 173.209C90.7413 171.873 101.445 169.406 107.746 16
8.309C114.048 167.212 115.706 169.628 125.756 168.974C135.807 168.32 136.596 16
6.257 144.316 164.679C152.035 163.101 144.913 165.381 154.936 163.427C164.96 16
1.473 160.055 161.502 165.735 159.557C171.415 157.612 170.313 158.806 176.402 15
5.829C182.491 152.851 183.571 151.071 187.663 148.89C191.754 146.708 196.592 14
5.087 203.05 140.512C209.508 135.937 220.721 127.385 229.395 116.47C238.069 105.
554 241.366 101.23 245.788 92.0708C250.211 82.9121 259.039 52.2399 260.674 45.64
65C262.309 39.053 266.162 22.9495 267.144 20.5762C269.019 16.1067 271.416 16.157
7 272.018 16.1753C272.62 16.1928 274.742 15.8917 275.902 19.587C276.711 22.1309
276.654 89.7212 278.753 92.297C280.853 94.8728 282.851 94.1065 283.353 92.132C28
3.854 90.1575 285.921 63.1811 286.684 60.395C287.447 57.609 289.047 58.1385 289.
554 59.3138C290.062 60.4891 289.491 117.552 291.802 121.969C294.113 126.386 296.
692 121.794 297.362 120.459C298.031 119.124 298.036 115.211 299.043 115.175C300.
05 115.139 300.855 113.515 302.535 116.028C304.216 118.54 302.76 130.4 304.416 1
32.513C306.073 134.625 306.934 134.31 307.863 132.105C308.793 129.9 308.574 115.
958 310.665 113.995C312.757 112.032 323.473 108.292 330.468 97.8768C337.464 87.4
617 342.08 68.5623 343.341 63.7333C344.603 58.9043 348.944 30.7115 352.138 28.12
2C355.332 25.5324 357.574 26.9487 358.373 29.2092C359.172 31.4697 357.79 39.991
359.703 42.2311C361.616 44.4712 363.296 42.1022 363.799 40.1961C364.303 38.29 36
7.13 27.8874 370.017 24.8491C373.332 21.1933 376.322 17.2494 378.95 13.0633C380.
855 9.94283 384.864 1.31305 387.859 0.481685" stroke="#56190F" stroke-miterlimit
="10"/>
</svg>
```

(C) Step 3: Tanvi exported each path's coordinate vector from Figma as a standalone file, which Jay imported back into his D3 code.

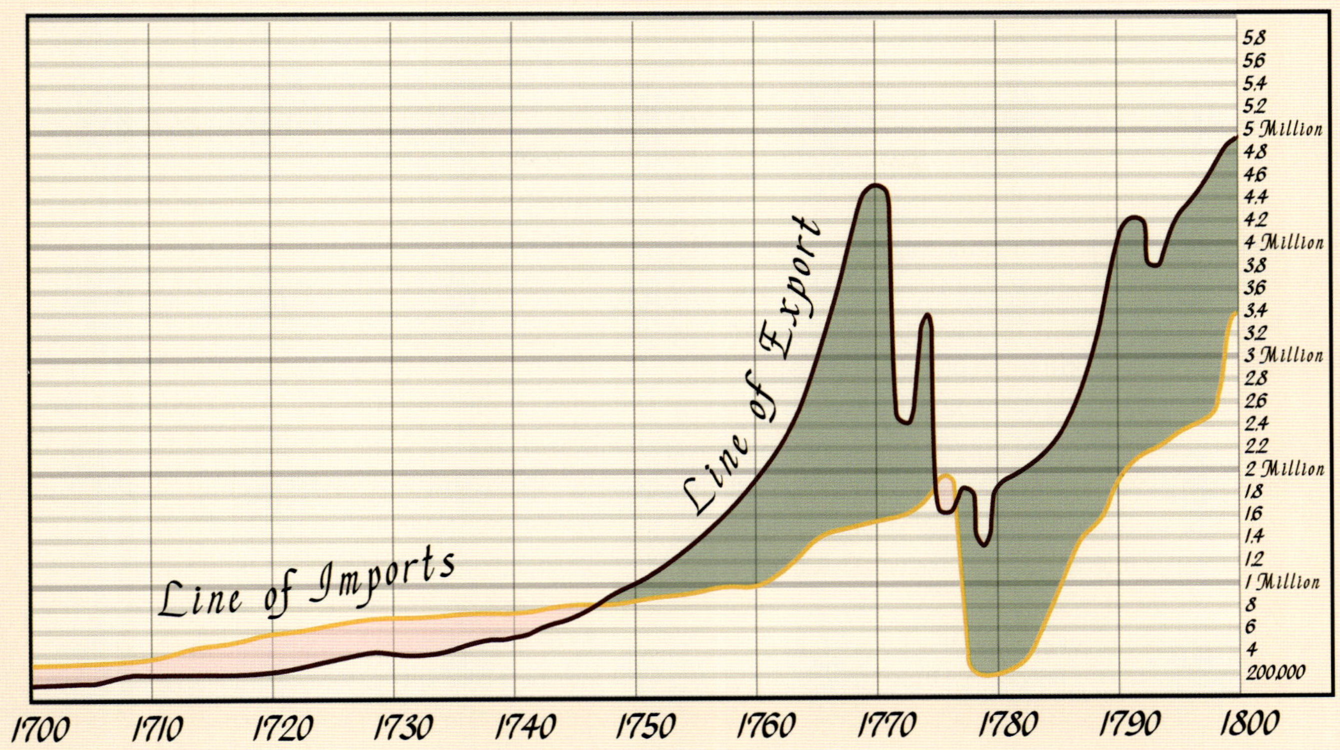

(D) The final recreation of "Exports & Imports to and from all North-America" that we created for our set of illustrations. Prototype by Anna Mola. Implementation by Jay Varner.

impression" prompted by his charts was paramount. The big picture view of his subject, gained even through a passing glance, outweighed his fidelity to the associated data. The tools that Playfair employed to create his charts enabled him to advance this epistemological claim. But today's tools encode a different view: of the fundamental importance of basing a visualization on the data it represents.

This is the dominant epistemology of visualization in the present moment. In most ways, it represents a positive development in the project of producing more precise and accurate knowledge. But there are other forms of knowledge that software libraries like D3 do not allow us to encode. Just as we must contextualize and interpret Playfair's original charts so as to surface their politics and the circumstances that contributed to their creation, so too must we remind ourselves to contextualize and interpret the range of visualizations that we encounter in the world today—as well as the range of tools we use to create them. This is what will lead not only to knowledge that is more accurate but—to return again to the words of William Playfair—knowledge that is more complete.

Reframing the Power of Playfair's Charts

Throughout his life, Playfair longed to be recognized for his innovations. In 1787, one year after the initial publication of the *Commercial and Political Atlas*, he authored a fictional dialogue between two of his heroes: Benjamin Franklin, whom, presumably, he admired for his ingenuity; and Joseph II, Holy Roman Emperor, whom, presumably, he admired for his power. Here, also, Playfair could not resist indulging a fiction of his own: an endorsement, on the part of Franklin, of his method of visual display. In the dialogue, Playfair writes in the voice of Franklin: "I have begun to practice the mode here, and it throws light on the state of our accounts, as if by inspiration, one minute giving a much clearer idea of the matter, than whole days and weeks without this simple invention."[51]

The reality, of course, was that it would take decades if not centuries for Playfair's "simple invention" to achieve the level of recognition that he craved. Playfair's charts were largely forgotten until the 1860s, when, according to the influential twentieth-century economist John Maynard Keynes, they were borrowed by his own economic forebear, William Stanley Jevons, to illustrate a statistical atlas of his own (FIGURES 2.18 AND 2.19).[52] Even then, the charts themselves remained out of sight. Jevons never published his atlas; the only visual record is four sketches now held in his personal archive. But a draft of Jevons's text was evidently circulated among the British statistical elite, where it—and, by extension, Playfair—won praise from none other than the statistician and eugenicist Karl Pearson, who, in a series of lectures on "The Geometry of Statistics," credited Playfair as the "father" of the field we call visualization today.[53]

For Edward Tufte, himself often credited as a father of the field, Playfair's charts epitomize the "graphical excellence" that Tufte has devoted his own life's work to promoting: a visual style characterized by the qualities of "clarity, precision, and efficiency," conveyed through a sparse, minimal aesthetic and an intentional absence of "chartjunk."[56] In the context of Playfair's own view about the value of visualization, it seems fitting that the enduring value of his charts has nothing to do with the underlying data, and everything to do with the "form and manner" in which they were created.

But by elevating Playfair's charts to this level of prominence, without also elevating anything about the context of their creation, we risk conflating the "simple and complete view" that Playfair sought to engender through the design of his charts with a simple and complete view of visualization itself. As this chapter has sought to underscore, the act of visualizing data is never simple, nor is any resultant image ever fully complete. The invisible politics of Playfair's charts, surfaced through his writing, should serve as a reminder—just as enduring as his images—to always ask questions about what we are looking at, as well as what we are not. What we now know to be missing from Playfair's charts can push us forward. It can open us up to the idea that knowledge itself is never complete. Our collective task thus becomes to continually seek out *more* knowledge, knowledge that can fill in gaps. This is how we work toward the most objective picture of what we know.[57]

This is, once again, why I find so much significance in the error engraved in "Exports & Imports to and from all North-America." With a slip of the hand that took less than a second, and that resulted in a line a hair-width wide, we are nonetheless provided with a window into a range of contexts—personal, political, and philosophical—that informed the creation of Playfair's chart. The chapters that follow continue to expand this contextual frame. In chapter 3, we will further explore the colonial

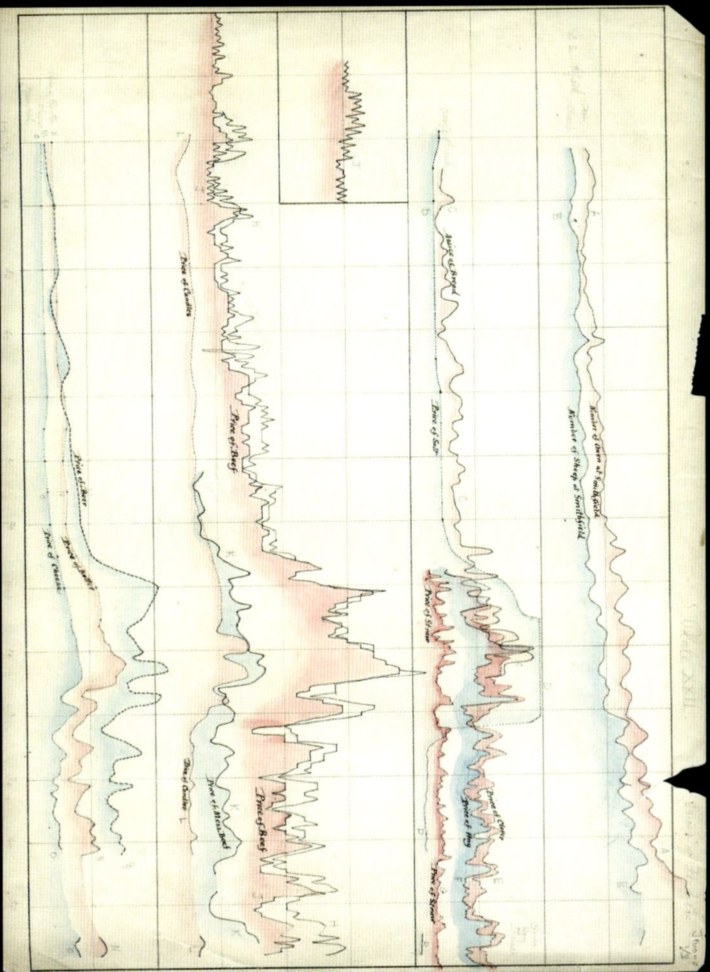

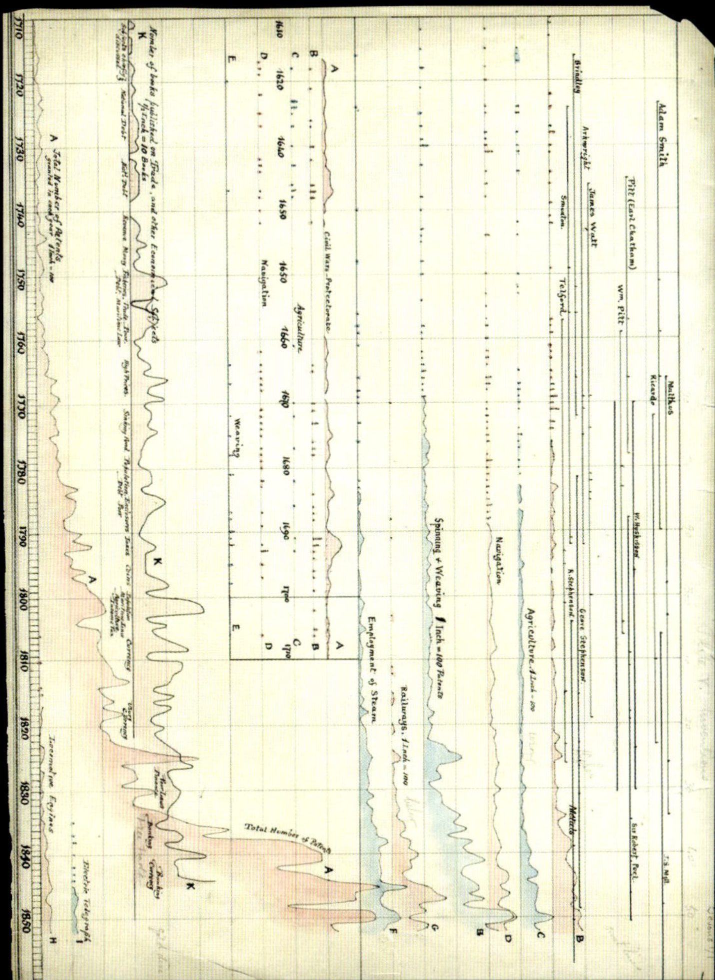

FIGURES 2.18 AND 2.19
Two sketches by William Stanley Jevons, ca. 1860–1861, are all that remain of the charts he created for his statistical atlas. They were deposited in his archive by John Maynard Keynes. Images courtesy of the Marshall Library of Economics, University of Cambridge.

context that, as we have already begun to learn, casts a long shadow on the development of modern visualization techniques. In chapter 4, we will begin to explore alternative epistemological frameworks that suggest additional ways in which visualization can produce knowledge, ways that exceed the scope of Playfair's "simple and complete" view. Here, however, we close with the claim that Playfair's charts remain crucial to the field—but not, or not only, for the reasons they have been celebrated to date. Playfair's charts, framed by the conditions that gave rise to them, and with their epistemological assumptions exposed, show how much space remains open for the field to continue to explore. For if the initial argument of *Data by Design* was of the importance of context, this argument expands into an even more powerful claim with the example of Playfair's charts. It is that the depth of our knowledge about the history of data visualization scopes our present vision of the field. More than that, by enriching our knowledge of this history, we can envision more expansive possibilities for its future.

Takeaways

FOR VIEWERS

Reject visualization's aura of objectivity

Ask for what purposes each chart has been designed

Consider what is rendered visible and what is not

Recognize the decisions that guide visualization work

FOR DESIGNERS

Acknowledge your decision-making power

Question the impulse to adhere to the default

Consider the capabilities and constraints of your chosen tools

Ask who will view your visualization and what they will need to know

CHAPTER 3

PEOPLE

Shanawdithit and Our Place in the Colonial Frame

Maps, like all visualizations, tell stories.
How can we learn to see the lines of power that they draw?

One morning in March 1819, at the first break of dawn, a small group of Beothuk people—the Indigenous inhabitants of the island now more commonly known as Newfoundland—were awakened by the sound of intruders.[1] The group had been asleep at their winter camp on the north side of Beothuk Lake, a long and narrow body of water at the island's center, when a half dozen or so British settlers surrounded them.[2] While the settlers' intentions were not yet known, the Beothuk had cause for alarm. Every previous encounter with the British had ended in destruction and death. This encounter would soon result in the same.

The Beothuk had been navigating European incursions for centuries. Some speculate that the mention of "scraelings" in the Norse sagas refers to ancestors of the Beothuk, which would date a first encounter to the eleventh century.[3] A second phase of more sustained contact began shortly after the Italian explorer John Cabot's initial visit to the island, in 1497, and persisted for the next two hundred years.[4] During this time, says environmental humanities scholar Fiona Polack, "fishing crews from Spain, Portugal, France, and Britain would spend the summer months catching and processing cod before returning home for the winter."[5] The seasonal nature of these stays granted the Beothuk "periods of unimpeded access to valuable materials, such as metal objects left in unattended fishing stations, and reduced the need for them to interact directly with the invaders."[6] Polack also documents how this arrangement—to which, of course, the Beothuk had no choice but to consent—began to strain as "increasing numbers of people from the British isles began to settle permanently on the island and compete directly with the Beothuk for resources."[7] It was this competition for resources, compounded over centuries, that in no small part led the British to Beothuk Lake that day.

But there were other, more direct reasons: several months earlier, on September 18, 1818, a group of Beothuk had surprised the same British settlers as they were preparing for a trip to market. Hidden in a canoe under a wharf at Lower Sandy Point, in the Bay of Exploits—north and east of Beothuk Lake, where its waters met the sea—the Beothuk waited for the "dense darkness" of night and then absconded with a boat carrying the season's catch of salmon, and possibly some furs.[8] This "theft and act of destruction" provided the rationale for John Peyton Jr., the owner of the boat (whose personal narrative serves as the source of the direct quotations in this paragraph) to request formal authorization from the governor of colonial Newfoundland to "search for his stolen property and, if possible, try and capture one of the Indians alive."[9]

Those at the winter camp did not know about the kidnapping authorization when they awoke that morning in March. But within minutes, the settlers' goal became clear. When the Beothuk fled to the woods, one woman, Demasduit, fell behind and was immediately set upon by the British. She "pointed out to the white men her full breasts to show that she had a child, and pointed up to the heavens to implore them, in God's mercy, to allow her to return to her child," but they "took hold of her," recalled John Paul, a Mi'kmaq-Innu man whose grandfather had been alive (but not present) at the time of the original events.[10] Demasduit's partner, a man named Nonosabasut, "came to her aid," but Peyton shot and killed him. Two days later, the child of Demasduit and Nonosabasut died as well—likely the result of starvation. One other young woman, Shanawdithit, who was then around 17, bore witness to it all.[11]

Shanawdithit's hand-drawn maps, which constitute the core of this chapter, offer documentation of these events in visual form. But in exploring her maps, our goal is not to argue for the utility of Beothuk cartographic techniques for the field of data visualization, nor to analyze the broader theories of knowledge production that they reflect.[12] Rather, the argument we present in this chapter is about the colonial context that gave rise to the maps' creation, and about how that context—what we call here the *colonial frame*—must be considered alongside any new knowledge that the maps bring to light.

This is the reason we are breaking the mold of the previous chapters by presenting you with a diagram before we present the original maps (FIGURE 3.1). This diagram was designed by Tanvi, with input from Lauren, as a way to give this colonial frame visual presence. The years lead out from the center, beginning with the first documented encounter between the Beothuk and British groups. But the time scale we employ is not linear; instead, the layers mark the major events that culminated in the creation of the three primary sources that we rely upon in this chapter. The shaded section at the top of the diagram displays the series of events that culminated in Shanawdithit's maps entering the colonial archive; the section in the middle right indicates those of John Peyton Jr.'s narrative; and the section at the bottom right indicates those of John Paul's oral history. Around them is open space. While additional areas of the diagram will be filled in as this chapter unfolds, some areas will remain open. The unfilled space is there to remind us of our distance from the past, and of what we—the project team, and therefore you our readers—can never know.

Our focus on our team's perspective is intentional. At various points in this project—in the preface, for example, and in our visualization of the Voyages Database that appears in chapter 1—we have been explicit about acknowledging the positions from which our work takes place. Here is another place where these positions matter, since research involving sources that document Indigenous cultures should be guided by the relationships among those sources, the people who steward their legacy, and the research team.[13] Our project does not contain any Indigenous team members, and while Tanvi is from India, which has its own colonial history, and other team members call other countries home, our team is firmly based in the United States—on Mvskoke, Lenape, Tongva land.[14] As such, our work cannot by definition be decolonial, as it is not led by Indigenous people and does not involve the repatriation of Indigenous culture or land.[15]

Instead, we aspire to an anticolonial stance.[16] Having reflected on our team's shared position over the course of creating this chapter, we have come to see our role—indeed, our responsibility—as not to argue for the utility of Shanawdithit's mapping techniques for the field of data visualization, but rather to illuminate their place in the long history of extracting Indigenous knowledge for colonial gain.

This history is a violent one, as you no doubt already know. The additional details about Shanawdithit and her maps that follow, which involve yet more instances of violence and harm, underscore how the maps at the center of this chapter cannot be separated from the inherent violence of colonialism.[17] While the history of visualization often elides such violence—think back to Playfair's "simple" view of British trade data, which hides the human sources of its profits and losses—this chapter asks what it would mean to acknowledge this violence as a substrate of so many of our current visualization techniques. We do this not to dismiss all visualization as colonial. Rather, we do so to underscore a broader claim. This has to do with how the dominant view of the value of visualization—that is, following Playfair, its ability to distill insight from complex data such that knowledge can easily and efficiently emerge—sits uneasily close to the colonial practice of extracting knowledge from its source.

As at previous moments in this project that have brought us to uncomfortable points, the response to this assertion is, we hope, not to close this book and walk away. Rather, we hope it will serve as an invitation to consider how we all might design future visualizations, as well as reposition ourselves with respect to existing ones, in ways that enable the *creation* of knowledge without the *extraction* of knowledge. This consideration entails an attention to the lives behind the data, as we have explored in chapter 1, as well as to the forms of insight that visualizations are most often designed to promote (or not promote, as the case may be), as explored in chapter 2. As we will see in this chapter, the colonial frame that surrounds Shanawdithit's maps prompts us to consider yet more: our relationships with the people who provide us with data and with those represented in it, and our responsibilities toward the knowledge that we, together, create.

Visualization as Extraction

With the importance of relationships and responsibilities in mind, we now return to the story of Shanawdithit and her maps as it emerges from the colonial archive. As it turns out, this story passes directly through Demasduit, the young mother, and her own eventual fate. Demasduit was taken first to the fishing village of Twillingate and then in the spring, after the ice had cleared, to the colony of St. John's. She made several attempts to escape her captors.[18] At some point during this time, she contracted tuberculosis. She succumbed to the disease less than a year later, while aboard a boat that was intended to return her to her family, the British ultimately achieving a deadly version of their goal of retribution.[19]

Shanawdithit was present the day that Demasduit's body was returned to Beothuk Lake, and she participated in Demasduit's burial ceremony, held over the course of several months the next spring. But the British would not learn of Shanawdithit until April 1823, when Shanawdithit was herself captured, along with her mother and her sister. The three women had been heading "to the seacoast in search of mussels to subsist on," following another winter in which food had been scarce and illness had been plentiful. A different group of British settlers—furriers, this time—came across them.[20] Concluding that it had become too difficult to

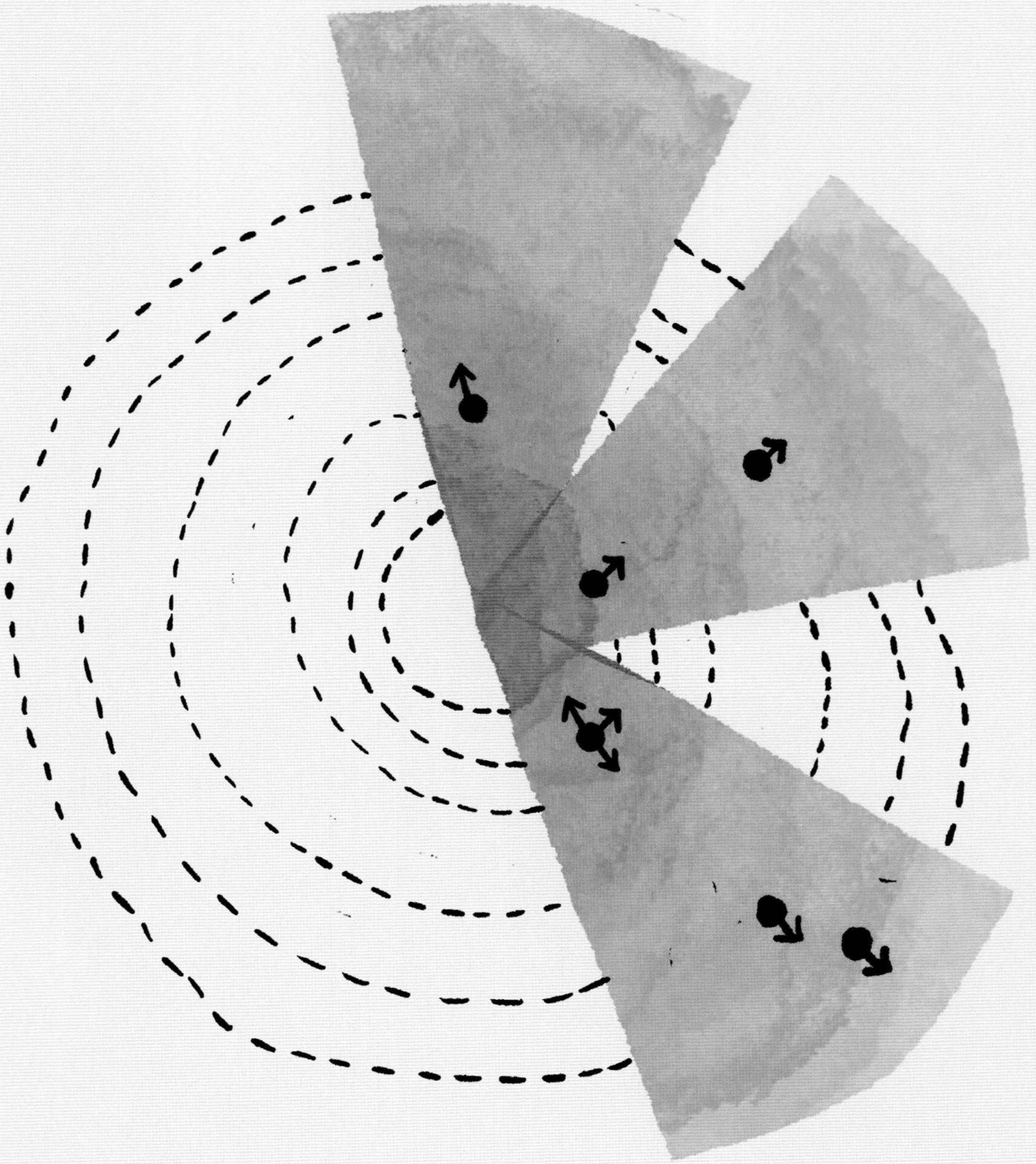

YEAR	DESCRIPTION
1818	"Theft" of Peyton's boat
1819	Attack on winter camp and capture of Demasduit
1819 - 1914	Grandfather tells John Paul his account
1829	Shanawdithit creates her maps
1871	Peyton relates his narrative
1914	John Paul's narrative is recorded

HOW TO READ
Innermost ring corresponds to the first year listed on left. Successive rings correspond to subsequent years listed.

LEGEND

- Speck
- Howley
- Shanawdithit
- All sources
- Unrecorded history
- Recorded history
- Primary source

FIGURE 3.1
The three primary sources referenced in this chapter: Shanawdithit's maps, John Peyton's narrative, and John Paul's account, emphasizing the decades-long chain of events that culminated in the creation of each. Diagram and key by Tanvi Sharma.

continue to keep themselves alive, according to another British account, Shanawdithit and her kin "allowed themselves in despair to be quietly captured."[21]

Like Demsaduit before them, the three women were taken to Twillingate, where they were held captive in the home of none other than John Peyton Jr., whose "stolen" goods had been the impetus for it all. Shanawdithit's mother and sister soon died, also of tuberculosis. But Shanawdithit persevered. For five years, she was forced to work for Peyton as a domestic servant, before she too fell ill. Following Demasduit's final path, Shanawdithit was then brought to St. John's, where she spent six of the final weeks of her life in the home of William Epps Cormack.[22] Cormack, the Newfoundland-born son of Scottish settlers who had earned early fame for his natural history of the island's interior, was the one to supply Shanawdithit with "paper and pencils of various colours," and who through some combination of enticement or coercion—we can never know—prompted her to create her maps.

In the final weeks of her life, Shanawdithit drew five maps, each documenting a different area of Beothuk Lake and the events connected to it. She also created five additional drawings that documented additional aspects of Beothuk tradition, culture, and daily life. The map on the next page, and which anchors this chapter, is the second in a roughly chronological sequence, spanning several decades (FIGURE 3.2).[23] It presents a composite picture of the series of encounters between the Beothuk and the British that culminated in Demasduit's capture and eventual death. In this and all of the maps, time is anchored by place.

No II
2 different scenes & times
~~The taking of~~
Captain Buchan's visit

The Taking of Mary March.

on the North side of the Lake

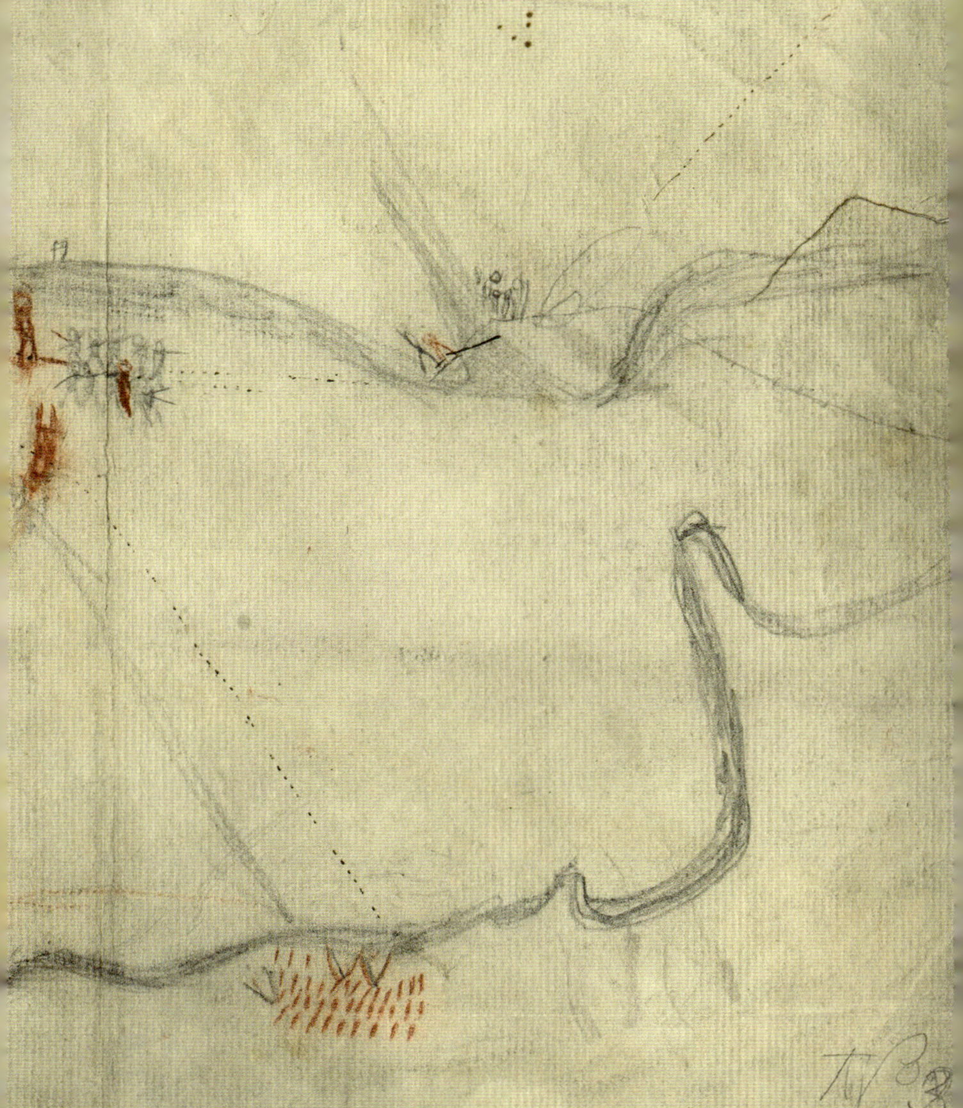

~~ry March~~

1810-11

~~15 16~~ at the South side of the Lake

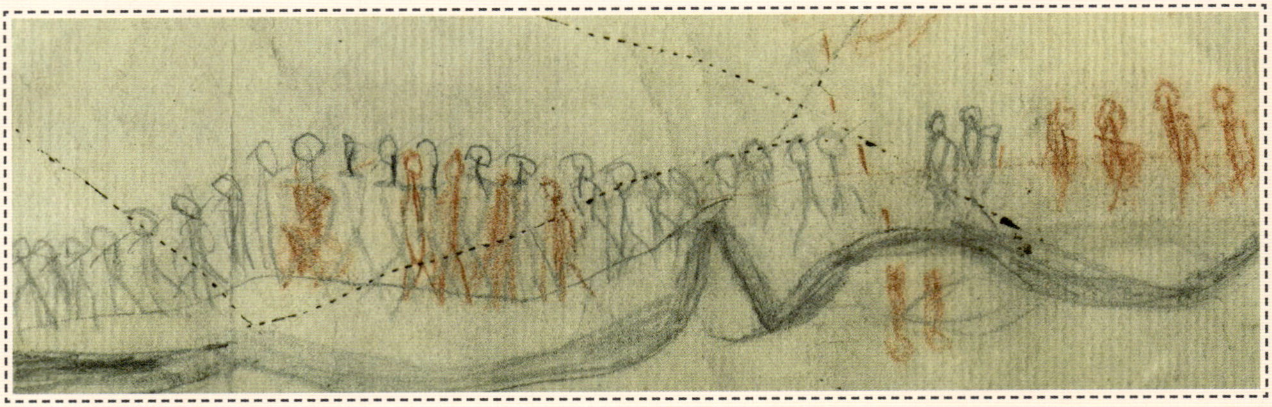

FIGURE 3.3
A closer view of the lower half of the map, depicting events that took place in the winter of 1810–1811.

The lower half of the map depicts an earlier encounter between the two groups, which took place on the south side of the lake in the winter of 1810–1811 (FIGURE 3.3). Thirty or so figures appear along the bank of the river. The figures drawn in red are Beothuk. Those in black are the members of the British party, led by a Scottish naval officer named David Buchan. They are pictured after their initial meeting, which was enabled by Mi'kmaq and Innu guides.[24]

The figures on the right side of the group—two black figures and four red ones—likely stand for two marines and four Beothuk whose distrust of the British (the result of several prior instances of kidnapping and murder) would result in the preemptive killing of the two marines the next day.[25] The two red figures oriented in the opposite direction may be the two Beothuk who briefly traveled with the British back to their camp before they were "told by signals to give chase," as John Paul reports.

FIGURE 3.4
Shanawdithit's depiction of her people as data. The 37 marks stand for each of the 37 inhabitants in the winter camp that year.

Off to the east—the right side of the original map—is the Beothuk winter camp (FIGURE 3.4). Three triangles stand for the three dwellings, called mamateeks, which housed the group. Thirty-seven marks stand for each of the 37 inhabitants in the winter camp that year. The dotted lines on the map correspond to paths taken across the frozen lake. A convention employed by many Indigenous cultures across Turtle Island (as the North American continent is often called by Indigenous groups today), these paths would seem to represent both established foot trails and the actual trajectories that culminated in the events depicted.[26] The lines serve a narrative purpose as well as a geographical one; they connect the series of events depicted as well as the two sides of the lake.

FIGURE 3.5
The first set of mamateeks on the north side of Beothuk Lake.

On the north side, at the center of the shoreline, we also see several mamateeks—two drawn in black and a third drawn in red (FIGURE 3.5). The red color and additional detail may indicate that this mamateek is the one covered in the sail that was stolen from Peyton's boat, in the episode described at the outset of this chapter.

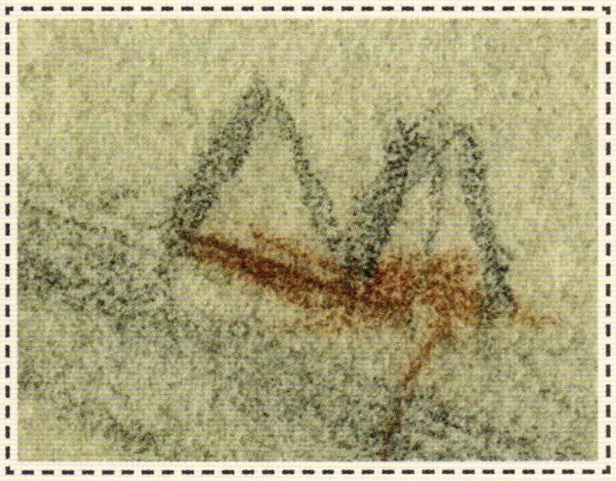

FIGURE 3.6
The second set of mamateeks on the north side of the lake.

A second set of mamateeks are positioned to the west of the winter camp. These may be the two mamateeks to which the Beothuk fled after the deadly encounter with Buchan's men, but this is not certain (FIGURE 3.6).[27]

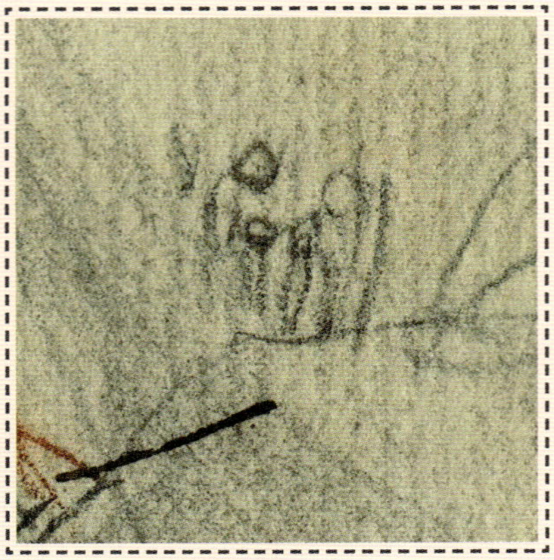

FIGURE 3.7
A closer view of Peyton's men.

As for Demasduit's capture, we see several phases of the events superimposed. Viewed chronologically, we first see several settlers to the east, drawn in black, whom we can infer from Peyton's narrative, and which Shanawdithit confirms, are some of Peyton's men who have hidden themselves in order to surveil the Beothuk camp before their morning attack (FIGURE 3.7).[28]

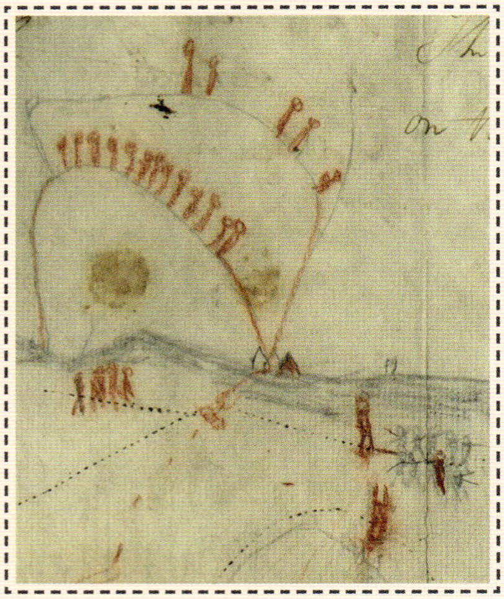

FIGURE 3.8
The scene of the attack. In the upper left, the inhabitants of the Beothuk winter camp, in red, are pictured as they sought safety in the woods. In the lower right is the actual scene of the attack.

In the center of the map we see several groups of red figures pictured along various footpaths; these, we might conclude, are the inhabitants of the winter camp who sought safety in the woods upon being attacked (FIGURE 3.8).

On the frozen lake is another group of figures (also see FIGURE 3.8). This cluster is composed of six figures in black and one in red, presumably Demasduit in the initial moment of her capture. To the left of that group is a large red figure, likely Nonosabasut depicted in the act of defending his wife. Just below the group is another red figure on the ground. Cormack claims that this figure represents Nonosabasut after being shot and killed. But Shanawdithit insists that two men were killed that day—the second being Nonosabasut's brother, who also came to Demasduit's aid.[29]

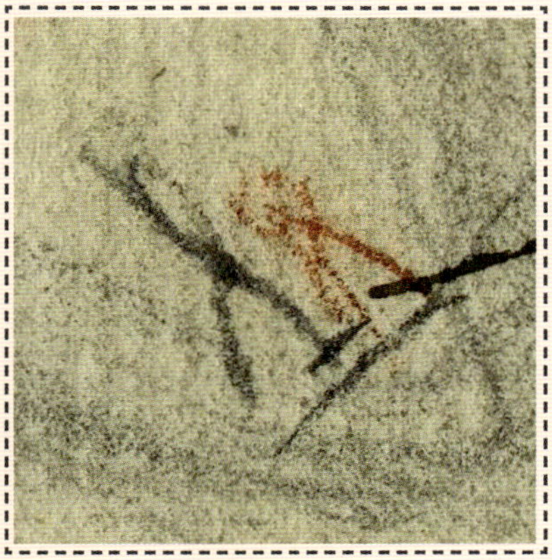

FIGURE 3.9
A closer view of the final phase of Demasduit's capture. Here she is being led away, perhaps by Peyton himself.

Positioned between this tragic scene and the initial surveillance of the Beothuk camp is a pair of figures, one red and one black, which has been interpreted as Demasduit and one of her captors—perhaps Peyton himself (FIGURE 3.9).[30] The man is leading her away from the home that she would never again visit alive.

→

Already, the inextricability of the maps' creation from the larger colonial project should be quite clear. But Cormack's words lay the extractive nature of this project bare: "I keep her pretty busily employed in drawing historical representations of everything that suggests itself relating to her tribe, which I find is the best and readiest way of gathering information from her," as he wrote in a letter to the Bishop of Nova Scotia in January 1829.[31] Cormack's sense of entitlement to Shanawdihit's knowledge is here apparent.

But Cormack's entitlement is also documented on the map itself, in the form of the textual annotations, which were penned not by Shanawdithit but by Cormack, likely at the same time that Shanawdithit set her own lines to the page. Cormack's handwriting encircles Shanawdithit's image, registering not just the "information" he sought to extract from her but, more profoundly, the power that he held over her as her captor. This power also colors the information presented on the map.

In fact Cormack's direct extraction of Shanawdithit's knowledge was only the first layer of how her knowledge has been mined.[32] In the early twentieth century, a British government official and geographer named James P. Howley redrew Shanawdithit's maps for inclusion in his own book, *The Beothucks or Red Indians: The Aboriginal Inhabitants of Newfoundland*. We display this version of the map later in the chaper. Howley's book is also the first place that Peyton's narrative appears. In addition to certain aesthetic decisions, such as smoothing out Shanawdithit's shading of the riverbanks, which had the effect of erasing the individual pencil strokes that more directly link Shanawdithit to the creation of her map, Howley also edited and rewrote Cormack's annotations, removing the erroneous words that Cormack first recorded and then crossed out.

This editorial decision underscores Howley's own sense of entitlement to Shanawdithit's knowledge, and his view of it as ethnographic information that could be easily severed from its source. This observation mirrors the broader critique of early ethnography as it emerged as an academic practice in the late nineteenth and early twentieth centuries in Europe and the United States.[33] But we might push this further still, observing that the act of converting information into data—a required step in the visualization process—shares uncomfortable similarities with this extractive process, in that the act of datafication very often also becomes one of severing information from its source.

With this in mind, our project team made the decision not to follow down Howley's path. While we could have easily converted Shanawdithit's information into GIS data and plotted it on a new map, we did not want to further the extractive process that was set in motion with Cormack—and, arguably, centuries before. Nevertheless, we are still implicated in this process. While we consciously prioritized Indigenous sources in constructing our account of the original map, just as we sought to use contemporary Indigenous scholarship to guide our interpretation, our model of slow and detailed explanation—the same that appears on the previous pages—is not dissimilar to the approach that Cormack and Howley both employed: of atomizing the map and clarifying the significance of its various parts.

This observation leads to a more general claim. Today, we still presume that the goal of visualization should be to clarify, and to enable deeper exploration only if required. The well-known "visual information-seeking mantra," also mentioned in chapter 2, exemplifies this approach.[34] But this approach leaves no room to consider how the process of clarifying the significance of the data runs the risk of further distancing the data from those who created it, or how enabling deeper exploration very often involves the transfer of explanatory power from those who created (or are represented in) the original data to ourselves. The point here, again, is not that we should never seek to visualize data. Rather, it is to ask how we might undertake this work more responsibly—that is, with more attention to, respect for, and engagement with, the people whose knowledge enables it.

We are by no means the first visualization researchers to ask these questions, nor are we the first to explore possible answers. Centuries of experiencing the effects of knowledge extraction, as well as a blatant disregard for responsibility, have motivated a range of Indigenous scholars to develop principles for maintaining Indigenous data sovereignty and governance.[35] In the past several years, several teams of students and scholars, as well as members of nonprofit groups, have sought to develop a parallel set of Indigenous visualization practices that similarly elevate questions of relationality and responsibility, as well as access and control.[36] Inspired by these efforts, and seeking to align them with our project team's anticolonial aims, we revised our initial diagram of sources with a goal of clarifying our own position with respect to the colonial frame (FIGURE 3.10).

In this version of the diagram, we add Cormack and Howley, along with ourselves, accentuating the layers of mediation that separate us from Shanawdithit's firsthand knowledge of the original events. Here we are guided by Seneca literary scholar Mishuana Goeman, who emphasizes the importance of resisting the "utopian" yet ultimately impossible goal of recovery.[37] No magnitude of desire or strength of effort, Goeman explains, can gain us access to "an original and pure point in history" before which all can be known, so that historical erasures can be remedied and injustices rectified.[38] Shanawdithit's maps exemplify this claim. Her direct knowledge is irrecoverable, and even if we could approximate it, the principles of Indigenous data sovereignty tell us that it is not ours to own.

As visualization designers, we cannot change our reliance on data; it remains the substrate of all that we do. But what we can change is our awareness of our position with respect to the data we use and the visualizations we create. This, according to Goeman, is the most generative form of knowledge that we can pursue: a better understanding of "the relationships set forth during colonialism that continue to mark us today."[39] Indeed, when we think of data only through the (literal) marks and channels that represent it, we lose sight of how—and by whom—that data came to be.[40] When we enter into a visualization project without sufficient regard for the data's provenance, we often fail to recognize what knowledge may have already been lost in the process of separating that data from its source.[41] It also becomes all the more difficult to consider any responsibility we might have to the people who created the data, the people whose data our visualizations represent, and the people who interact with our visualizations in their final form.

Indigenous Cartography within the Colonial Frame

The circumstances that surround the creation of Shanawdithit's maps make it clear that we cannot view them as unmediated expressions of her worldview. Yet it is also clear that, despite Shanawdithit's captivity and Cormack's role in prompting the creation of the maps, Shanawdithit was able to incorporate many of her own ideas into the maps' designs. This is evident in the fact that many of her design decisions were aligned with

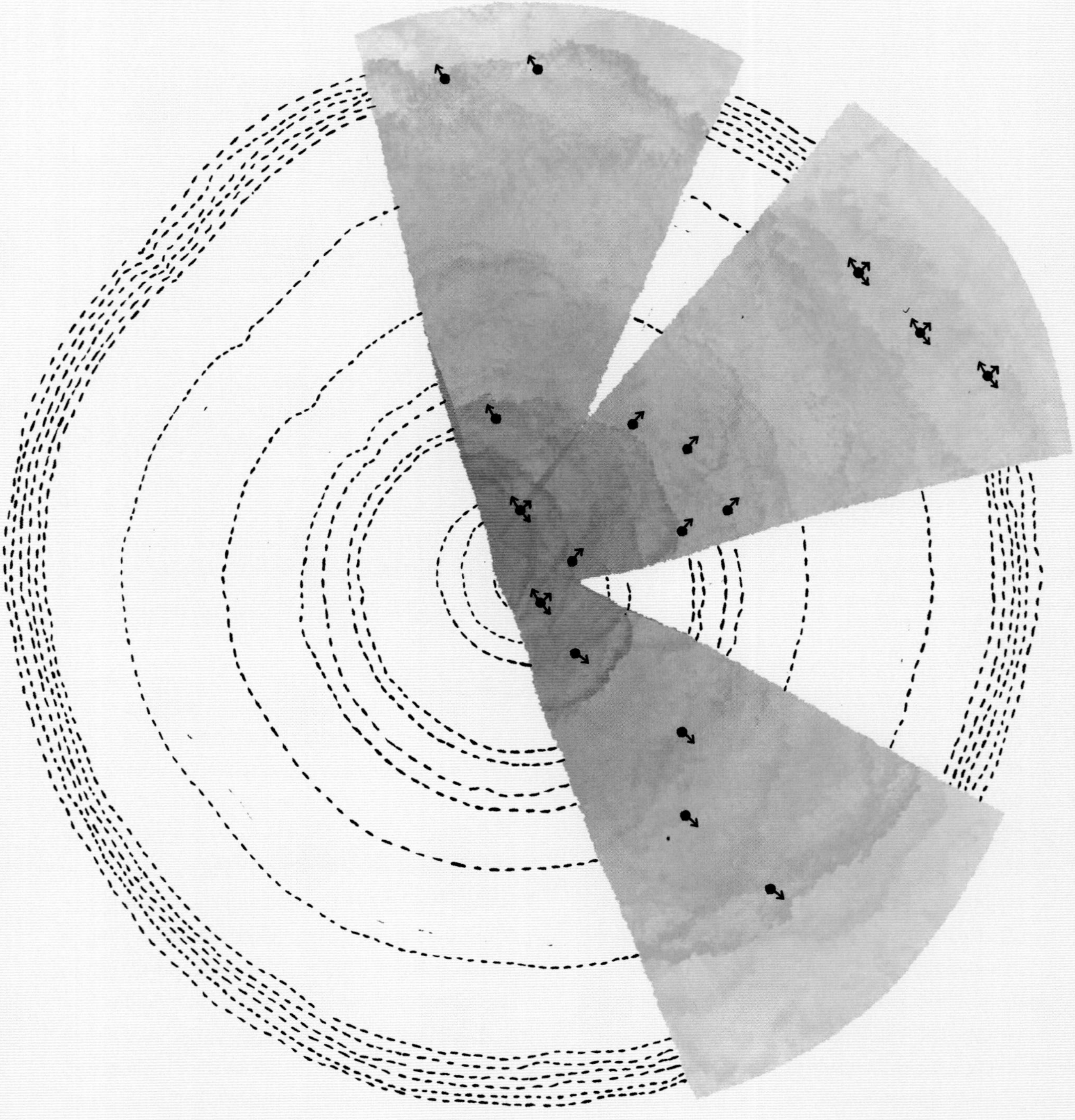

YEAR	DESCRIPTION
1810-1811	Beothuk encounter David Buchan and his men
1818	“Theft” of Peyton’s boat
1819	Attack on winter camp and capture of Demasduit
1819-1914	Grandfather tells his account to John Paul
1829	Shanawdithit creates her maps
1829	Cormack annotates Shanawdithit’s maps
1871	Peyton relates his narrative to Howley
1914	Speck records John Paul’s narrative
1915	Howley redraws Shanawdithit’s maps
1915	Howley publishes his book
1922	Speck publishes his book
2008	Speck’s book is scanned and uploaded to Internet Archive
2021	Lauren begins research on this chapter
2022	Tanvi begins designing visualizations for this chapter
2023	Lauren visits The Rooms in St. John’s, Newfoundland
2024	DxD team receives approval to include map in chapter
2026	Chapter is complete

HOW TO READ

Innermost ring corresponds to the first year listed on left. Successive rings correspond to subsequent years listed.

LEGEND

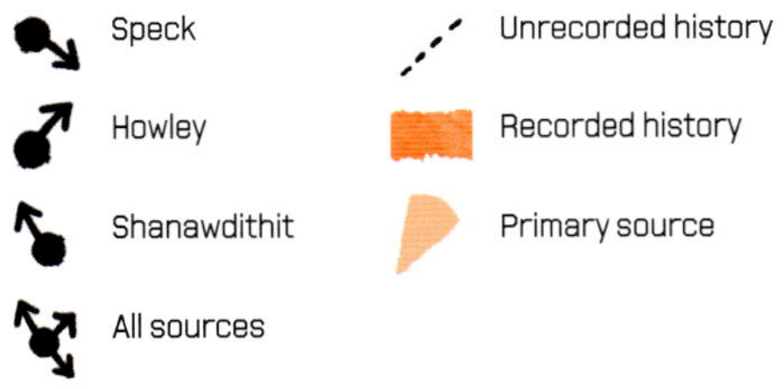

FIGURE 3.10
An expanded diagram (with key) of the sources and people, including ourselves, who have contributed to the account offered in this chapter. Diagram and key by Tanvi Sharma.

mapping practices then common among a number of Indigenous cultures across Turtle Island. Unpacking these practices, and the elements of the map of Beothuk Lake that illustrate them, can help to attune us—if not to grant us full access—to ways of knowing outside of the colonial frame.

Before moving forward, it is important to make clear that there is no single or exhaustive set of such practices. As Indigenous cartographers Margaret Wickens Pearce (Potawatomi) and Renee Pualani Louis (Kanaka ʻŌiwi) explain, Indigenous mapping practices span cultures, continents, and genres, ranging from "Hawaiian performative cartographies to Navajo verbal maps and sand paintings and the Nuwuvi Salt Song Trail." They also span multiple forms and "may be gestural, chanted, or inscribed in stone, wood, wall, tattoo, leaf, or paper." Such "maps," a term Pearce and Louis use as shorthand and we borrow here, are equally varied in the ends they are enlisted for: "to assess taxes, guide a pilgrim, connect the realms of the sacred and profane, or navigate beyond the horizon."[42] What binds these examples together, for Pearce and Louis, as for other scholars of Indigenous cartography, is how they are understood as part of a larger *process* of knowledge-making, rather than as a definitive source of what is *there*.[43] This process is premised on relationships among people as well as places, relationships that continue to acquire meaning as they unfold.

The relational basis of Indigenous cartography is perhaps most visible in how, across this range of genres and forms, many maps express temporal rather than spatial points of view. We see this foregrounding of a temporal perspective emerge in Shanawdithit's decision to depict a series of events, which transpired over decades, in the single place of Beothuk Lake. Cormack's difficulty in determining what it was, precisely, that Shanawdithit had pictured on the page underscores its divergence from the spatial perspective that was (and remains) characteristic of colonial maps. Cormack crosses out one of his earlier incorrect labels, "The Taking of Mary March," which he had first positioned on the south side of the lake, and rewrites in a more accurate location on the north side.[44] He also adds a clarifying note at the top left of the map, just below the reference number he has provided: "2 different Scenes & times." The note is underlined for emphasis. It appears that Cormack himself required this note in order to remind himself of what was depicted, even as the link between the two scenes was (presumably) self-evident in Shanawdithit's mind. Underscoring this point, in Howley's redrawing of the map this uncertainty is fully removed. There are no crossed-out labels. Instead, Howley presents an unequivocal declaration of what Cormack's own worldview made it so difficult for him to see.

Indeed, Shanawdithit's worldview diverged sharply from Cormack's—and, for that matter, from Howley's as well. Whereas Cormack and Howley saw her map as self-contained, Shanawdithit very clearly saw her map as only one piece of a larger system. This is indicated by her decision to include human figures on the map. The figures exemplify what Pearce characterizes as an emphasis on place *as it is experienced*, "as opposed to the Western convention of depicting space as universal, homogenized, and devoid of human experience."[45] Along similar lines, geographer Laura Harjo (Mvskoke) theorizes the Mvskoke conception of the world as a "kin-space-time constellation," which she explicitly contrasts with Cartesian mapping. We might further connect this idea to Edward Tufte's famed formulation, with respect to Charles Minard, of the "space-time-story graphic."[46] Viewing Minard's flow map alongside

Howley's recreation of Shanawdithit's map is startlingly revealing, as the map is both more closely connected to Minard's and more distinct than might initially appear (FIGURES 3.11 AND 3.12).

While the two maps appear to visually rhyme, human figures are wholly absent from Minard's work. It follows, then, that Tufte's formulation of the "space-time-story graphic" does not include kin. In his analysis of Shanawdithit's maps, geographer Matthew Sparke probes the significance of the maps' human dimensions, albeit not through the lens of the "kin-space-time constellation" that Harjo names. Sparke notes how even the symbolic components of the map, such as the paths across the lake, push back against Western orthodoxies of space and scale. By depicting "the uneven possibilities of travel by foot across uneven landscape," Sparke suggests, Shanawdithit incorporates an embodied dimension into the elements of the map that would otherwise be interpreted only for their geographical information that they encode.[47] More pointedly, Shanawdithit's maps make it impossible for their viewers to conceive of the land without the people—the Beothuk—who had first inhabited it.[48]

Before moving forward, it is important to underscore several additional features of the maps. First, they are incredibly spatially precise. Howley is among several settlers who comment on the maps' "extraordinary minuteness of topographical detail."[49] Second, Shanawdithit was not simply drawing her land and her people; she was actually drawing herself. In the map at the center of this chapter, Shanawdithit appears in multiple places and in multiple forms: in the form of data as one of the 37 tick marks on the south side of the lake, and again on the north side as one of the figures in red that sought shelter in the woods. While she may have been recording "information" about her people for Cormack, to return to his words, she was also testifying to the events of her own life. It follows, then, that this map is also a document of Shanawdithit's "survivance," to enlist a term coined by Chippewa scholar Gerald Vizenor; in continued unfolding of colonial violence, he argues, survival constitutes an act of resistance in and of itself.[50]

With that said, the violence that surrounds—and is recorded in—this map's creation ensures that it can never be upheld as a triumph alone. For even if it might exemplify a "kin-space-time constellation," it also vivifies the violence that is the reason it was even set to the page. An additional biographical detail accentuates this point. The art historian Nicholas Chare, who has written on Shanawdithit's maps through the lens of trauma studies, locates in a note written by Cormack the otherwise unremarked-upon fact that Shanawdithit "received two gunshot wounds at two different times, from shots fired at the band she was with by the English people at Exploits," and that "one wound was that [of] a slug or buck shot thro[ugh] the palm of her hand."[51] While it is unknown which hand Shanawdithit employed to create her drawings, "it may well have been the hand she sketched with," Chare suggests.[52] The wound and the scar it left on her skin—one which Cormack reports that he saw—thus serve as a visceral reminder of how Shanawdithit's maps were a direct output of colonial violence, the very same that led to the destruction of her culture and the death of her kin.

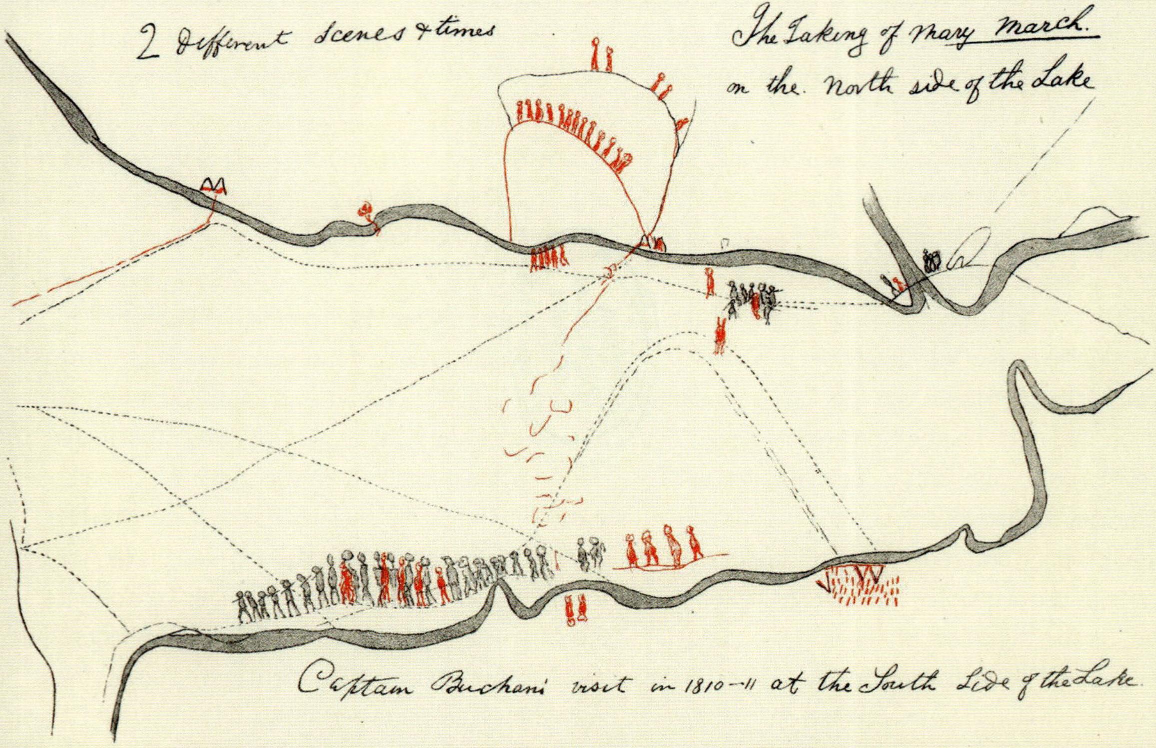
2 different scenes & times
The Taking of Mary March.
on the North side of the Lake
Captain Buchans visit in 1810–11 at the South Side of the Lake

pour la bibliothèque impériale

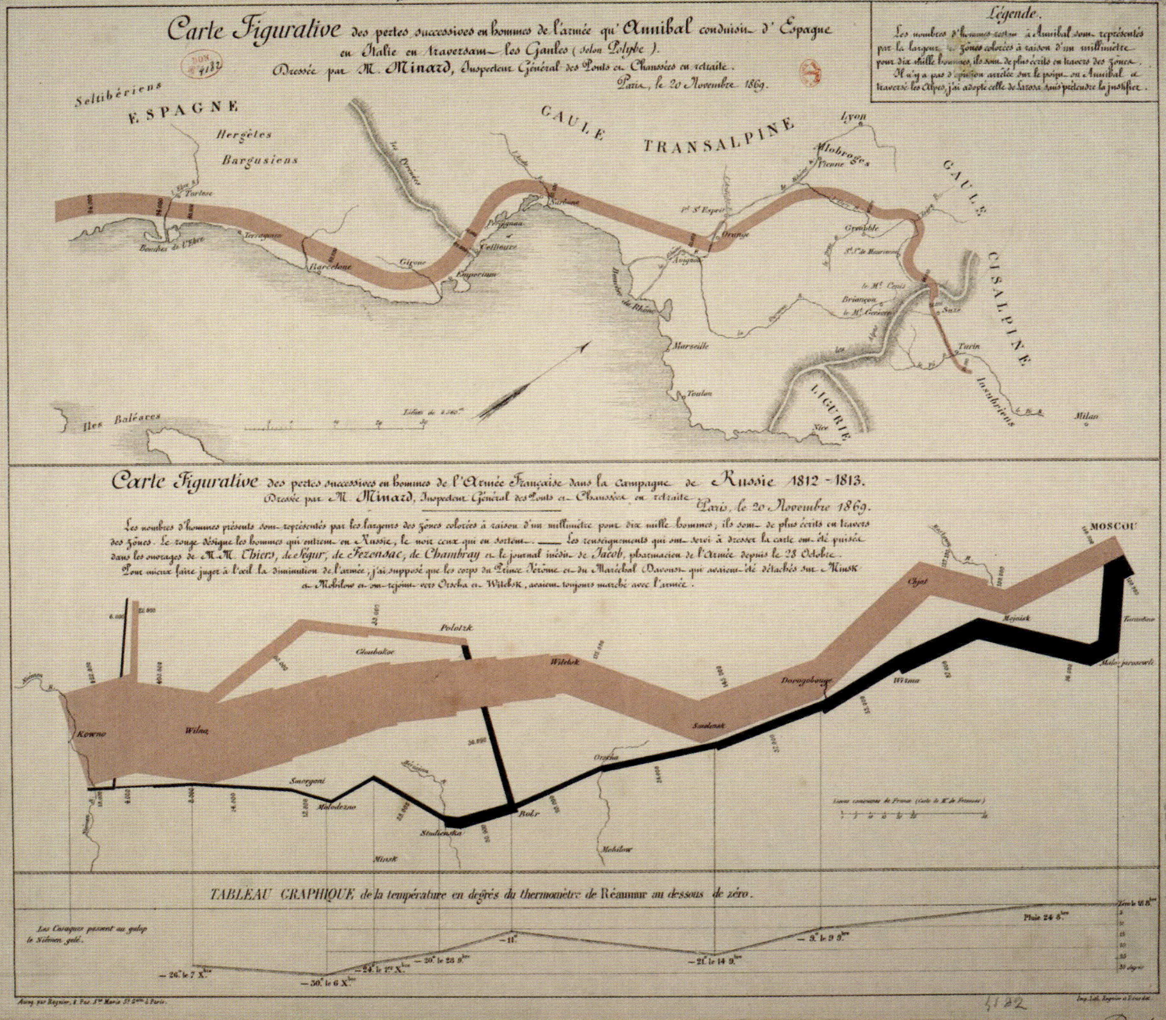

Maps and the Production of Colonial Power

Maps can literally create nations and dismantle others—a lesson that most Indigenous inhabitants of Turtle Island had learned well before the encounter between the Beothuk and Peyton and his men.[53] A resonant example is the so-called "Walking Purchase," which dates to 1737, nearly a century before Shanawdithit set her maps to paper, when the Lenape leader Teedyuscung agreed to sell a parcel of land to the Penn family (of the then-colony of Pennsylvania) that was bounded by the distance that a man could walk in a day and a half. After the treaty was signed, the Penns' agent cleared a trail through the land and hired three of the fastest runners he knew to run along it, resulting in the Lenape ceding a swath of land twice as long as was initially envisioned. In response to the "fraud"—Teedyuscung's own characterization in his report to colonial officials—he subsequently "insisted on drawing his own map to delineate [the Lenape] territory and solidify their rights."[54]

Or consider the end result of a seemingly innocuous encounter between Ac ko mok ki, a Siksika leader, and a surveyor for the Hudson's Bay Company named Peter Fidler, which took place at an outpost just east of what is more commonly known today as Alberta in February 1801. At the time, the outpost represented the frontier of colonial knowledge as well as settlement. When asked by Fidler about what lay further north and west, Ac ko mok ki traced in the snow—from memory—a map of more than 200,000 square miles of the continent, narrating the features of the map as he drew. Fidler then copied the map onto paper "reduced ¼ from the original," annotated it with the information he'd heard Ac ko mok ki speak aloud, and then sent the map back to the headquarters of the Hudson's Bay Company in London (FIGURE 3.13). Ac ko mok ki's knowledge was then incorporated into the map of the continent that the Hudson's Bay Company had been preparing, and which three years later would be used by Meriwether Lewis and William Clark to determine the route for their expedition to the west coast.[55] Their mission is widely recognized as authorizing the United States' future claims to the full width of the continent, and seeding the idea of "manifest destiny" that would validate US territorial expansion into the next century and beyond.

For the Indigenous knowledge that they consistently capture, and the dispossession they continually leave in their wake, such maps might be understood as weapons—weapons of map destruction, to adapt a phrase from Cathy O'Neil. They become tools to dismantle Indigenous sovereignty just as effectively as they consolidate the knowledge on which colonial power depends.[56]

A key example of how colonial maps consolidate knowledge, and further shape it, comes to us from a Connecticut-born woman by the name of Emma Hart Willard. An educator and activist—and student of data visualization, as will be further explored in chapter 4—Willard created nine maps to serve as the structuring images of her US history textbook (FIGURES 3.14–3.23). Willard's *History of the United States, or The Republic of America* was first published in 1828, the same year that Shanawdithit created her maps. Unlike Shanawdithit's maps, however, which have never seen wide circulation, Willard's textbook would be reprinted every year until the 1860s, when the US Civil War would require wholesale revision to the nation's origin story.

We have previously seen how the "simple and complete" picture staged by William Playfair in his import-export charts allowed his viewers to reject complexity and overlook potential harm. In Willard's

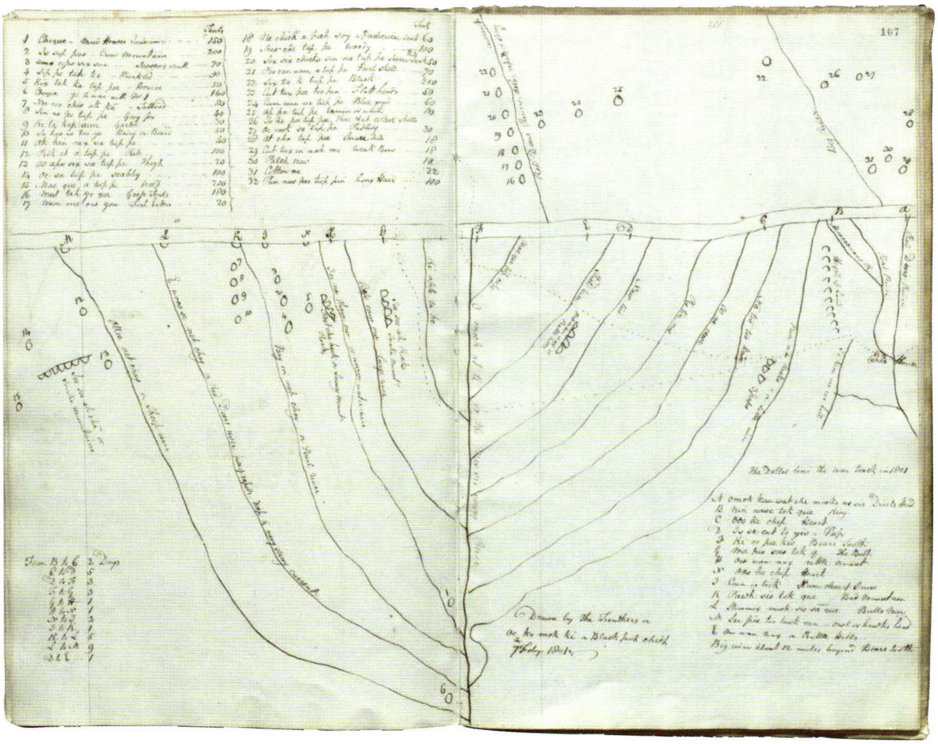

FIGURE 3.13
A copy of a map first drawn in the snow by Ac ko mok ki, a Siksika leader, in February 1801, which represented nearly 200,000 square miles of the North American continent. Image courtesy of the Hudson's Bay Company Archives, Archives of Manitoba. Digitized by the Hudson's Bay Company Archives.

maps, we find a similar simplification, here enlisted in the service of a story of nation formation. From the "First Map or Map of 1578," which depicts the routes taken by European explorers—including John Cabot to Newfoundland—to the "Second Map or Map of 1620," which depicts the colony of Virginia (along with an inset documenting the Pilgrims' landing at Plymouth Rock), and on to the final "Ninth Map or Map of 1826," which depicts the then-present day, Willard presents a "cumulative statement of nationhood," as historian Susan Schulten explains, one that enlists the consolidating power of the map in the service of a story of America's national emergence.[57] As Willard herself explains, her maps connect otherwise isolated historical "facts" in history and as a result "contribute much ... to the growth of wholesome national feeling."[58]

This "wholesome national feeling" for the United States was premised, not surprisingly, on the removal of all existing (Indigenous) nations from its own national origin story—as from the land itself. This removal was required to remove any competing claims to its narrative of divinely ordained flourishing and growth. It is notable that the only map in the textbook that features Indigenous peoples or nations in a place of prominence is the "Introductory Map" that appears before the map labeled "First Map." The map is subtitled "Locations and Wanderings of the Aboriginal Tribes," a title that rejects the possibility of Indigenous agency through its description as "wanderings" of what was, in truth, a combination of seasonal and forced migration.[59]

Willard places labels in the approximate locations of each Indigenous nation or tribe that she knew.

The color of the circles and the lines connecting them indicate affiliation and "migration," in Willard's terms, although a more accurate word would be displacement.

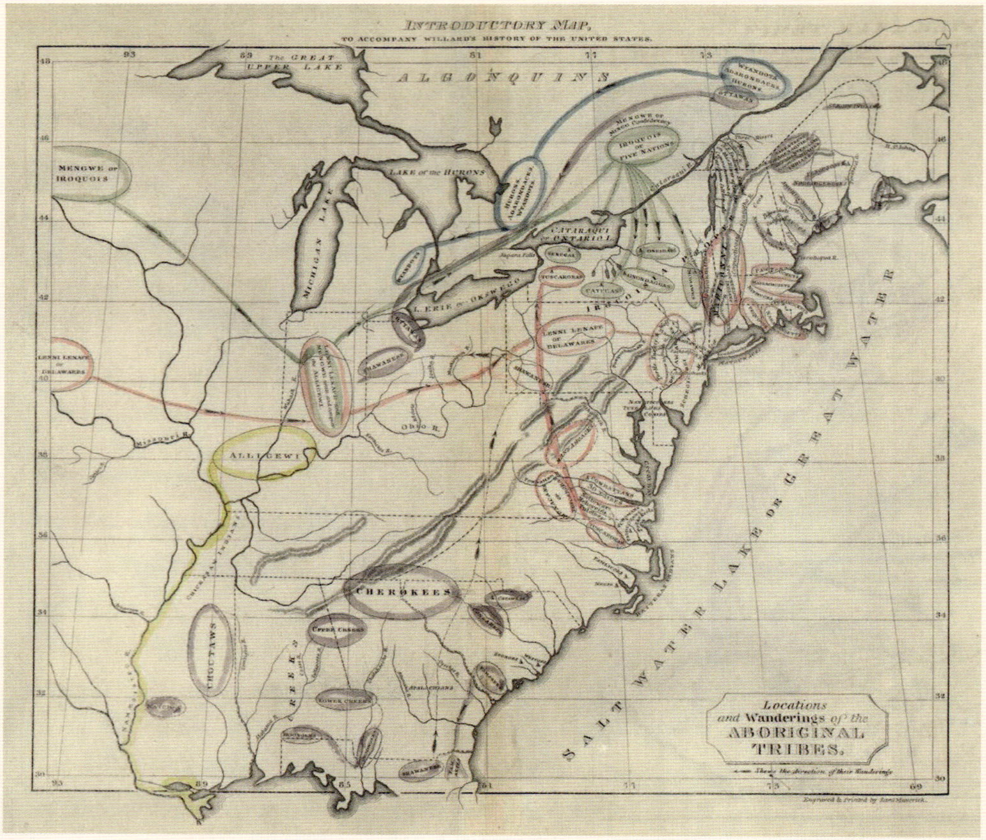

← Willard makes the clear choice to label certain geographic features with names intended to evoke an Indigenous worldview, such as "Salt water Lake or Great Water."

← The map is subtitled "Locations and Wanderings of the Aboriginal Tribes," rejecting the possibility of Indigenous agency by failing to recognize what was, in truth, a combination of seasonal and forced migration.

↑ Willard places the state borders of the not yet extant United States in the background of the map, presenting them as the literal ground truth on which Native peoples are only temporarily superimposed.

↑ Willard places labels in the approximate locations of each Indigenous nation or tribe that she knew. She also circles each of the tribe's names, with the size of the circle indicating its "size and relative influence."

FIGURES 3.14
The "Introductory Map" of Willard's *History of the United States, or The Republic of America* (1828), subtitled "Locations and Wanderings of the Aboriginal Tribes." Courtesy of the David Rumsey Map Collection, Cartography Associates.

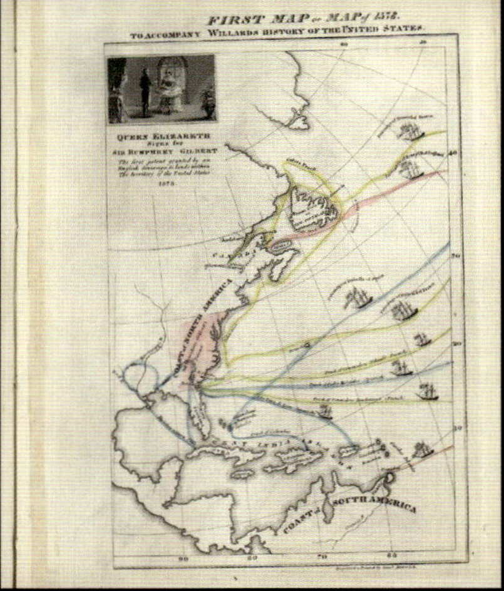

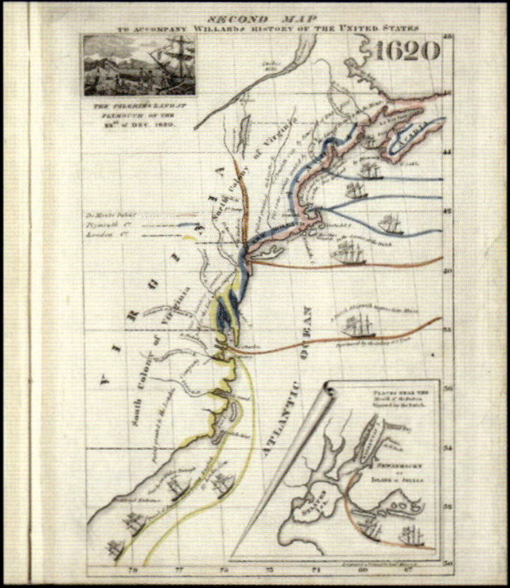

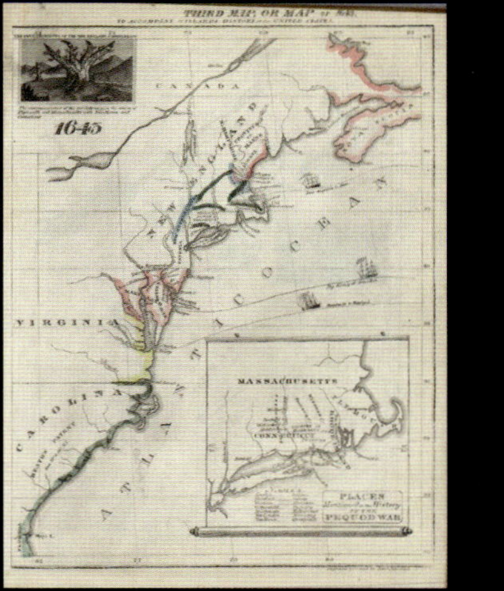

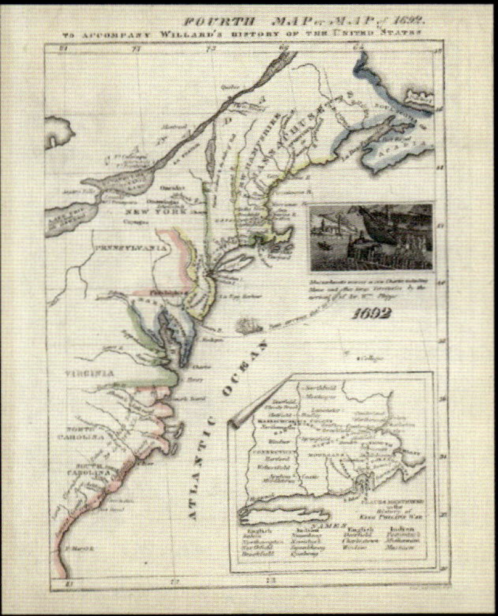

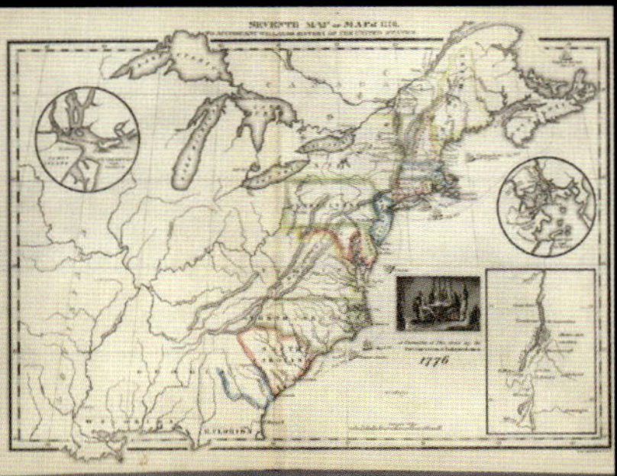

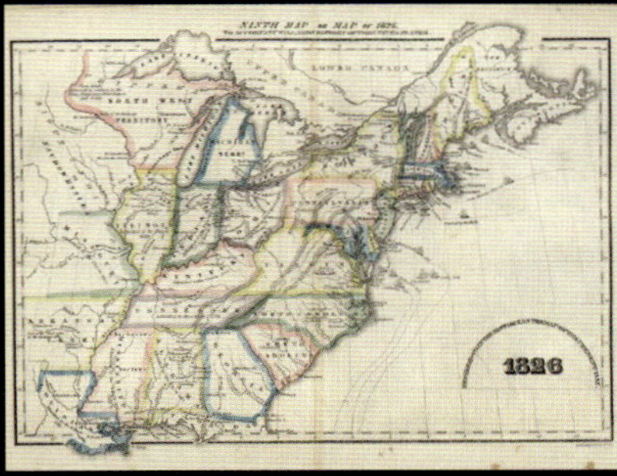

FIGURES 3.15–3.23
The nine numbered maps that accompany Emma Willard's *History of the United States, or The Republic of America* (1828), spanning the years between 1578 and 1826. Images courtesy of the David Rumsey Map Collection, Cartography Associates.

She also circles each of the tribe's names, with the size of the circle indicating its "size and relative influence." The color of the circles and the lines connecting them indicate affiliation and "migration," in Willard's terms, although a more accurate word would be displacement. Yet the perspective inhabited by the map is contradictory. Willard makes the clear choice to label certain geographic features with names intended to evoke an Indigenous worldview, such as "Salt water Lake or Great Water," which she explains in the accompanying chapter are two names given by the Delaware people to the Atlantic Ocean at various times.

Yet Willard also chooses to present this map as "introductory," rather than accord it the position of "First Map." That title is reserved for the map depicting the voyages of the European explorers, as previously discussed. More pointedly, Willard never again employs the same color and label combination to designate Indigenous sovereignty; after the first map, color is reserved only for British and other European voyages, then settlements, then colonies, and eventually US states. This design choice "reinforced the contemporary assumption that Native Americans existed in a timeless space prior to human history," Susan Schulten explains.[60] Nations such as the Beothuk are not granted a place in the future of North America, only its past.[61]

This view is confirmed when considering Willard's work as an example of the "thematic map" genre. Such maps can be analyzed in terms of the layers of data that they visualize, and the designer's choices about how to order them.[62] Generally, whatever data is plotted as the bottom layer of the map is presumed to be stable and true—the "base data" over which new layers can be added, and through which new insight can emerge. Thus, as historian and cartographer Bill Rankin has observed, it accords whatever dataset is placed at the bottom the status of incontrovertible fact.[63] In the case of "Locations and Wanderings of the Aboriginal Tribes," Willard places the state borders of the not yet extant United States in the background of the map, presenting them as the literal ground truth on which Native peoples are only temporarily superimposed.

With our eyes attuned to the layers of Willard's map and the claims implied by each, it is worth returning to Shanawdithit's map once more to consider the parallel claims implied by its "kin-space-time constellation." In rejecting the distinction between foreground and background, and by presenting people, place, and time in a single visual plane, Shanawdithit insists on her cultural as well as geographic authority. By contrast to Willard, who strategically deploys temporal data in order to impose her own story onto the land, Shanawdithit employs time to unify the many stories that connect people to the land, across past, present, and future. Viewing her map from a settler perspective, and more than two centuries removed, we—the contributors to this chapter—cannot know the exact stories that order those relations. But we can recognize the additional stories that have shaped our relationship to the map ever since (and, in truth, even before) it was set to the page. These are signaled by Cormack's annotations, and point to the relations between colonizer and colonized, and between knowledge and knower. These are not "good relations," but they are necessary to acknowledge and to understand, because these relations—to recall the words of Mishauna Goeman quoted in this chapter's introduction—are what "continue to mark us today."

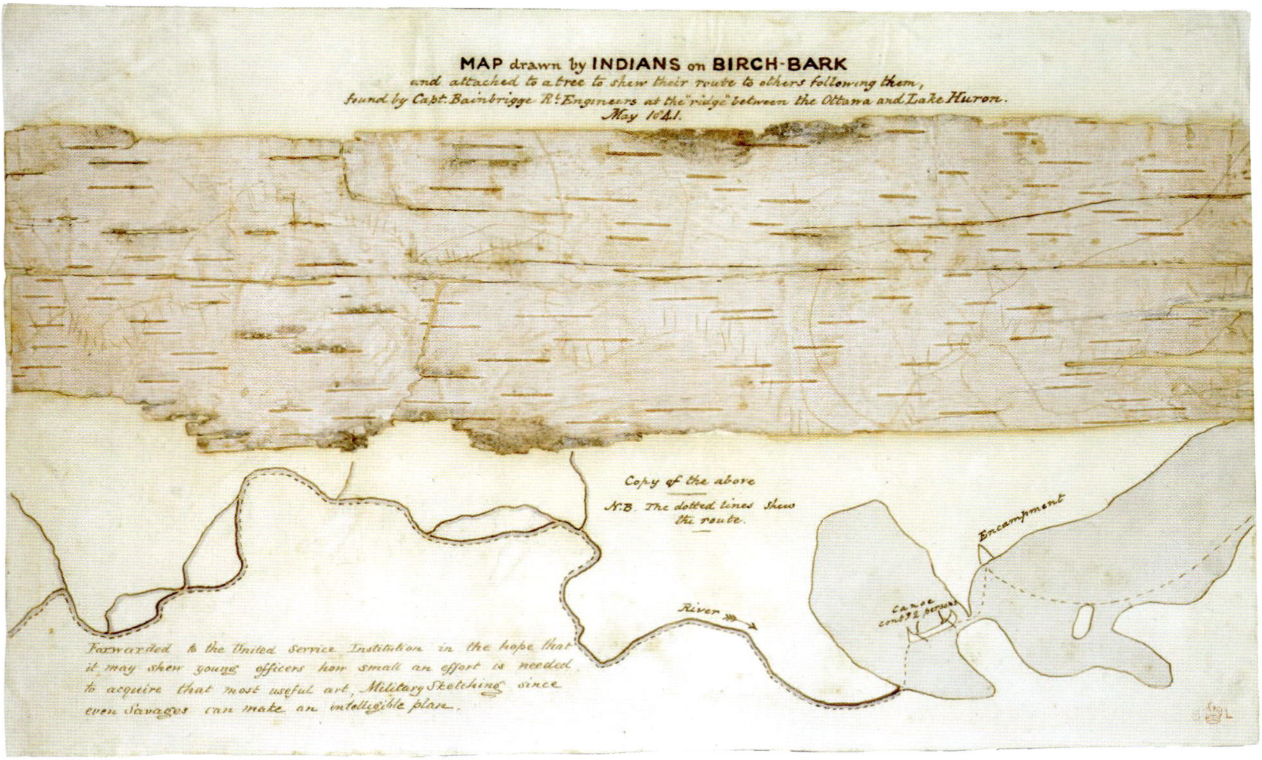

FIGURE 3.24
An Ojibwe *kikaigon*, or directional map, which was found pinned to a tree by a British surveyor, Captain Bainbrigge. Bainbrigge reproduces the map below, adding his own notes and comments. Courtesy of the British Library.

The Birch Bark Map and the Lookout Tree

About a decade after Shanawdithit set her maps to paper, just over a thousand miles to the west, two men, likely Ojibwe, stood atop a ridge between Lake Huron and the Ottawa River. They were a day ahead of the rest of their group, and wanted to inform those behind them of the progress of their trip and the route they were soon to take. So one or both of the men did something they had done countless times before. They etched a kikaigon, or directional map, into a piece of birch bark and attached it to a tree. The map depicts the men's path across Lake Huron, including the location of the campsite where they had spent the night, as well as the canoe they were using to travel. Two vertical lines drawn inside the canoe, each attached to an oar, represent the men themselves. The map suggests that they planned to continue down the Ottawa River, perhaps all the way to Montreal, another 400 miles away.[64]

We do not know whether the other members of the group ever saw the map. We do know, however, that it was seen by one Captain Bainbrigge, of the Royal Engineers. We know this because Bainbrigge took the map off the tree and affixed it to a larger sheet of rag paper. Below the original, he drew his own copy

FIGURE 3.25
In *Birch Bark* (2022), contemporary artist Sarah Sense (Chitimacha/Choctaw) employs Native weaving techniques as a way to interrupt the process of knowledge extraction that is documented in the original 1841 map and redirect its meaning-making force.
© 2022 Sarah Sense. Image courtesy of the artist.

of the map, replacing the vertical hash marks used to indicate the men's route with more constrained dotted lines, but keeping the icons used to indicate the camp, the canoe, and the men. He added an indication of the direction of the river and several other annotations, as well as a pejorative note.[65] He then sent the map back to England, where it eventually arrived at the British Library. As a result, it has earned distinction as the oldest known example of a birch bark map to have been preserved (FIGURE 3.24).[66]

Yet Bainbrigge's preservation of the map flattens it—and not only in a literal sense. By removing it from the time, place, and people for whom its insights were intended, Bainbrigge removes much of its meaning. His annotations, even more than Cormack's, impose his own assessment of its value. And because the original map and Bainbrigge's copy are literally on the same page, this assessment is impossible for contemporary viewers to ignore. We can no longer employ the map to produce the full range of insights for which it was initially designed. But these limitations offer a new set of insights for us today: Bainbrigge's colonialist arrogance—his presumption that he could and should use the map for his own purposes—reminds us of the necessary limits of our own efforts to seek knowledge with visualization, which are often (but not always) the result of the colonial frame.

But contemporary artist Sarah Sense (Chitimacha/Choctaw) demonstrates how such presumption can still be contested and further transformed. In her digital artwork *Birch Bark,* she employs her own ancestral basket-weaving techniques as a way to interrupt the unidirectional process of knowledge extraction that the Bainbrigge map records (FIGURE 3.25). Her own process "re-Indigenizes" the map, as she explains in the accompanying artist's statement, redirecting some of the map's meaning-making force.[67] Here, the "warp" of the image—what in basket-weaving practice is viewed as the more passive layer, since it serves as the basket's base—is a reproduction of the Bainbrigge map. Woven through the map, in patterns derived from Sense's own Chitimacha and Choctaw heritage, is a photograph of the land. Considered as a thematic map, Sense's choice to make the land the more active "weft," the layer controlling the pattern that is produced, contests the authority of the map that serves as its base.

Our project team returned to Sense's image multiple times when considering how we might similarly infuse Shanawdithit's map with new meaning. But ultimately we recalled the lessons of this chapter, reminding ourselves that Shanawdithit's map and the knowledge it encodes was not ours to further dissect. With this in mind, we turned away from the map and back toward the archive that set the story we tell here in motion. We asked ourselves how we might employ our team's perspective, and our skills, to contest the authority of the colonial archive from within.

One particular passage seemed to hold the key to this effort. It is a lengthy footnote that appears in Peyton's narrative (here reproduced in its entirety):

> Mr. Peyton afterwards learned from the woman Shanawdithit, the full particulars of the manner in which his boat was stolen. She was present all the time and knew every incident connected with this event. As Mr. P. rightly conjectured, it appears the Indians were watching all his movements very closely. There was a high wooded ridge behind his house, which from its peculiar outline had been named Canoe Hill. It bore some resemblance to a canoe turned bottom up. One tall birch tree on the summit of this ridge, (still standing at the time of my first visit in 1871), was pointed out by

> Shanawdithit as the lookout from whence the Indians observed Peyton's movements, during several days preceding the depredation. She also informed him, that when he paid his last visit of inspection to the long wharf before the taking of the boat, that the Indians were actually hidden in their canoe beneath the wharf, but kept so perfectly motionless, that in the dense darkness he did not observe their presence.[68]

Here in this footnote, perhaps deliberately relegated to the bottom of the page, is the suggestion of a version of the events that contests the authority of Peyton's account. In this version, Shanawdithit serves as the authoritative source of knowledge, since she "was present all the time and knew every incident connected with this event." Peyton, meanwhile, only learns "the full particulars" from Shanawdithit after the fact. With this evidence of an earlier phase of surveillance, one far more sustained, in which the Beothuk "observed Peyton's movements" for "several days preceding the depredation" from the tree, it is the Beothuk—and not the British—who are in control.[69]

The footnote also draws out a second theme. I have previously discussed how Shanawdithit's maps are mediated documents, inseparable from the colonial violence that produced them. But so too is Peyton's narrative. It was filtered not only through his eyes but through his memory, recorded late in his life, in 1871—nearly a half-century after the original events transpired—by none other than James Howley. It was then rewritten by Howley for the publication of his own book, *The Beothucks or Red Indians,* which was published by Cambridge University Press in 1915. Howley was likely the author of the footnote, and any additional edits he might have made to Peyton's version of the events—as he did to Shanawdithit's—will remain forever unknown.

FIGURE 3.26
A photograph of a lookout tree at Beothuk Point, taken in 1914 by the American anthropologist Frank Speck. Digitized by Margy Adams.

SPECK—BEOTHUK AND MICMAC PL. II

VIEW ACROSS RED INDIAN LAKE FROM RED INDIAN POINT, TAKEN FROM NEAR THE TOP OF "LOOKOUT TREE," SHOWING VIEW OF THE WATER WHERE CARIBOU SWAM BY

FIGURE 3.27
Speck's photograph of the view across Beothuk Lake, taken from "near the top" of the lookout tree. Digitized by Margy Adams.

But his is not the only narrative that has been filtered through memory (or power or time). John Paul's account, which we last considered in our analysis of the events depicted in Shanawdithit's map, was also recorded in the twentieth century, by an American anthropologist named Frank Speck, who published it in his own book on the Beothuk and Mi'kmaq in 1922.[70] Speck's book, interestingly, also contains a series of photographs, which he took during his own visit to Newfoundland in the summer of 1914. One of these photographs is of a lookout tree (FIGURE 3.26).

This lookout tree is not the same as the "tall birch tree" that Shanawdithit told Peyton about, as Speck labels it as being located at Red Indian (now Beothuk) Point and identifies it as a "large white spruce."[71] But it remains resonant nonetheless, especially since he provides a second photo: a view from the top (FIGURE 3.27). For Speck, the view from the top of the tree suggests a window in an unmediated past, a sense of what the Beothuk themselves might have seen.[72] "I climbed [the tree] to experience the sensation of observing these wastes"—an archaic term for uncultivated land—"from the vantage point of the ancients," he explains. Needless to say, his presumption that he could "experience the sensation" of Shanawdithit and her people by, one hundred years later, climbing the same tree, is just that—a presumption. But what if we understood Speck's view from the lookout tree differently? In terms of his own "vantage point," and ours?

When I first encountered this photo, in a low-res scan of Speck's book made available online via the Internet Archive, I wasn't entirely sure what it showed. The emphasis on the view across the lake and the land in the distance had resulted in the foreground becoming out of focus. I speculated, but could not be certain, that at the bottom of the photo were the lower branches of the spruce reflected in the water. Even the center of the image seemed blurry in contrast to the crispness of the ripples of the water that appear, in focus, on the lower right. It was not until I myself traveled to St. John's to view Shanawdithit's maps where they are currently housed, in the cartographic archives of The Rooms, the provincial museum of Newfoundland and Labrador, that I realized what Speck's photo had captured: a covering

YEAR	DESCRIPTION
*	Birch tree grows from seed
1810 - 1811	Beothuk encounter David Buchan and his men
1818	Beothuk observe Peyton and his men from lookout tree (birch)
1818	"Theft" of Peyton's boat
1819	Attack on winter camp and capture of Demasduit
1819 - 1914	Grandfather tells his account to John Paul
1823 - 1829	Shanawdithit tells Peyton about lookout tree (birch)
1829	Shanawdithit creates her maps
1829	Cormack annotates Shanawdithit's maps
1871	Peyton relates his narrative to Howley
1914	Speck records John Paul's narrative
1914	Speck photographs lookout tree (spruce) at Beothuk Point
1914	Speck climbs lookout tree (spruce) at Beothuk Point
1915	Howley redraws Shanawdithit's maps
1915	Howley publishes his book
1922	Speck publishes his book
2008	Speck's book scanned and uploaded to Internet Archive
2021	Lauren begins research on this chapter
2021	Lauren sees Speck's photos of lookout tree (spruce)
2022	Tanvi begins designing visualizations for this chapter
2023	Lauren visits The Rooms in St. John's, Newfoundland
2023	Lauren sees clouds across lake in St. John's
2024	DxD team receives approval to include map in chapter
2025	DxD team interprets Speck's photos for this chapter
2026	Chapter is complete

HOW TO READ

Innermost ring is the first year listed on left. Successive rings correspond to subsequent years listed.

LEGEND

- Speck
- Howley
- Shanawdithit
- All sources
- All of history

FIGURE 3.28
A final diagram of the people and sources that have contributed to the account offered in this chapter, taking the power and presence of the Beothuk into account. Design by Tanvi Sharma.

of low, dense clouds. St. John's is nearly fully surrounded by the Atlantic Ocean, after all, and the water contributes to near-constant moisture in the air.[73] The clouds in the photograph may well have been the "vantage point of the ancients" that Speck thought he could experience by climbing the tree. But I believe that Speck captured even more: a metaphor for the mediated nature of his own vantage point, and, in turn,of ours today.[74]

Over one hundred years after Speck's visit to the Beothuk winter camp, and another hundred years after the British attack and the devastation that ensued, we would seem to have an unprecedented ability to capture the conditions of the present, just as we do to understand the complexities of the past. From the ever-increasing array of sophisticated mapping and visualization tools, to the always expanding capabilities of digital photography, now coupled with AI, it is tempting to assume that our potential insights are limited only by our ability to imagine how we might fill in missing data, leading to knowledge previously cloaked by metaphorical (or actual) clouds. But as this chapter has sought to show, we can learn as much from the clouds themselves as we can from the knowledge they conceal. Once we learn to see what separates us from the sources of information that we rely upon in our work, we can begin the work of repair: of identifying the additional people whose perspectives might complementement our own; the relationships we might begin to (or might better) cultivate in order to ensure true collective benefit; and the responsibilities we carry—to the people who will seek insight from our work in the future, to those we work with in the present, and to those who created our sources of data in the past.

We thus close with a final diagram of the sources of this chapter, recentered around the lookout tree (FIGURE 3.28).[75] It remains an incomplete picture, but its incompleteness—like Speck's clouds over Beothuk Lake—is intended as a guide: of how we can pursue more complete knowledge just as we recognize what we must leave for others to explore.

We will soon move on to consider alternative visual forms and epistemological frameworks, those that reject the unidirectional path of knowledge production in favor of a more balanced exchange of ideas (chapter 4). We will also consider examples of how more conventional visual forms can push back against extractive aims by centering otherwise overlooked perspectives, and by including methods that might otherwise be set to the side (chapter 5). In short, we will move from lessons of limits to those of possibility. But the colonial context that explicitly frames Shanawdithit's work, and implicitly frames so much more, should remain the context for the next chapters as well. And along with this context, we hope, is a lesson of intention and care—care for how we choose to frame future knowledge, as well as for our relationships with the people, past and present, who have enabled our work—and whose knowledge we seek to steward and share.

Takeaways

FOR VIEWERS

Resist the extraction of knowledge from its source

Consider your relation to the colonial frame

Recognize the people who enable your knowledge

Accept when that knowledge is not yours to own

FOR DESIGNERS

Examine your responsibility to your data and its stewards

Aim to keep your data more connected to its source

Balance clarity with context and connection

Allow for the possibility of not visualizing at all

CHAPTER 4

KNOWLEDGE

Elizabeth Palmer Peabody and the Value of Epistemology

Visualization can produce knowledge in many ways.
What perspectives do we gain when we expand our view of the past?

Elizabeth Palmer Peabody was born within a year or two of Shanawdithit, in 1804.[1] But her life circumstances—stability of food and family, membership in the Boston intellectual elite, and status as US-born settler woman—enabled her to live for nearly a century, over 65 years longer than Shanawdithit, who would die of tuberculosis several weeks after creating her maps. By the time that Peabody died, in 1894 at the age of 89, she had left her mark on a number of fields. The bookstore that she ran out of her home at 13 West Street, just off Boston Common, functioned as the de facto salon of the transcendentalist movement. She edited and published the first version of Henry David Thoreau's essay on civil disobedience. She can also be credited with starting the first kindergarten in the United States, which opened on the other side of the Common, at 15 Pinckney Street, in 1860. This was the same year that Abraham Lincoln would be elected President, South Carolina would secede from the Union, and the stage for the US Civil War would be near fully set.[2]

But even just a few years before, this national crisis seemed far from inevitable—especially to Peabody, whose own antislavery consciousness had yet to fully coalesce—and so Peabody left her bookstore in the hands of a caretaker and set off to ride the rails. She traveled as far north as Rochester, New York; as far west as Louisville, Kentucky; and as far south as Richmond, Virginia, all to promote a pair of history textbooks she had recently written, *The Polish-American System of Chronology* (1850) and *Chronological History of the United States* (1856). In the textbooks, Peabody described a new method of visual learning—the Polish-American system, she came to call it—a method with data visualization at its core (FIGURES 4.1–4.4).

Peabody's contributions to the development of data visualization are rarely recognized at the same level as her other achievements. But as this chapter will show, Peabody offered a transformative view of the uses and value of visualization—one that challenges the standard account of the field's emergence, as well as the epistemological assumptions that, as a result, have become embedded in best practices today. As further evidence of this claim, consider how Peabody boarded the train that would take her on her national tour with not one but two versions of her charts: the first, the textbooks that she hoped to sell; and the second, a set of rolled-up set of charts between five and six feet long. These were large-scale replicas of the data visualizations that appeared in the books. Peabody identified

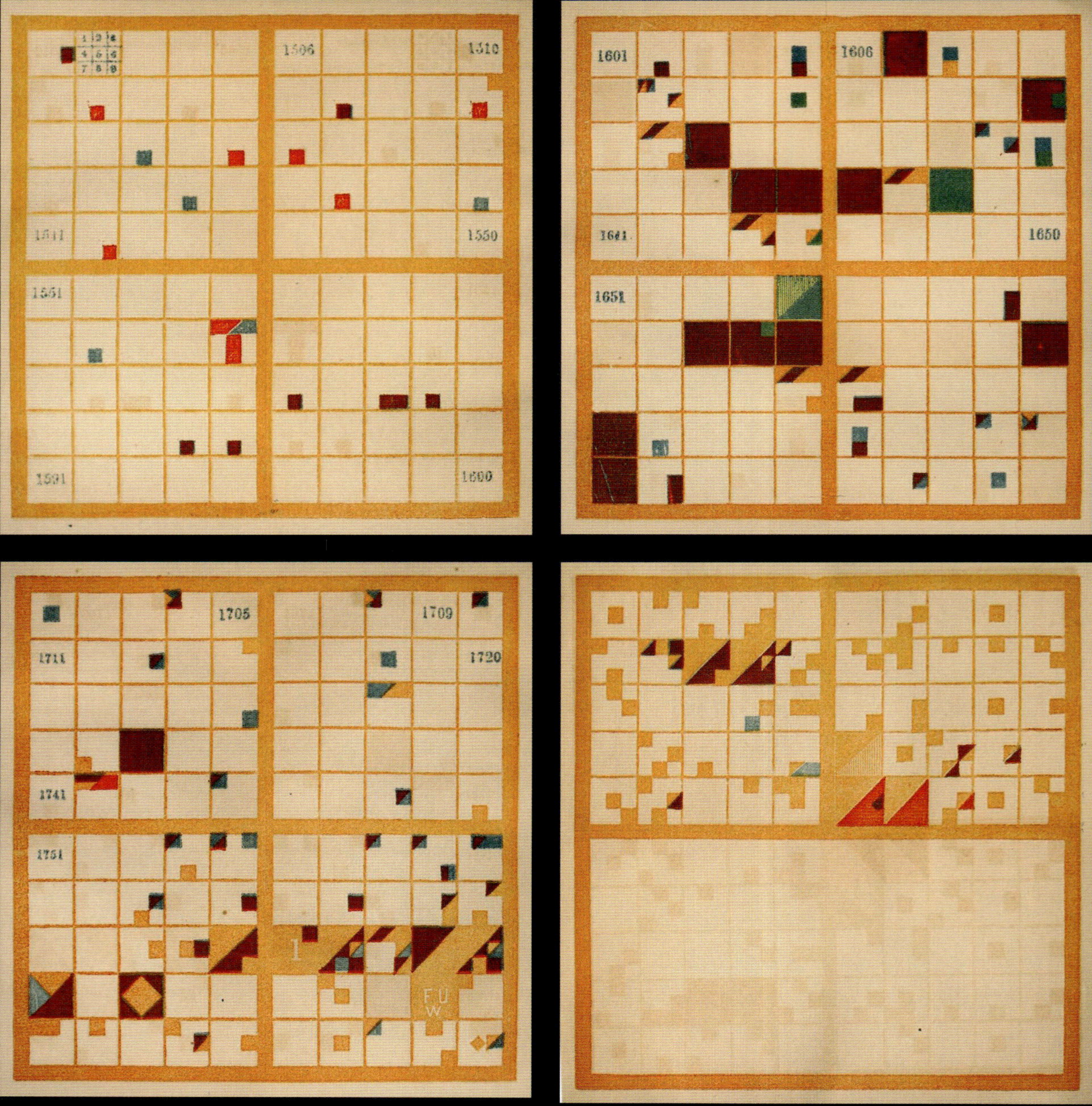

FIGURES 4.1–4.4
The four chronological charts included in Elizabeth Palmer Peabody's *Chronological History of the United States* (1856), which display the significant events of the 1500s, 1600s, 1700s, and 1800s. Images courtesy of the Internet Archive. Digitized by the Library of Congress.

the charts, with their vibrant geometric abstraction, as key components of her visual epistemology. The "painted centuries," as she sometimes also described them, were "intended to do for the science of history what maps do for that of geography; and ... [make it] easy to lay the foundations of historical knowledge in the minds of the young."[3]

Like Thomas Clarkson, William Playfair, and the other early proponents of data visualization whom we've encountered thus far, Peabody recognized the power of visual perception as a means of crystallizing the insights that lead to new knowledge. But she wielded this power in a very different way. She designed her charts to be abstract rather than indexical or concrete; to promote sustained reflection rather than immediate or intuitive understanding. And she did so with a clear goal in mind: to provoke a unique imaginative response in each viewer. Peabody's charts thus offer an alternative to the dominant view of how visualization leads to new knowledge—of the "single view" advanced explicitly by Clarkson and Playfair, and implicitly by Cormack in his request that Shanawdithit produce her maps.

As this chapter will argue, Peabody's charts model a feminist epistemology *avant la lettre*—that is to say, before the term "feminist" carried much of the meaning that it has today.[4] I make this claim not because Peabody herself was a woman, although this chapter does explore some of the reasons for the gender skew of the history of the field. Rather, by repositioning the source of the insights that lead to new knowledge from the visualization itself to its interplay with the viewer, Peabody flattens the hierarchy that more commonly structures the relationship between the designer of a visualization and those who perceive (or interact with) it. As we will learn in this chapter, this is a fundamentally feminist move.

In addition, in their strategic use of abstraction and the involvement of the whole body, Peabody's charts challenge several additional hierarchies that remain entrenched in visualization design—particularly those that place purportedly "objective" representational strategies above those that encourage more affective and embodied ways of knowing. In their rejection of the belief that every phenomenon in the world can be easily captured and rendered visible by data, Peabody's charts attune us to the multiple ways by which visualization can lead to new knowledge, as well as to the multiple ways in which we might understand the nature of knowledge itself.

These are not small contributions. Yet there is still more we might learn from Peabody and her charts. If there is an overarching lesson of this chapter, it is that Peabody's charts are but one set of examples of the many missing forms—and missing figures—that might be incorporated into the story we tell about the emergence of modern data visualization. Recall the argument of the introduction, where we were reminded that our knowledge is never complete. Peabody's charts provide a compelling example of how we might enrich our sense of our field's history, but the truth is that there are many more that remain to be explored. Indeed, if Peabody's charts expand our awareness of possibilities for the future, they should also expand our awareness of what remains unknown about the past—what is still waiting for us to come to see.

The Value of Relational Knowledge

Peabody's method of visualizing events of historical significance was inspired by a system developed in Poland in the 1820s and '30s by the educator Antoni Jażwiński (FIGURE 4.5), and popularized in subsequent decades by the military general (and erstwhile math

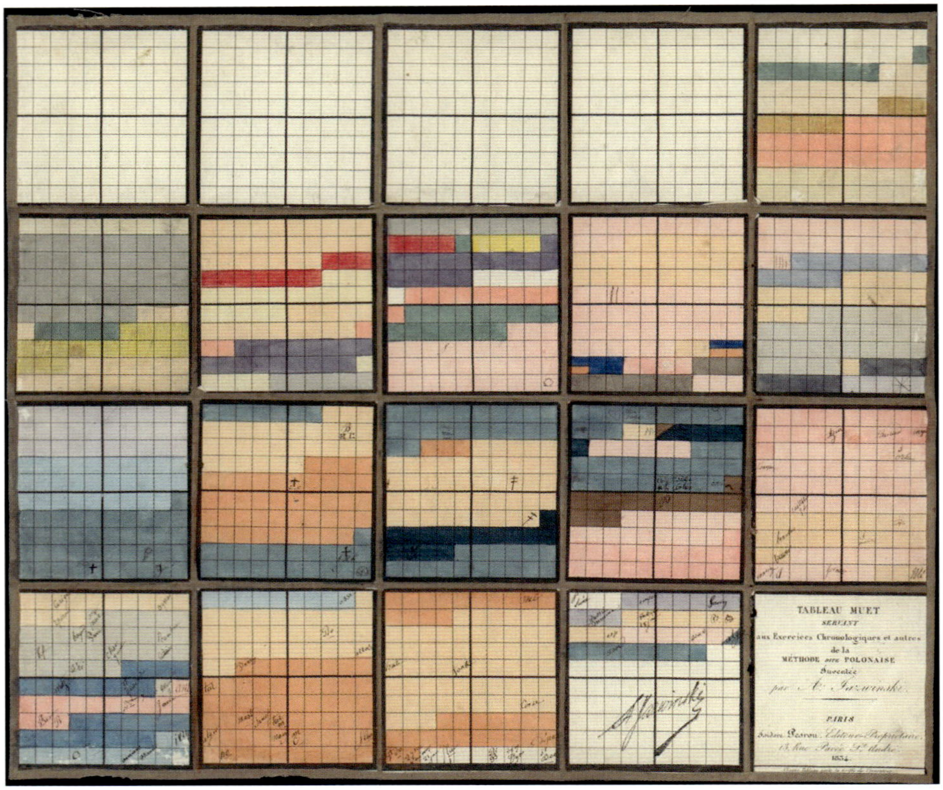

FIGURE 4.5
An example of Antoni Jażwiński's "chronological exercise" as published in his 1834 *Tableau muet servant aux exercises chronologiques et autres de la méthode dite polonaise inventée par A. Jazwinski*. Image courtesy of the David Rumsey Map Collection, Cartography Associates.

teacher) Józef Bem.[5] The Polish system, as it was called, aimed to visually represent events in time through a grid overlaid with shapes and colors. In *Cartographies of Time: A History of the Timeline*, Daniel Rosenberg and Anthony Grafton describe how the system "swept across Europe and North America" in the middle decades of the nineteenth century.[6] But Peabody first encountered it by chance: through a traveling lecturer who briefly boarded with her family on West Street. The boarder, a man named Joseph Podbielski, had come from Poland with copies of Bem's charts, which he intended to promote on a lecture tour of the United States. While he soon departed the family's residence, Peabody remained "captivated" by the charts, according to one of her biographers, Bruce Ronda.[7] She went on to devote several years of study to the Polish system, culminating with the development of her own modified version: the Polish-American system that prompted her national tour.

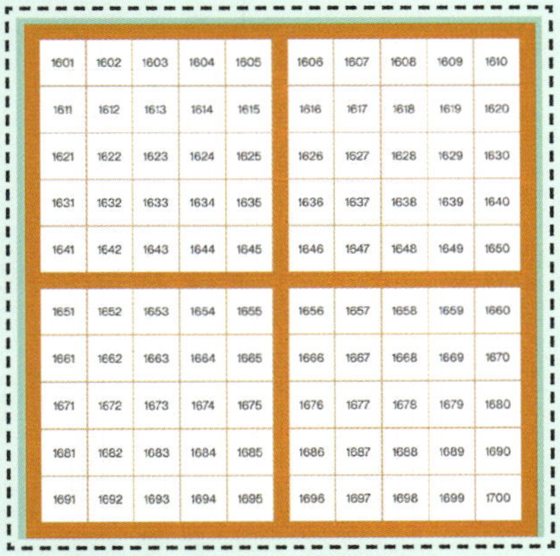

FIGURE 4.6
Peabody's numbered grid, borrowed from Józef Bem.

Peabody's system borrows from Bem the idea of a numbered grid, with each year of a given century marked out in its own square. They are arranged in rows of ten—one row for each decade (**FIGURE 4.6**). The thick orange lines that we might interpret as horizontal and vertical axes do not serve that purpose; they are intended only as "landmarks to the eye," Peabody explains.[8]

FIGURE 4.7
The subdivisions of each year-square.

Each "year-square" (Peabody's term) is then subdivided into nine interior squares, with each location in the year-square corresponding to a particular type of historical event (**FIGURE 4.7**).[9] In Peabody's system, as in Bem's, the top left corner of the year-square—labeled 1 in this diagram—is the location for battles, sieges, and beginnings of wars. The top middle—labeled 2—is the location for conquests, annexations, and unions. The top right—labeled 3—is the location for losses and disasters, and so on.

FIGURE 4.8 A color-coded year-square, here associating the bottom center square, "deeds," with England. In this case, we see the attempted English settlement of Nashua land.

The events are also color-coded, indicating the various countries involved in a particular event (FIGURE 4.8). On this point, Peabody makes special note that she employs "a somewhat different, and, as it seems to me, a more expressive distribution of colors" than the original system.[10] In Peabody's version, Spain is red, England is purple (now faded to a rich burgundy), France is blue, Holland is "yellowish-green," and "Americans," a term she specifies as including both "aboriginals" and "citizens of the United States," are orange.[11]

FIGURE 4.9 A solid-color year-square, here indicating the settlement of Jamestown by the British.

When an entire year is shaded a single color, it indicates an event of such magnitude or complexity that the other events in that year literally pale in comparison. For example, the square for 1607 is entirely burgundy, marking the settlement of Jamestown, the first permanent British colony in the Americas (FIGURE 4.9).

FIGURE 4.10 The square depicting the arrival of the first enslaved people being brought to American shores, which Peabody dates to 1620.

In 1620, we see two events (FIGURE 4.10). More significant, for Peabody, is the settlement of Plymouth by the Pilgrims, as it creates a "C" shape around a small green (formerly "yellowish-green") square. The latter marks the first enslaved Africans being brought to Jamestown, so shaded because they arrived aboard a Dutch ship. (This event is now recognized as taking place in 1619; Peabody was off by one year.) Peabody was on the side of abolition though by no means its most radical proponent, especially at that point in her life; the year-square's outsized ratio of burgundy to green reflects Peabody's awareness of—if not an urgency about—the need to end slavery in the United States.

FIGURE 4.11
The death of Pocahontas, as visualized on Peabody's chart of the seventeenth century.

Elsewhere on the charts, Peabody uses diagonals to show events that involve two or more nations. In 1622, we see a representation of the series of deadly conflicts between the Powhatan people and the Jamestown settlers (FIGURE 4.11). First, in the position that indicates battles, sieges, and beginnings of war, we see the attack by the Powhatan on the Jamestown colony that John Smith documented in his history of Virginia. In her accompanying list of events, Peabody labels this the "Jamestown Massacre." In the top left of the square is orange, the color for "Americans," here representing the Powhatan and their position as instigators of this particular battle. In the bottom right is burgundy, representing the British settlers, in their position as besieged. In the next square over—the position that indicates conquests, annexations, and unions—the colors are reversed, indicating retaliation by the British on the Powhatan. In the textbook, this is labeled "Assault on Powhatan Settlements." Peabody shades the final square in the row, for losses and disasters, entirely orange. The corresponding text reads "Indians are conquered."

In the same year, we see another orange square in the bottom right, the position representing "deaths of remarkable individuals." Peabody tells us this represents the death of Pocahontas, the daughter of Chief Powhatan himself.[12] Returning to the overall view of the century, that orange square at once represents the power of commemoration and the insufficiency of a colored box to stand in for a human life, however saturated with color—and meaning—that box may be.

→

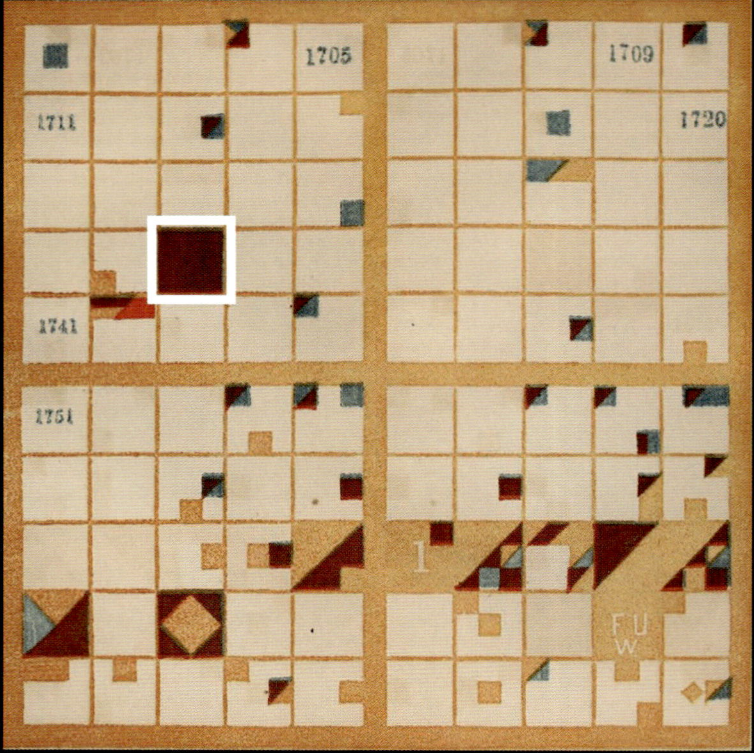

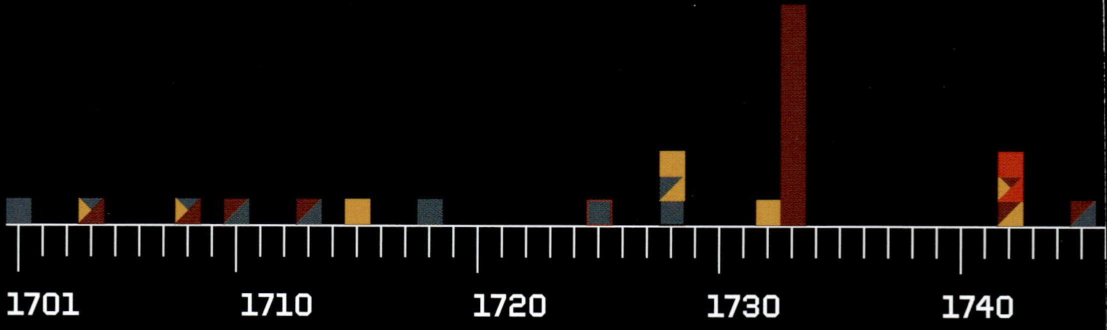

Visualizing Time

The Peabody Chart as Bar Graph

1500s 1600s 1700s 1800s

France Americas England Spain

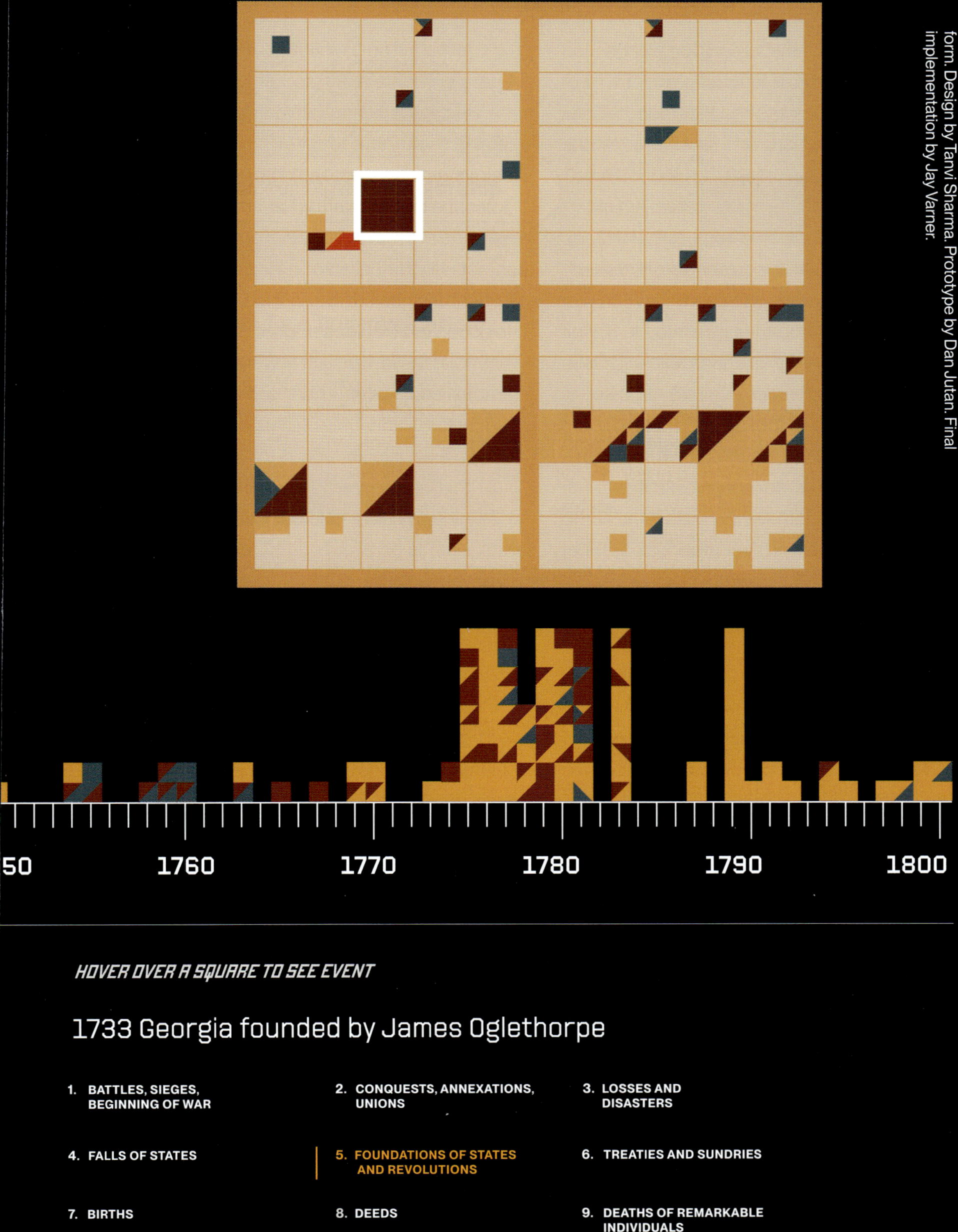

FIGURE 4.12
An interactive version of Peabody's chart in which event squares can be cross-referenced with their description. Below, the same event squares are visualized in timeline form. Design by Tanvi Sharma. Prototype by Dan Jutan. Final implementation by Jay Varner.

In *The Polish-American System of Chronology*, Peabody covers a tremendous temporal expanse: the period between 2500 BCE and what was then the present (1849 CE). But Peabody also saw the need for a textbook that focused exclusively on the United States. And so, shortly after the release of *The Polish-American System*, she began working on the book that would be published, in 1856, as *Chronological History of the United States*. This textbook contained the four full-color plates displayed at the start of this chapter; one for each of the centuries since the first European colonists set foot on Native land. As Peabody envisioned it, the basic exercise was to read a chapter of the textbook, which contained a narrative account of the events of a single century, and then work through a list of those same events that came at the conclusion of the chapter, matching each item with its visual representation on the corresponding chart.

Chronology—the study of events in time—is not, of course, the same as historiography—the study of how history is written. But in Peabody's mind, the one led to the other: "If you have the dates here [on the charts] represented perfectly by heart," she explains to her imagined students in the introduction to *Chronological History*, "events are so connected in the narrative of history."[13] In keeping with the leading pedagogical theories of the day, which emphasized mental recall, students were expected to commit the charts to memory.[14] But Peabody's approach diverged from the rote memorization that characterized most mid-nineteenth-century classrooms; her ultimate aim was for each student's mental picture of past events to prompt a richer—and, crucially, a personally constructed—narrative of history. This narrative would, in turn, help guide each student's future actions. As young citizens of the world's first representative democracy, on whose participation the future of the nation would soon depend, an understanding of the "origin and consequences" of past events would help guide their "own inevitable action" as each student became a "creator of its events" in adulthood, or so Peabody believed.[15]

Far from an antiquated line of thinking, Peabody's belief in the catalyzing effects of chronology remains deeply embedded in US culture today. As a prominent example, one might consider the efforts of *New York Times* journalist Nikole Hannah-Jones to replace 1776 with 1619—the year the first enslaved Africans actually arrived at Jamestown—as the starting point for the history of the United States. The goal of this revised origin point, as Hannah-Jones explains in the visual feature that introduces the "1619 Project," is to "reframe the country's history by placing the consequences of slavery and the contributions of black Americans at the very center of our national narrative."[16] Ideally, this recentered narrative would—in a view endorsed by the *Times* editorial board—then prepare US citizens of all races "for a more just future."[17]

But it is not only the narrative of the nation's founding that can benefit from a recentering of the enduring costs of its legacy of slavery, or of the contributions made by its Black citizens. In *Dear Science and Other Stories*, Black studies scholar Katherine McKittrick takes on the project not of history but of science itself, explaining how an account that centers Black people, Black life, and Blackness more broadly can reveal the "asymmetrically connected knowledge systems" that structure modern scientific inquiry.[18] For McKittrick, an awareness of the range of related yet unequally weighted sources of knowledge is what enables her own vision of a liberatory Black science to unfold. But it also offers lessons to scholars outside the field of Black studies—including white scholars, such as myself, who cannot claim to know Blackness firsthand—about the "asymmetrically connected knowledge systems" that structure all of our work.

As should be clear, all of our ways of knowing—about science, about history, or about phenomena in the world—are shaped by asymmetrical power, and data visualization is no exception. We have already seen this in theory in chapter 1, with feminist philosopher Donna Haraway's articulation of the "god trick," and in practice in chapter 3 with the colonial frame surrounding Shanawdithit's maps, among other examples discussed thus far.[19] The question that McKittrick's formulation of relational knowledge now allows us to answer is: *what is to be done?* For our awareness of the asymmetrical power associated with visualization does not diminish the validity of what it has allowed us to learn to this point, nor does it foreclose any future knowledge it might enable. On the contrary, a more nuanced understanding of the relations among what McKittrick calls "knowledge systems" and what I have called "epistemologies" is, as I see it, what will lead to the most complete knowledge that we can obtain.[20]

With a renewed sense of the value of multiple ways of knowing, and of what can be gained by understanding the (at times unequal) relations between them, we might return to Peabody's chronological charts with an observation that is both painfully obvious and quite profound: very few viewers—either in the twenty-first century or the nineteenth—can intuit the significance of the events encoded in the charts without first taking the time to learn how they work. This lengthy and difficult learning process might be viewed as a liability by those who champion the ease and efficiency of visualization, or by those who maintain that data visualization is best deployed to amplify already existing ways of seeing. In defense of this view, consider what happens if we alter

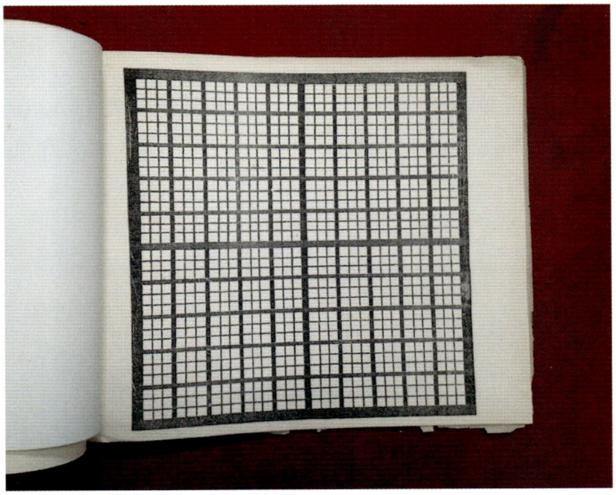

FIGURE 4.13
A blank chart from *Blank Centuries for Monographs of History* (1870), the workbook that Peabody printed to accompany her history textbooks. Courtesy of the American Antiquarian Society. Photo by Lauren Klein.

the representation of the events encoded in Peabody's chart of the seventeenth century, making it a timeline rather than a grid (FIGURE 4.12). The progression of events over the course of the century, culminating in a series of wars and rebellions in the late 1680s and early 1690s, is much more immediately legible than the top-to-bottom, left-to-right layout of her grid.

But for Peabody, her charts' near-total abstraction was precisely the point. She designed her charts to appeal to the senses directly, to provide what she called "outlines to the eye."[21] Her hope was that, with the outline of history provided—and, crucially, only the outline—each viewer could fill in the missing parts of the story themselves.[22] The result would be not only a proliferation of historical narratives, originating in the mind of each viewer, but, very crucially, narratives that reflected their own interpretation of the chart. This aspect of the Polish-American system separated it from most other visualization strategies of its day. It not only affirmed each interpretation of each image as a source of knowledge, but also established individual interpretation—shaped by deliberate pedagogy and informed by historical expertise—as the ground on which future decisions might be based.

The Epistemological Stakes of Interaction

Peabody was a lifelong proponent of what might be described today as participatory learning. In the early 1860s, as the nation became increasingly consumed by the Civil War, Peabody found personal comfort in a recommitment to early childhood education. Her kindergarten—as mentioned, the first of its kind in the United States—which she opened with her sister, Mary Tyler Mann, on Pinckney Street, served as a proving ground for her innovative pedagogical ideas. With her sister, she also published a series of essays documenting the theories they enacted in the classroom.[23] And on her own, Peabody continued to iterate on the teaching and learning materials emerging from the Polish-American system.

In 1870, Peabody began printing workbooks with sheets of blank charts—"blank centuries," as she called them—so that students could themselves create the visualizations that they would then study (FIGURE 4.13). In the interactive version of this book, we adapted this participatory process for the web (FIGURE 4.14).

It is not necessarily a negative that you are only able to read about this exercise, rather than participate in it; the difficultly level is high. Evidently, this was also true for nineteenth-century students. Peabody's nephew, Julian Hawthorne, who served as her first test subject, recalled that she "labored during some years to teach me all the leading dates of human history," but that he nevertheless remained "most inapt and grievous" throughout the process.[24] Surprisingly abundant archival evidence confirms this account. The American Antiquarian Society, the Library Company of Philadelphia, Yale's Beinecke Library, and Princeton's Special Collections all contain multiple student copies of Peabody's workbooks. Each tends to follow a similar pattern: a page or two of grids filled out in earnest; then, in the next few pages, a series of attempts abandoned halfway; and then, for the remainder of the workbook, an obvious shift in purpose—either the grids are entirely blank, indicating the exercise abandoned, or they are consumed by doodles and drawings, becoming a canvas for pattern and play (FIGURES 4.15–4.22).

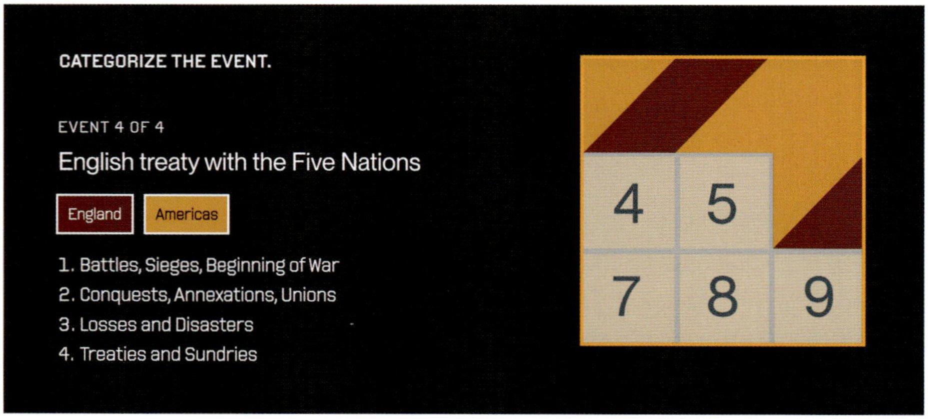

FIGURE 4.14
An interactive quiz designed by the *Data by Design* project team as a way to reimagine Peabody's workbook process in digital form.

Peabody would have likely been dismayed by such blatant disregard for her intentions; but it is also plausible that she would have taken at least partial pleasure in these students' reimagined uses of her grids. After all, the difficulty of the Polish-American system was both a liability of the form and part of the point. Peabody first developed her method at a time when the nation's future seemed to hang in balance. The second half of the 1840s had brought an increased awareness of the nation's growing sectarianism, as well as its range of social ills—albeit with a (mostly) optimistic view about the potential of its governing structures to address these challenges. But as the 1850s unfolded, especially after the passage of the Fugitive Slave Act galvanized those in the North—including Peabody, who had previously sought to sit out of the political fray—the magnitude of these challenges became increasingly pronounced. Over those years, Peabody came to recognize that the moral obligation of ending slavery, not to mention a resolution to the underlying issue of sectarianism (which was her original concern) would require difficult political maneuvering. She understood that any successful resolution of these conflicts would require considering all possible paths forward, as well as time for sustained effort and space for deep thought.

Peabody's goal with the Polish-American system was to create a framework, equal parts pedagogical and epistemological, through which this difficult thinking could take place. "The old world is covered with bad institutions which men have created, very often with positively good intentions, but on false notions, or, at least, without large and profound ideas," she explains in the preface to the *Chronological History*. "Whether the new world shall estimate and sift out these evils, or repeat these mistakes, depends on young Americans, who are now sitting in schoolrooms all over the country, unconscious of their powers and consequent responsibilities."[25] Her hope was that the act of creating the chronological charts, rather than simply studying them, would prompt both self-reflection and new ideas. Put another way: Peabody hoped that by prompting her students to create new narratives of the past, they would also imagine alternative possible futures.

The historical and political context that motivated Peabody to publish her textbooks is crucial for understanding both her ideas about the uses of visualization and the form that her visualizations take. We have already seen several other examples of this. In chapter 2, we learned how Playfair's time series charts were born of a desire to communicate a big-picture view to "men of high rank" who did not have time to concern themselves with the details. In chapter 3, we considered how Emma Willard's choice to put the outlines of US states in the background of her map of Indigenous nations accorded the United States the status of established fact, rather than its still-contested reality. To probe the convergence of Peabody's epistemology with her pedagogy and political ideology, we will turn to one of Willard's later charts, which helpfully puts its own ideology right on the surface (FIGURE 4.23).

Willard designed her *Temple of Time* in 1846 after nearly two decades of her own continual experimentation with ways of visualizing history. She explored the affordances of maps, as discussed in the previous chapter, and then a set of visualizations inspired by Alexander von Humboldt (who was himself inspired by the Colombian naturalist Francisco José de Caldas) (FIGURES 4.24–4.27), and then the idea of a memory palace, as exemplified by *Temple of Time*.[26]

In the chart, Willard depicts past centuries as the pillars that support the titular temple. Here, the nineteenth century is represented as an unfinished column,

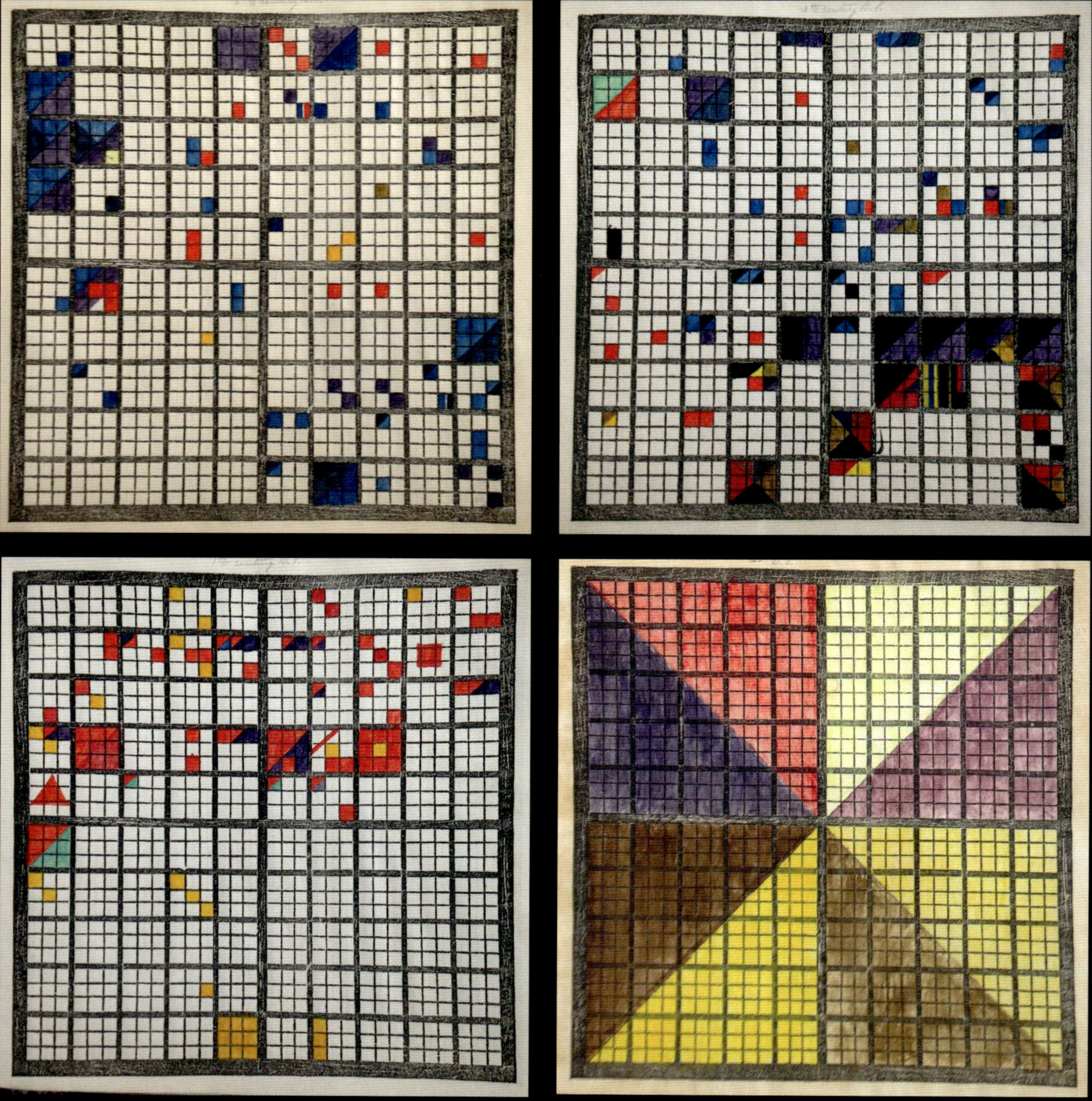

FIGURES 4.15–4.22
Images of student-created charts from a copy of *Blank Centuries for Monographs of History* housed at the American Antiquarian Society. Courtesy of the American Antiquarian Society. Photos by Lauren Klein.

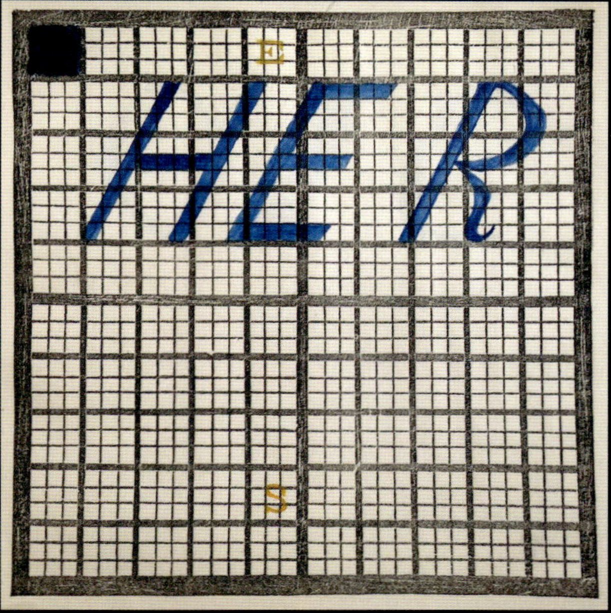
E
HER
S

D
OIC
G
A
T

J
AGE
J
I
E
A
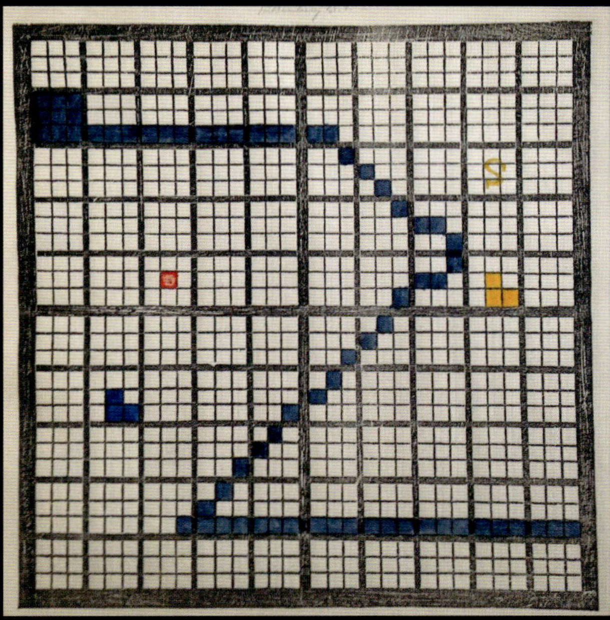

not yet stable enough to support the weight of the past. On the ceiling of the temple, Willard catalogues key figures from each prior century, a revealing who's who of statesmen, philosophers, discoverers, and poets. On the floor of the temple, she lays out the developments of major nation-states, their paths drawn as rivers subjected—like water levels—to the expansions and contractions of state power over time. Consistent with the narrative enacted by her maps, Willard places the path of the United States front and center. This river flows directly toward us, the viewers. The message is not subtle: we will be enveloped by the expanding influence of the United States in the years to come.[27]

While Willard's charts sharply diverged from Peabody's in terms of visual form, Peabody nevertheless identified them as a major source of inspiration, crediting Willard with creating "the most ingenious chart ever besides [her own] invented."[28] As for Willard in turn, her stated influences included Joseph Priestley's "New Chart of History," from 1769, discussed in the introduction; and Playfair's *Commercial and Political Atlas*, discussed in chapter 2, which she had managed to learn about—perhaps from Humboldt—in spite of their still-niche circulation. According to historian Susan Schulten, Willard appreciated the efficiency of Playfair's charts, but felt that they gave "little sense of the *dimension*, such as the relative importance of periods or the subjective experience of time."[29] It is still safe to say today that Playfair's import-export charts do not provide a sense of the subjective experience of time, but they remain just as interpretable as they were at the time of their making—one of Playfair's primary aims, as you might recall.

By contrast, the more abstract impression first conveyed by Peabody's chronological grids requires additional interpretation. Her configuration of her data not according to the Cartesian grid (as with Playfair) but instead as a visual text, designed to be read from left to right, top to bottom, underscores her interpretive intent; Peabody designed her charts to serve as the basis for future knowledge, rather than that knowledge's definitive source. Here we begin to see how epistemologies that extend beyond the positivist approaches that have long dominated the field can also generate knowledge.[30] We can also begin to see the possibilities for visualization strategies that invite in outside knowledge—those that allow for the possibility that viewers of visualizations, as well as designers of them, may bring their own expertise.[31]

With the addition of the workbooks, which enabled students to create their own charts, Peabody underscores her belief in this more participatory form of knowledge production. Her requirement that her students create the charts challenges the typical hierarchy of knowledge-making, in which the designer of a visualization is inherently above those who view or otherwise interact with it. According to Peabody's pedagogical process, it is the student who is authorized to *both* create *and* interpret the image, rather than the charts' original designer—in this case, Peabody herself.

If there is a single message that is communicated by Peabody's charts, it is that their meaning is not fixed. Rather, they impress upon the viewer a sense of responsibility—first for encoding the past into an assemblage of colors, shapes, and positions; and then for using that assemblage of the past to help design their own future course of action.

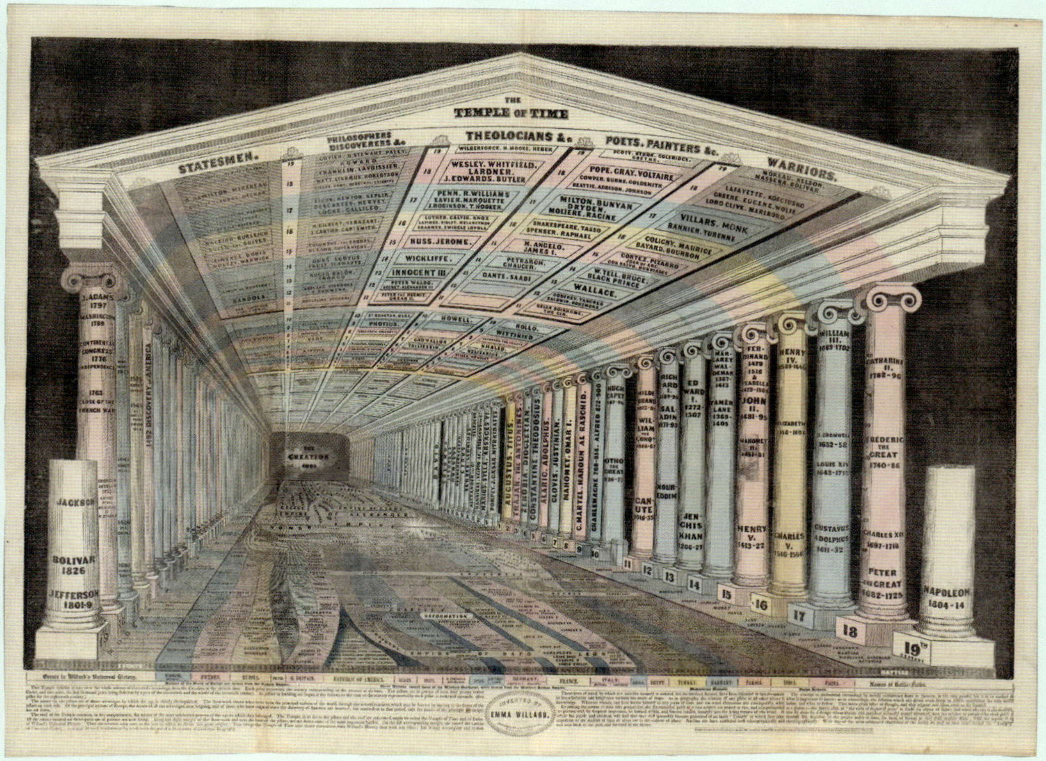

← On the ceiling of the temple, Willard catalogues key figures from each prior century, a revealing who's who of statesmen, philosophers, discoverers, and poets.

← Willard depicts past centuries as the pillars that support the temple, their colors connecting them to the figures named on the ceiling above.

← The nineteenth century is represented as an unfinished column, not yet ready to support the weight of the past.

↑ On the floor of the temple, Willard lays out the developments of major nation-states, their paths drawn as rivers subjected to the expansions and contractions of state power over time.

↑ Consistent with the narrative enacted by her maps, Willard places the path of the United States front and center. The message is not subtle: viewers will be enveloped by the expanding influence of the United States in the years to come.

FIGURE 4.23
Emma Willard's perspectival *Temple of Time* (1846). Image courtesy of the David Rumsey Map Collection, Cartography Associates.

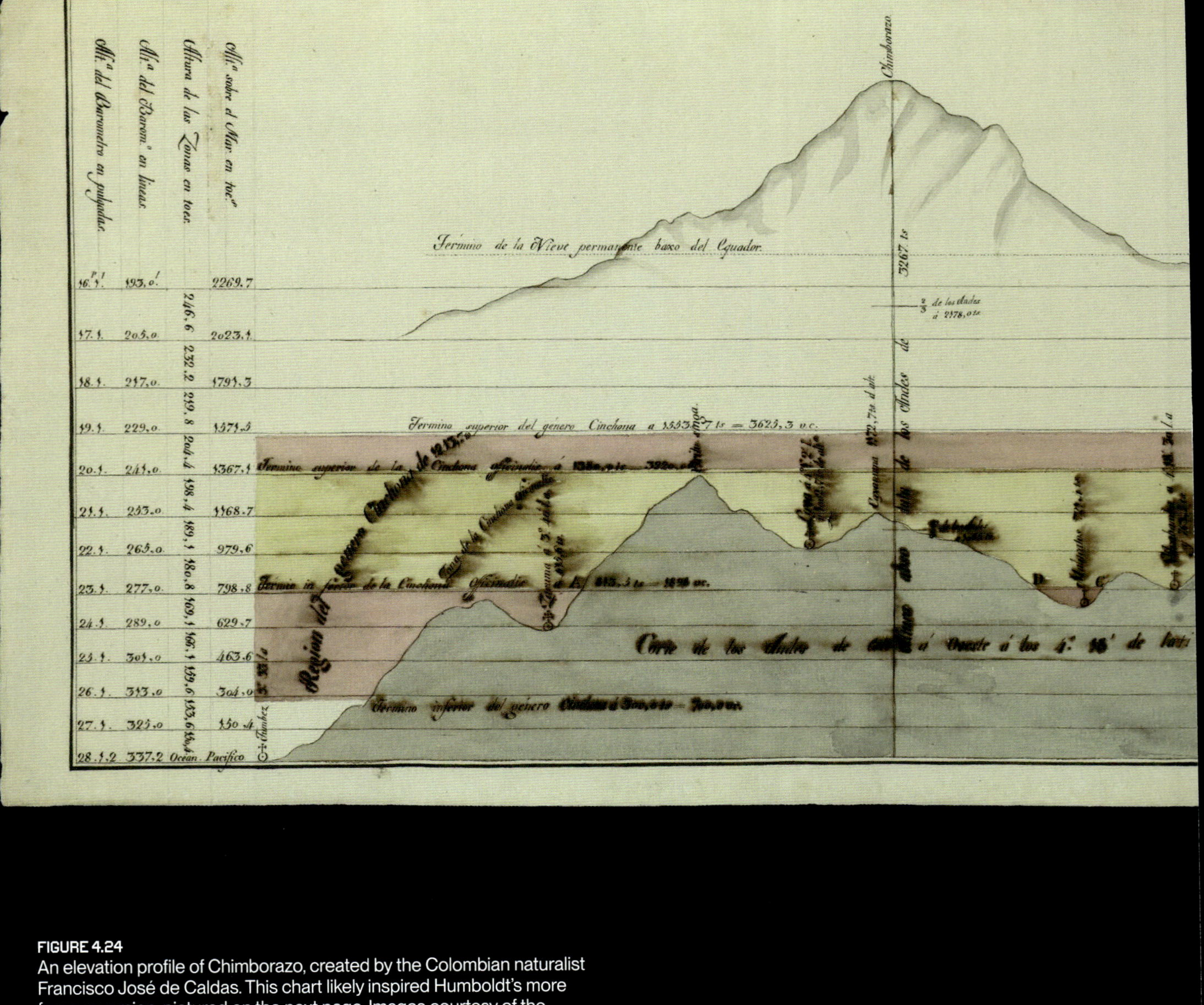

FIGURE 4.24
An elevation profile of Chimborazo, created by the Colombian naturalist Francisco José de Caldas. This chart likely inspired Humboldt's more famous version, pictured on the next page. Images courtesy of the Archivo del Real Jardín Botánico, CSIC, Madrid, AJB, Div. III, M529.
© Real Jardín Botánico, CSIC.

N. 2.

Nivelacion de las Quinas en gral, y en particular de la de Loxa, ó de la Cinchona officinalis

por Fran.co José de Caldas.

Geographie der Pflanzen in den Tropen-Ländern;
ein Naturgemälde der Anden,
gegründet auf Beobachtungen und Messungen, welche vom 10ten Grade nördlicher bis zum 10ten Grade südlicher Breite angestellt worden sind, in den Jahren 1799 bis 1803.
von ALEXANDER VON HUMBOLDT und A. G. BONPLAND.

Top of Chimborazo 21000 feet from the level of the Sea. — Cotopaxi 18,876 Feet.
Plate 1.
Altitude
Feet.
21000
18000
15000
12000
9000
6000
3000
Sea.
Latitude.
Degrees.
90
80
70
60
50
40
30
20
10
0 Equator.
Perpetual Snow.
Region of Lichens.
Lichens.
Region of Grasses.
Region of Wheat & Rye
Region of the Wine Grape.
Region of Figs and Olives.
Region of Palms, Spices and Sugar Cane.
Miss Lee del.
Illman & Pilbrow Sc.
Progress of vegetation at different elevations, corresponding with its progress in different Latitudes.

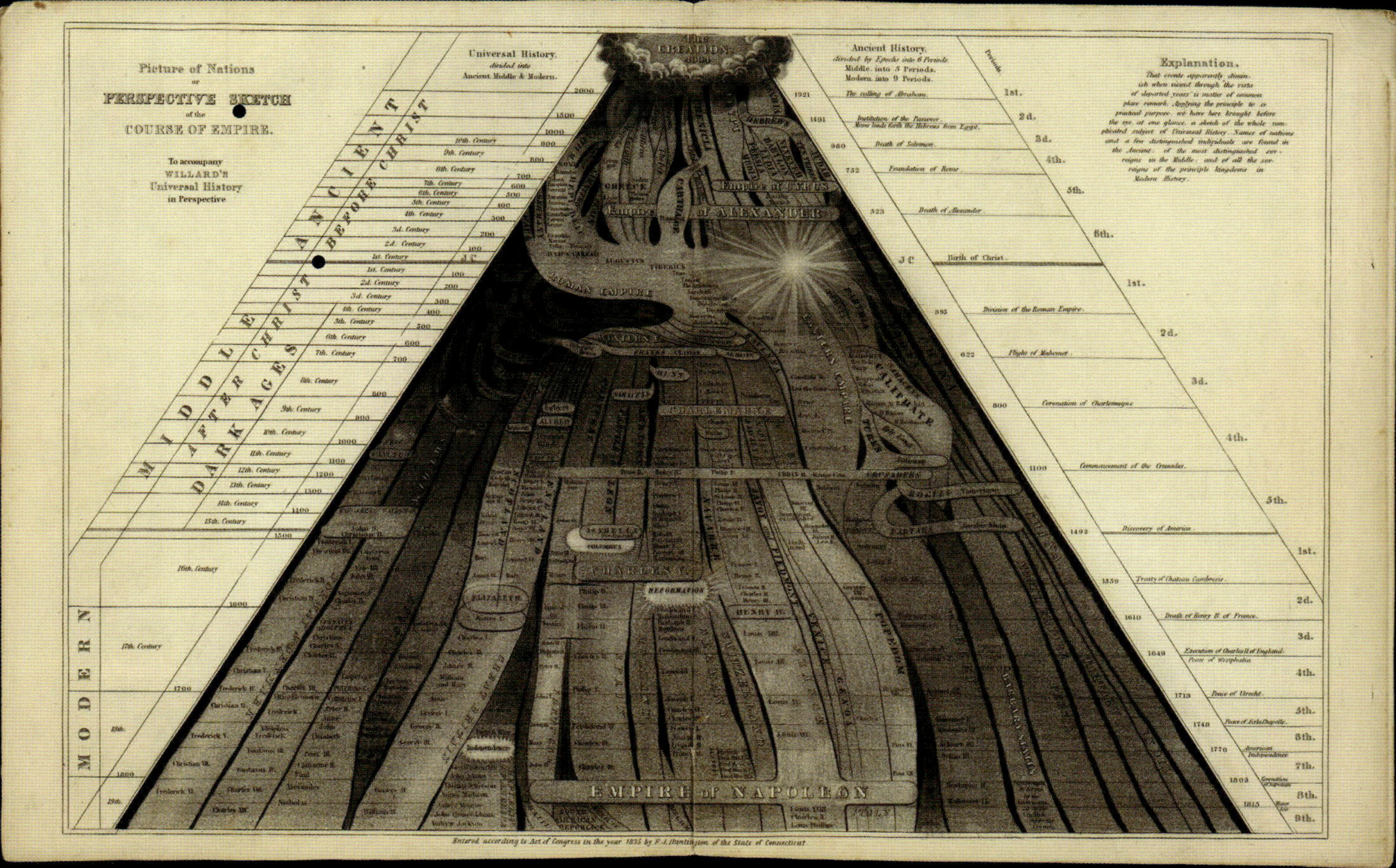

FIGURE 4.27
Willard's "Picture of Nations or Perspective Sketch of the Course of Empire" (1836), which adopts the triangular form of Humboldt's Chimborazo diagram in order to represent the rise and fall of nations. Image courtesy of the David Rumsey Map Collection, Cartography Associates.

FIGURES 4.25 AND 4.26
OPPOSITE, TOP: Humboldt's famed diagram of the vegetation of Mount Chimborazo, created with Aimé Jacques Alexandre Bonpland. Image courtesy of Wikimedia Commons. Digitized by Leibniz-Institut für Länderkunde, Leipzig.

OPPOSITE, BOTTOM:
Illustrator Thirza Lee's recreation of Humboldt's "Top of Chimborazo" for *Familiar Lectures on Botany*, an 1829 botany textbook for young women written by Willard's sister, Almira Hart Lincoln Phelps. Image courtesy of the Library Company of Philadelphia.

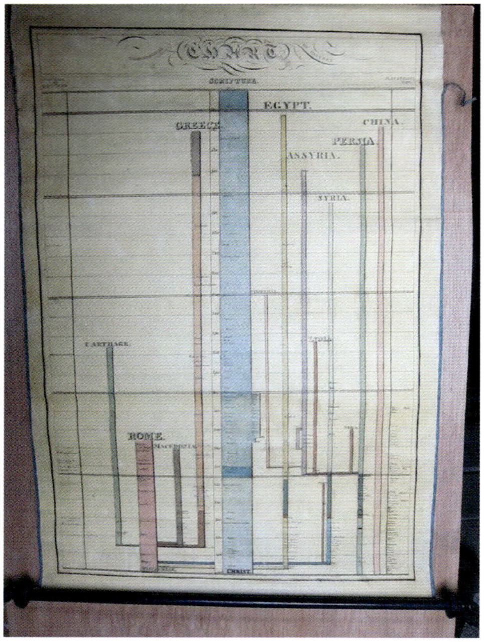

FIGURES 4.28 AND 4.29
RIGHT: A vertical timeline created by Anne Laura Clarke as it was displayed on her traveling lecture tour. **BELOW:** A second chart as it was encountered by Granville Ganter in a Clarke family attic. Here the first chart is visible on the left. Photos courtesy of Granville Ganter.

The Gendered Archive of Data Visualization

As part of her participatory pedagogy, Peabody also created larger "painted centuries," or "mural charts" as she also sometimes described them, in order to guide classroom discussion.[32] These were the charts she traveled with on her national promotional tour, and by all accounts they were dazzling—triangles and squares of crimson, ochre, and forest green, set against a sharp black grid. In her version of a sales pitch, Peabody would "lay [a] chart down on the floor" and invite her would-be textbook adopters to sit around it and contemplate the colors and patterns that enveloped them.[33]

While the experience of sitting on the floor to learn together, rather than in rows of wooden desks, was—to most adults—quite shocking, the pedagogical impact of this more collective and embodied form of learning was, as best we can gather, nothing short of transformative. "I have never known a system which placed the events of the history of all nations before the mind with such clearness, so little confusion, and so much permanency," wrote Eliphalet Nott, then president of Union College, who participated in one of Peabody's teaching demonstrations.[34] Anticipating a decidedly twenty-first-century view of the value of embodied learning, Peabody staged an encounter with the data that involved the whole body. This is akin to what Kelly Dobson has described (and enacted) in a contemporary context as "data visceralization," the idea of making data legible to the whole body, not just the eyes.[35] Indeed, if visualization is to continue to offer "richer understandings that enable researchers to ask bolder questions," as visualization researcher Ben Shneiderman asserts, then—far from a fringe example—the mural chart seems to represent an early apotheosis.[36]

Until one of Peabody's mural charts was acquired, digitized, and put online by the David Rumsey Map Collection in 2024, most scholars—including me—believed that none of Peabody's mural charts had been preserved.[37] We were unsure even of their basic features.[38] Peabody's nephew Julian Hawthorne's recollection of the "huge, colored charts" which "hung on the walls of our sitting room" tantalized our project team, nearly two hundred years later, with the most detailed description of what they would have been like to perceive.[39] I doubted that I would ever see one in person, especially since a colleague who'd been researching another nineteenth-century woman chart-maker, Anne Laura Clarke, sent me a photo of how he'd found Clarke's own large-scale charts: brittle and water-damaged in her sister's family's attic (FIGURES 4.28 AND 4.29).[40] Peabody's parents had only rented their West Street home; there would be no attic discovery for me.

But Peabody's embodied pedagogy seemed, to me, so primed for the present that I could not let the idea of the mural charts go. In my research lab, working with multiple cohorts of students over multiple years, we sought to reimagine the mural charts for today.[41] Where Peabody used a stick to point to specific events on her chart, we wanted our users to be able to engage their bodies more directly; our "Floor Chart," as we came to call it, needed to respond to touch. To do this, we first created a three-foot-by-three-foot matrix made of strips of copper tape.[42] The matrix worked like a computer keyboard, with columns and rows of conductive material—in this case, the copper tape—separated by a foam spacer. When a person pressed on a square of the grid, the two layers of copper tape touch each other, creating a connection. Above the touch matrix, a layer of quilted fabric, also created by members of the lab, held in place strips of individually

FIGURE 4.30A AND B
RIGHT: A diagram of the layers of the floor chart.
FROM BOTTOM TO TOP: the foam base, the bottom half of the copper touch interface, the foam spacer, the top half of the copper touch interface, the foam cover, the quilted base, the LED strips, the soft diffusion layer, and the quilted top. Diagram by Tanvi Sharma.
BELOW: A look inside the floor chart, with the touch interface exposed. Floor chart design, prototype, and implementation by Angela Vujic, Marninder Japra, Courtney Allen, Alexandra Beall, Gwendolyn McGuire, Sarah Schoemann, Svyateslav Kucheryavykh, and Noah Sutter. Photo by Margy Adams.

addressable LEDs.[43] The result was a grid that could be toggled (or preprogrammed) to display the events listed in the *Chronological History*—or any others that have transpired in the intervening years. In this way, users could engage with the past using their hands (FIGURE 4.30).

The Floor Chart project required expertise in a range of domains, from electronics prototyping to signal processing to circuit board design. It also required a truly tremendous amount of physical labor. Each yard-long strip of copper tape needed to be perfectly aligned, lest a small misalignment at one end result in a significant gap at the other. Each of the nine hundred square-shaped holes of the spacer for the touch matrix was required to be cut out by hand, as laser-cutting would have released harmful toxins. Each electrical connection had to be soldered, tested, and then—in almost all cases—soldered again, to ensure that the circuit remained intact. As much as an exercise in physical fabrication—a delight to my Georgia Tech students—the project became an exercise in the physicality of work itself. It called us back to the focus that is required of any project that is made by human hands, and the resultant fatigue.[44]

The tedious, time-intensive nature of the Floor Chart project also called me back to another historical detail related to Peabody's original charts: the fact that she did not only demonstrate the charts as part of her sales pitch; as an additional incentive, she also promised an original mural chart to any teacher who purchased copies of her textbooks for their entire class. Writing to a friend in 1850, Peabody revealed that she was "aching from the fatigue of making Charts for the Schools who will take the book." The letter continues:

> Every school must have a mural chart—& there is but one way of making them (until they can be made by ten thousands) & that is by stencilling [*sic*]. ... I can do one a day. But I must sell them cheap. ... To day I worked 15 hours—only sitting down to take my meals—& so I have done all week—so much fatigue stupefies one—but as soon as it is adopted in a few towns I shall be able to hire someone to do this drudgery for me.[45]

While the digitized mural chart in the David Rumsey Map Collection appeared online as if by magic (FIGURE 4.31), letters like these help to attest to the intensity of the physical labor that was required to produce the original charts.[46] With its reference to the "stencilling" through which Peabody created her colorful symbols, as well as to its characterization of the tasks involved as "drudgery," the letter also exposes the gendered dimensions of Peabody's knowledge work.

It is not a coincidence that the fabrication of the mural charts involved what is plainly legible as "women's work," and that these same charts were scarcely preserved. (Evidently, the one extant mural chart surfaced only in 2021, when a listing for a lot of four "antique wall maps" from a private collection caught the eye of rare map dealer.)[47] That salvaged map notwithstanding, the status hierarchy of work aligns with the social hierarchy of gender that persists, still, in the United States. Work that is performed outside the home is valued, both culturally and monetarily, over work that is performed within it. Work that is perceived as more rigorous or more professional—like, for instance, the political economy that functioned as Playfair's primary trade—is valued, again, both culturally and monetarily, over work that is perceived as more intuitive or more domestic—like, for instance, the teaching that functioned as Peabody's main employment. Even within the

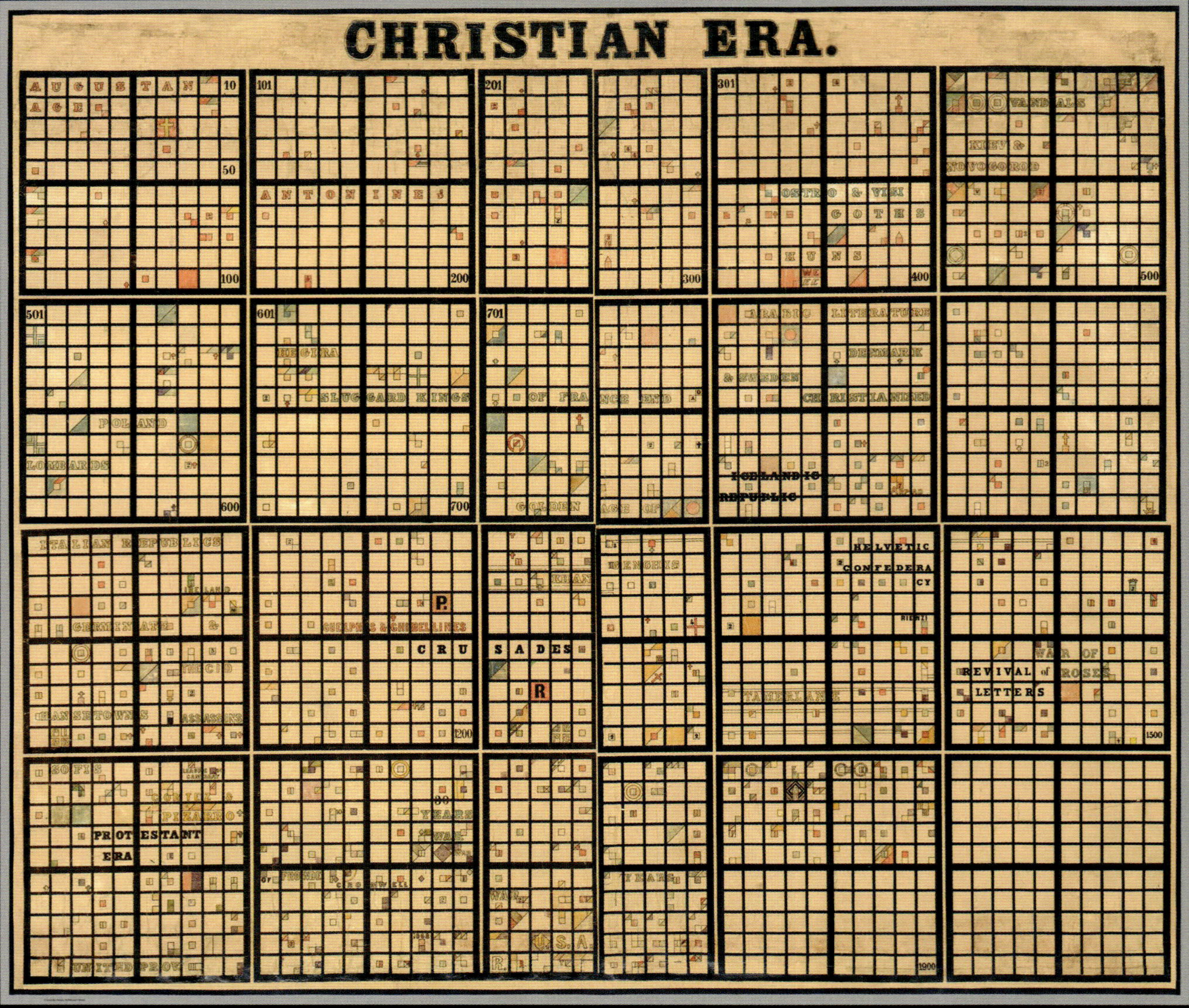

FIGURE 4.31
The only extant example of one of Peabody's mural charts. This, Christian Era," ca. 1850, was restored by Barry Ruderman and Alex Clausen of Ruderman Antique Maps in the early 2020s. Image courtesy of the David Rumsey Map Collection, Cartography Associates.

art world, creative work that is perceived as high art is valued above work perceived as craft.[48]

These gendered divisions of labor are among the primary reasons that Peabody's mural charts did not enter officially sanctioned archives at the time of their creation. They are also among the same reasons that her charts have not (yet) been centered in our typical telling of the rise of modern data visualization. But think about how Peabody's approach diverges from Playfair's, and how generative that divergence can be. It is not, after all, that Peabody did not recognize the power of visualization. In point of fact, she also sought to create an "instantaneous impression" with her charts, to recall the goal shared by Clarkson and Playfair. It is, rather, that Peabody saw the power of creating that impression—along with perceiving it—as an experience that could be *shared*. Who else are we missing when we fail to include charts like Peabody's in the stories we tell about the emergence of data visualization? And in these omissions, what possible future approaches to visualization design are we also not (yet) able to see?

It is now uncontroversial to assert that the work of women has been crucial to the development of most scholarly disciplines and professional fields, even if the value of that work is still often only recovered after the fact.[49] While the obvious turn, here, may be to the women "computers" who were the first computer programmers, I would like to retain a focus on the charts' formal features. Most contemporary viewers, when seeing Peabody's charts for the first time, observe that they look like paintings by Piet Mondrian, the famous Dutch modernist. To be sure, Peabody's charts do strongly resemble Mondrian's bold, colorful, geometric grids. But Peabody's own account of making the mural charts brings to mind a second point of reference, which is not painting but quilting—an art form, traditionally practiced by women, that has long been relegated to the world of "folk art" and craft.[50]

The area of Alabama known as Gee's Bend, a small, rural Black community, 35 miles south of Selma, traces its roots to a cotton plantation established there in the early nineteenth century. The quilts created in Gee's Bend have been valued by its residents for generations, both for their formal features and for the family histories that they encode (FIGURES 4.32 AND 4.33). (Not only specific patterns but also specific pieces of fabric represent specific stories and events.) But the quilts have only recently begun to be recognized by art historians as representing an alternate stream of modernist aesthetics.[51] Exhibitions at the Whitney Museum in New York in 2002, the Turner Contemporary in London in 2019, and others have proposed that the quilts "predate like-minded works by their more famous abstract art cousins."[52]

The life experiences of the Gee's Bend quilters—Black women who pursued their art while enslaved, and have continued that art through slavery's perpetual wake—could not be farther removed from those of Peabody, a white woman well established in Boston's intellectual elite. But the creative work of the Gee's Bend quilters, as distinguished historian Elsa Barkley Brown has shown, offers a model that might bring together these complementary perspectives. Drawing inspiration from their "polyrhythmic, 'nonsymmetrical,' and nonlinear" patterns, Barkley Brown advocates for a continual "pivoting" of what we think of as the center—that is, not decentering one perspective in favor of another but instead intentionally and continually shifting the focus from one perspective to the next.[53] The result of this strategy, which Barkley Brown devised in her own history classroom, is capacious and multifold: it

FIGURE 4.32
An early twentieth-century quilt from Gee's Bend by Rachel Carey George, entitled "Housetop—Sixteen-Block 'Half-Log Cabin' Variation Sashed with Feed Sacks," ca. 1935. © 2025 Estate of Rachel Carey George / Artists Rights Society (ARS), New York. Photo by Stephen Pitkin/Pitkin Studio. Courtesy of Souls Grown Deep Foundation.

FIGURE 4.33
A more contemporary quilt from Gee's Bend, "Housetop variation," designed by Mary Lee Bendolph in 1998 and quilted by Essie Bendolph Pettaway in 2001. © 2025 Estate of Mary Lee Bendolph / Artists Rights Society (ARS), New York. Photo by Stephen Pitkin/Pitkin Studio. Courtesy of Souls Grown Deep Foundation.

allows the artifacts and experiences under analysis to be understood in the context of their own creation, and it allows the students performing this interpretive work to "become the voices of authority in their own education."[54] Ultimately, Barkley Brown concludes that when done right, "the class is a quilt. It is precisely the contrast which organizes the whole and holds it together."[55]

What if we understood the history of visualization as a quilt? This is what Barkley Brown's pedagogical theory suggests, and what my own research group literally fabricated. So consider the surprise that greeted me, along with no small degree of pleasure, upon discovering that a quilt created by Loretta Pettway, one of the Gee's Bend quilters, graces the cover of Edward Tufte's most recent book, *Seeing with Fresh Eyes* (2020). More than merely a compelling image—"unorthodox, fresh, amazing," is how Tufte describes the quilt in the text—we must learn to see such quilts, like Peabody's charts, as systems of knowledge-making.[56] Even more, as McKittrick advises, we must understand them in relation to each other.

In this task, these two systems of knowledge-making helpfully converge. Their connection is very clearly asymmetrical, the result of differences in race and class, physical and economic mobility, social and cultural status, and more. But it is also forged across their shared methods. Both employ shape and color to represent and recall past events: the quilts in order to commemorate a community's ancestors and their stories; and the charts, as we have learned, in order to craft new narratives about the nation's defining historical events. Both also rely upon sense perception—and more specifically the tactile experiences of the body—to assimilate visual display into knowledge. Whether enveloping oneself in a quilt or gathering together around a mural chart, the result is a more embodied encounter with the object—and with the events of the past that the object seeks to commemorate and to convey.

Barkley Brown reminds us that the Gee's Bend quilts are "illustrative of a particular way of seeing, of ordering the world."[57] We might again extend this assertion to data visualization. Each image that we encounter reflects the designer's own way of seeing and ordering the world. This fact does not invalidate the insights that the image might prompt, or the knowledge those insights might lead toward. On the contrary, acknowledging the particular way that the designer sees and orders the world further enriches the knowledge their visualization helps bring to light. With a wider awareness of the multiple ways of seeing the world, a broader range of methods for ordering its data, and a more acute sense of the relations among them, we can expand the basis of what we presently know, and—as Peabody envisioned—open up new possibilities for future knowledge.

The Visualization Work to Come

Since the start of this book, I have argued for the additional inspiration we might find, and the knowledge we might gain, by expanding the history of data visualization to include a wider range of forms and figures. From Peabody and from the Gee's Bend quilters, we are pointed to the meaning-making capacity of geometric abstraction, and to the epistemological possibilities of more immersive—and more prolonged—visualization experiences. We are also pointed to the possibilities that emerge when we place visualization midway through the process of knowledge production, rather than at its end. As we have seen enacted—but, I would argue, not sufficiently theorized—in twentieth-century formulations of exploratory data analysis, this use of exploratory data visualization prompts yet more questions about who is authorized to produce knowledge, both in the design of visualizations and in their interpretation.

Here we might draw from Peabody's example of encouraging multiple perspectives, placed in relation to the Gee's Bend quilters, so as to articulate a broader claim: that as visualization designers, we must expand the range of sources—and the range of people—whom we enable to make knowledge claims. How can we better recognize the people who we are currently excluding? How might Peabody's efforts—as well as how they were circumscribed—help to elevate our own attempts to reckon with the relational nature of all that we know? We have not, after all, fully considered Peabody's relation to the colonial frame, and how her desire for national wholeness—not dissimilar to Emma Willard's, discussed in chapter 3—did not allow her to leave space for events or stories that might reflect opposing perspectives. For example, how might she have better distinguished those that might have been offered by members of Indigenous nations had they not been grouped into the same color category as their settler-colonial kin?[58]

As we move into the final historical chapter of this book, which centers on the charts that W. E. B. Du Bois and his students designed for the 1900 Paris Exposition, we will continue to explore the power

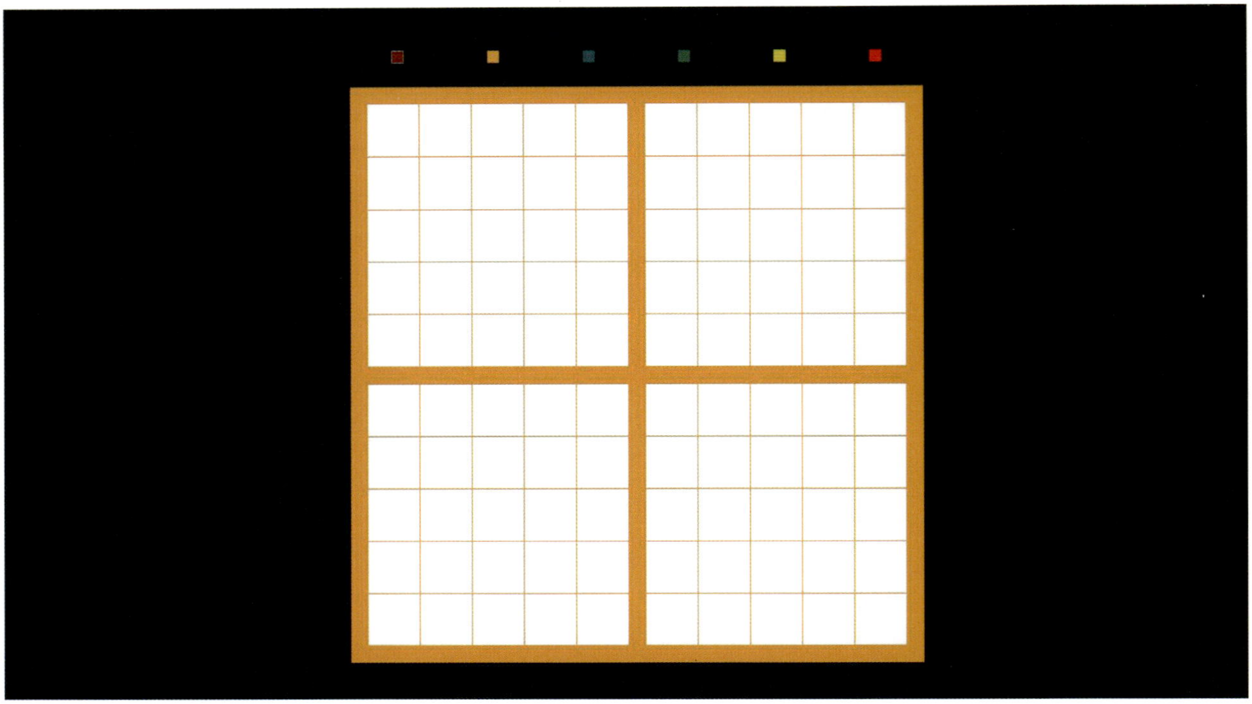

FIGURE 4.34
Inspired by Peabody's blank workbooks and her feminist epistemology, this grid is left open for you, the reader, to fill in.

of data visualization to support larger narratives: of peoples, of nations, and of the field. At the same time, we will consider how Du Bois himself grappled with the limits of what could be learned from visualization alone. With Du Bois and his students, we will also continue to probe questions of labor, asking how we might value the full range of labor that contributes to the creation of data visualizations, and how we might honor all of those we rely upon to perform this work. As we move forward in time, Peabody's "painted centuries"—her "knowledge system," as it might be termed—will become valuable once again, because it authorizes us to support multiple narratives rather than a single view. With epistemological frameworks that allow us to bring out these narratives and bring them together—and to do so with an awareness of the not-always-equal relations among them—we can acquire the most complete picture of what we know. In the end, the most liberatory version of Peabody's charts, as we seek to evoke in the final chart we have designed for this chapter (FIGURE 4.34), is that when presented with only the outlines of history, we might take it upon ourselves to color them in.

Takeaways

FOR VIEWERS

Remember that histories are shaped by power

Expand your ideas about what visualization includes

Recognize that you can create knowledge

Ask whose knowledge is missing from what you see

FOR DESIGNERS

Draw inspiration from multiple ways of knowing

Empower viewers in the process of knowledge production

Design for more complete knowledge, not total knowledge

Probe the history of visualization for future ideas

CHAPTER 5

CHANGE

Lula Iola Mack, W. E. B. Du Bois, and Visualization for Liberation

Visualization can bear powerful witness to oppression. How can we use charts to make change in the world?

The last week of December 1899 had been cold and exceptionally rainy.[1] But on the day after Christmas, when Lula Iola Mack crossed the "high and narrow iron bridge" from the women's dormitory to the main campus of Atlanta University, one of the nation's flagship Black colleges, to begin the winter term of her senior year, she likely walked with a confident step.[2] Mack was beginning the term as one of the top students in her class. She had already been awarded high honors after her sophomore year—the only member of her class to receive such recognition—and she was on track to repeat the feat at graduation.[3] She'd spent the previous term learning a range of state-of-the-art methods for data collection and analysis, so she likely approached the next course in her required sociology sequence feeling well-prepared.[4] She was to undertake a project involving "a general survey of social conditions," just as the course catalogue promised, and just as students had done the year before.[5]

Little did Mack know—nor, for that matter, did her professor—that the scope and scale of this project was about to dramatically change. For the past several months, a Washington, DC-based lawyer and government official by the name of Thomas J. Calloway had been leading a letter-writing campaign with a single goal: for the United States to stage an "Exhibit of American Negroes" at the 1900 Exposition Universelle, which was to be held in Paris the following spring. The exhibit, Calloway believed, should stand alongside the other celebratory displays of national might and ingenuity that were the main attraction of these turn-of-the-century world's fairs.[6]

On January 25, 1900—one month to the day after the winter term at Atlanta University had begun—Calloway received a financial commitment from the US federal government.[7] He tasked Daniel A. P. Murray, then the Assistant Librarian of Congress, with assembling a set of books by Black writers to put on display. He also reached out to a college friend to work through what the "social" component of the exhibit should include.[8] This friend was Mack's professor, none other than the esteemed sociologist, author, and activist W. E. B. Du Bois. And for the next three months, Mack, Du Bois, the three other students enrolled in his year-long sociology sequence, and an additional student—William Andrew Rogers—who had graduated the previous spring, threw themselves into the work that would culminate in a set of 63 (or 64) "plans, charts, and figures" for the Paris Exposition, as the event was also called.[9] Once installed in the southeast corner

EXHIBIT OF AMERICAN NEGROES AT THE PARIS EXPOSITION.

FIGURE 5.1
A photo of the “Exhibit of American Negroes” at the 1900 Exposition Universelle that appeared in the November 1900 issue of the *American Monthly Review of Reviews.* Photographer once known. Image courtesy of the Library of Congress, Prints and Photographs Division, LC-DIG-ppmsc-04826. Digitized by the Library of Congress.

← The chart on the far right of this row is the one believed by Jason Forrest to be the missing 64th chart, as it does not appear in the Library of Congress collection.

← In this row, the charts, photographs, and other artifacts were encased in swing-out vitrines.

← In this row, more charts and photographs are installed in picture frames.

← The top of the bookcase, at desktop height, is stacked with books, pamphlets, and other artifacts—including some of the charts created by Du Bois and his students—intended to be perused.

← The low L-shaped bookcase that held the books collected by Murray.

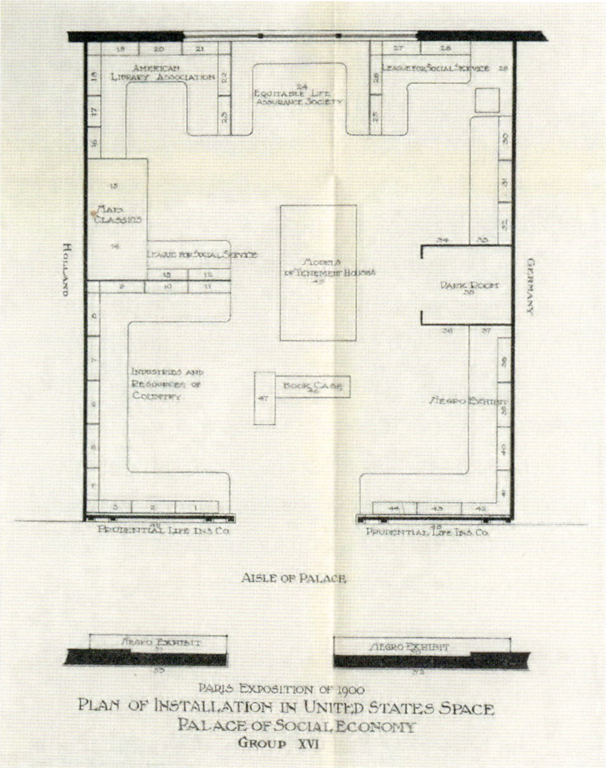

FIGURE 5.2
The floor plan of the Palace of Social Economy and Congresses; the exhibit was installed in the southeast corner of the US exhibition space. From "Report of the Commissioner-General for the United States to the International Universal Exposition, Paris, 1900," February 28, 1901. Courtesy of the Library of Congress.

of the Palace of Social Economy and Congresses, the charts would be viewed by over fifty million visitors from around the world who would travel to Paris over the course of the Exposition's seven-month run (FIGURE 5.1).[10]

Since the exhibit was positioned in a corner, there were only two full-length walls on which to display the charts (FIGURE 5.2). (The outside of a model darkroom that had also been constructed for the space provided a third half-length wall.) Photos of the exhibit show a low L-shaped bookcase with shelves lined end to end with the books collected by Murray. The top of the bookcase, at counter height, is stacked with more books, pamphlets, and other artifacts—including some of the charts created by Du Bois and his students—intended to be perused. Above the bookcase are more charts, along with photographs and additional artifacts, each encased in a picture frame or swing-out vitrine and installed in three dense rows. Viewed from any direction—left, right, up, down, or straight ahead—visitors could immediately perceive the bright colors and bold patterns of these "data portraits" of Black American life, as scholars Whitney Battle-Baptiste and Britt Rusert evocatively characterized the charts in their landmark 2017 book.[11] Recovered and reincorporated into the history of data visualization in the years since, the charts have captivated a new generation of data visualization designers and enthusiasts who see in them evidence of how visualization can be enlisted in the service of justice while also advancing the formal possibilities of the field.

And yet, when the story of the charts' creation is told, it generally tends to follow an arc much like those of William Playfair, Charles Minard, and other "founders" and "fathers" of the field. The result is that Du Bois is celebrated for his singular achievement, but we fail to see the bigger picture—or the additional people—that allow us to understand the full significance of the charts. To be sure, just like Playfair's line graphs, Minard's flow maps, and so many other deservedly iconic charts, the charts created for the Paris Exposition exemplify the insight-producing power of visualization. Through form and color, abstraction and scale, they expose socially significant patterns and trends, in Du Bois's own words, "at a glance."[12] In this case, they expose one trend in particular: the growth and progress of Black Americans in spite of the intensity of the "prejudice" (Du Bois's term for racism) that they still faced.[13]

But as a scholar who was then pioneering the idea of social scientific research, and—put plainly—as a Black man in the United States, Du Bois was as attuned to the limits of visualization as he was to its uses. His scholarly expertise and lived experience both pointed to the fact that visualization *on its own* could not hope to convey a complete picture of the progress of Black Americans to date, just as visualization *on its own* could not create the change that would be required to achieve true liberation.

This is why the final historical chapter of *Data by Design* is centered on Du Bois, his charts, and the students who helped to create them: to show how data visualization is best understood as one method, among many, that can be enlisted in the service of social change. Indeed, if this book began by considering the power of data visualization to change the mind of a king, it now concludes with an example of its power to change the minds of the people—that is to say, of a wide range of viewers, with a wide range of perspectives, who were not always aligned in their values or goals. As we will learn over the course of this chapter, Du Bois was required to recalibrate his own confidence in the power of data visualization, and data analysis more broadly, as a result of the racism—and, in one particular moment, outright racial terror—that he personally faced. But he could not, and did not, convey his call for change on his own. In the end, it is the example of Du Bois *through his students* that teaches us how to approach the main goals of this project: of striving for more complete knowledge, the pursuit of the truth, and a more just and equitable world.

As this chapter will confirm, the charts of the Paris Exposition can indeed set us on a path to a more liberatory practice of visualization design. But this path opens up only when we learn to see the charts in a certain way: as examples of how visualization must be placed in its social and political context, of how lived experience must enhance scholarly expertise, and of how we must remain open, ourselves, to additional methods of making knowledge. From this perspective, the charts also serve as a reminder of the power that each of us holds in choosing how to tell a story—about the charts designed by Du Bois and his students, about the broader history of data visualization, and through the charts that we ourselves design. This chapter thus seeks to affirm our responsibility as both designers and viewers, sources and subjects, and students and teachers—modeling how we can each, from our particular place in the world, pursue greater knowledge in the service of change.

Visualization and Countervisualization in the Paris Exposition Charts

The 63 (or 64) poster-sized charts that Mack, Du Bois, and the other Atlanta University students made for the Paris Exposition were organized into two distinct sets. The first, *The Georgia Negro: A Social Study*, consisted of 36 charts that visualized statistics related to the Black population of that state (FIGURES 5.3 AND 5.4). Scholars generally believe that William Andrew Rogers, the student from the year before, was responsible for much of the design and implementation of this set of charts.[14] (Notably, it is also Rogers's lettering, adapted into a typeface by Tré Seals of Vocal Type, that is used on the cover and chapter titles of this book.)[15] The second set, *A Series of Statistical Charts Illustrating the Condition of the Descendants of Former Slaves Now in Residence in the United States of America*, drew from several data sources, including the US Census, in order to put the Black population of the United States in national and international perspectives (FIGURE 5.5).[16] These charts

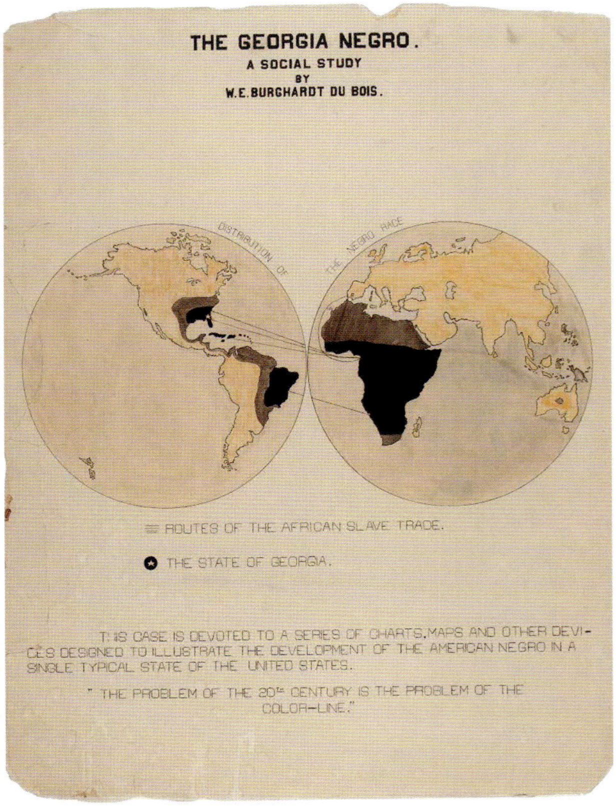

FIGURE 5.3
The introductory image of *The Georgia Negro: A Social Study*, the first set of charts included in the Paris Exposition. Image courtesy of the Library of Congress, Prints & Photographs Division, LC-DIG-ppmsca-33863.

are collectively attributed to "Negro Students of Atlanta University," but nowhere does Du Bois credit any of these students by name.[17]

For visualization researchers, focused on the charts themselves, this lack of direct attribution has made the identities of these students somewhat of a dead end. But for humanities scholars, particularly those trained in techniques for recovering meaning from fragmentary archives, this "archival silence," as it is often called, is where the real work begins.[18] The lists and ledger books, catalogues and records—the bureaucratic and household detritus that, for most people, can be safely set to the side—are, for these scholars, the main focus of their research. Certainly they've been been the main focus of mine.[19] In this particular case, it's the annual *Catalogue of the Officers and Students of Atlanta University*—the same that Mack would have consulted for information about her sociology course at the start of the term—that points to the likely contributors to this set of charts: Mack along with three others—Henry Napoleon Lee, Edward Lee Simon, and William George Westmoreland. Why might this be so? Because they are the students listed in the catalogue as the four members of the Atlanta University Class of 1900.[20]

Because Du Bois's year-long sociology course sequence was required of all Atlanta University seniors, we can safely assume that Mack, Lee, Simon, and Westmoreland took his winter-term class.[21] We also know from extensive scholarship on Du Bois—and, again, from the catalogue itself—that the organizing principle of Du Bois's pedagogy, and the sociology sequence in particular, was to involve his students in the research (and requisite skills of data analysis) required to produce public-facing work.[22] In point of fact, when Du Bois received the invitation to contribute to the Paris Exposition, he canceled all of his winter-term classes except for his sociology course.[23] Presumably he identified a direct alignment between the course's learning objectives and the work that creating the charts would entail. But perhaps more than that, Du Bois recognized from the start that, both practically and conceptually, the work of creating the charts could not be completed alone.

The title chart of the *Georgia Negro* series—the set commonly attributed to Rogers—is often posited as a key to unlocking the significance of the charts as a

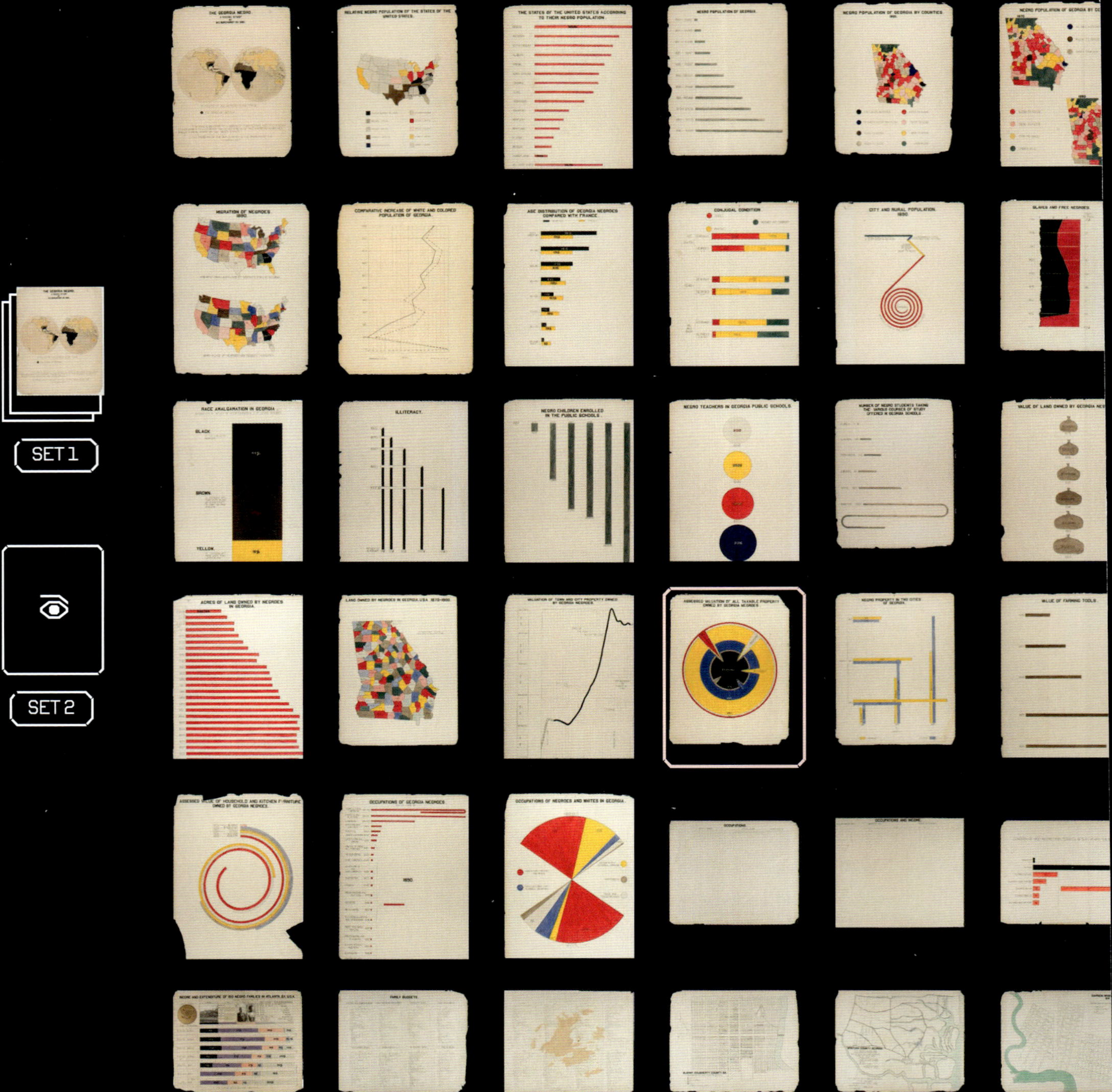
SET 1
SET 2

The 36 charts of *The Georgia Negro: A Social Study* that visualized statistics related to the Black population of that state. It is generally believed that William Andrew Rogers, a member of the Atlanta University Class of 1899, was responsible for much of the design and implementation of this set of charts. Design by Tanvi Sharma. Prototype by Zhou Fang. Implementation by Jay Varner.

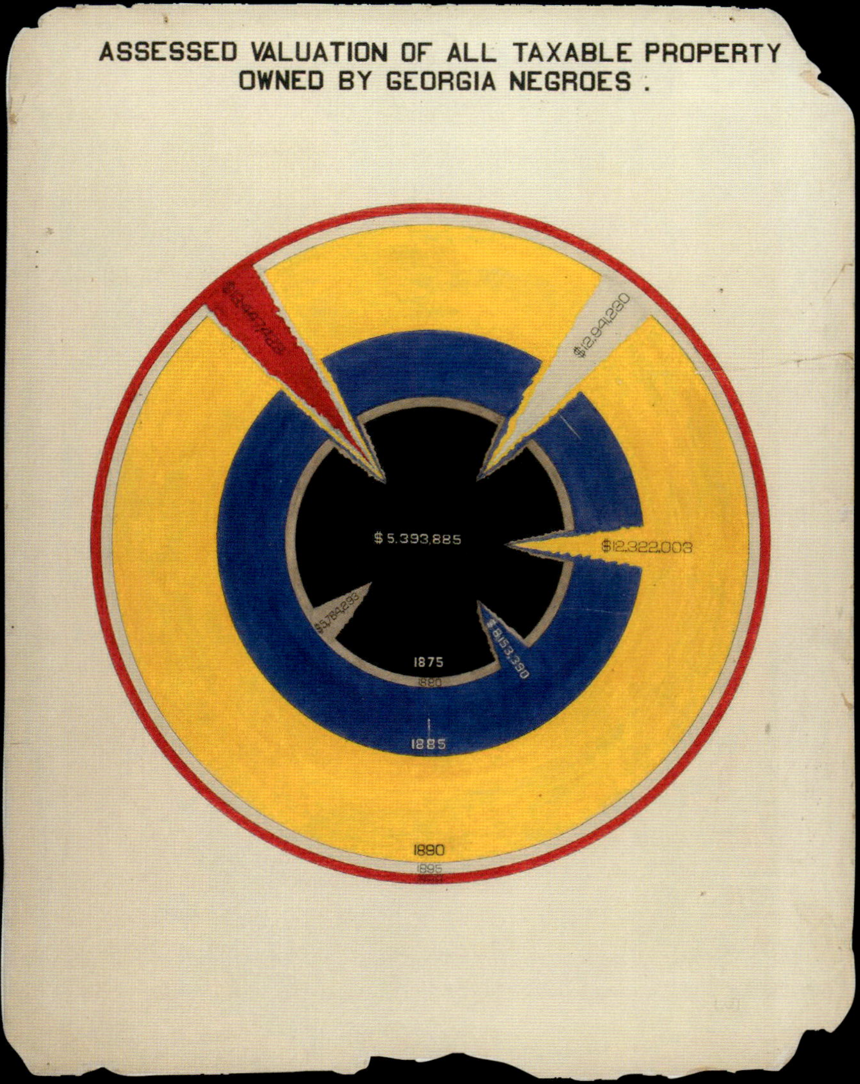

Assessed Valuation of All Taxable Property Owned by Georgia Negroes.

Image courtesy of the Library of Congress, Prints & Photographs Division, LC-DIG-ppmsca-33884.

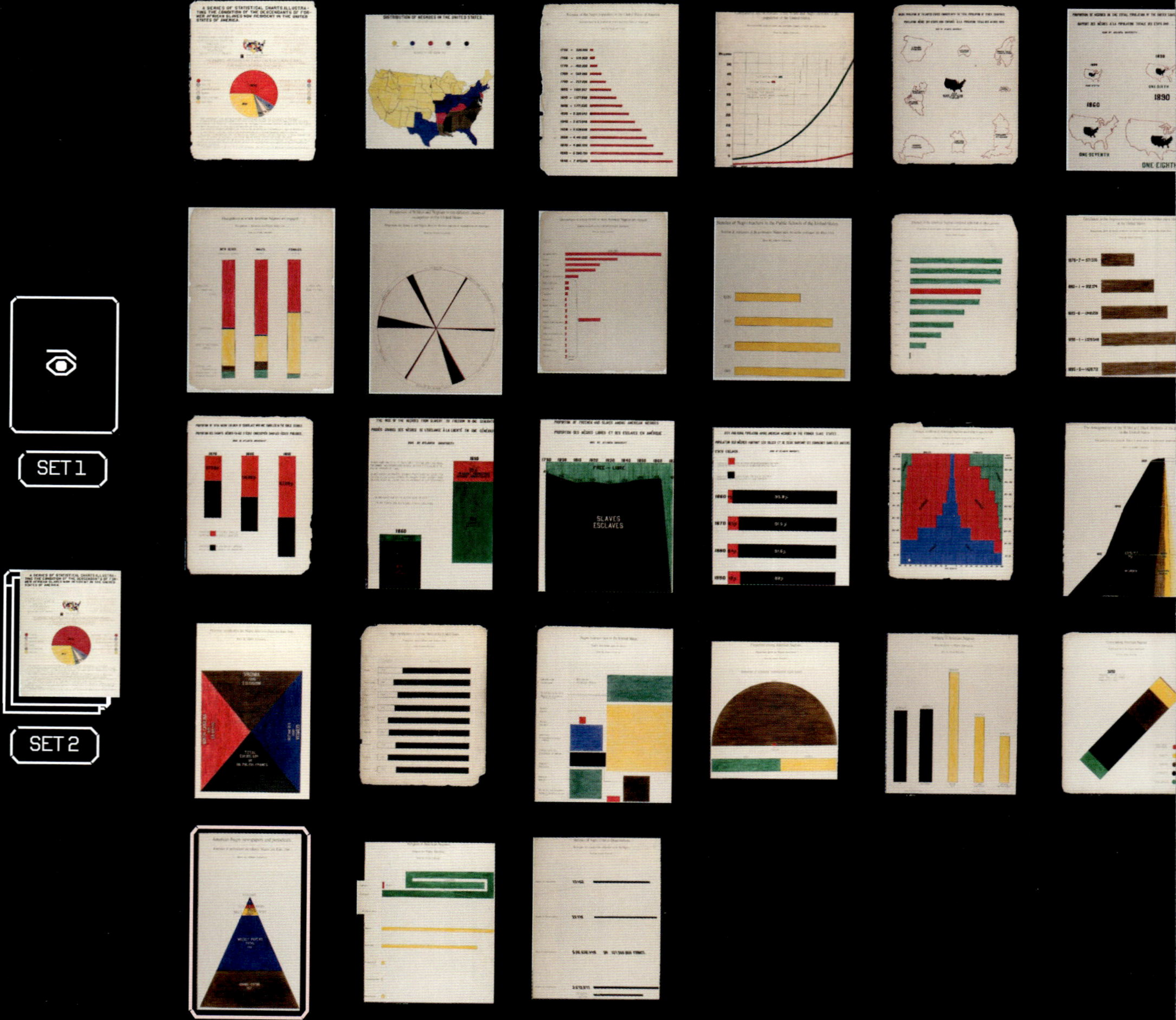
SET 1
SET 2
DISTRIBUTION OF NEGROES IN THE UNITED STATES.
1860
1890
ONE SEVENTH
ONE EIGHTH
FREE - LIBRE
SLAVES
ESCLAVES

FIGURE 5.5
The 27 extant charts in *A Series of Statistical Charts Illustrating the Condition of the Descendants of Former Slaves Now in Residence in the United States of America*, which drew from several data sources to put the Black population of the United States in national and international perspectives. These charts are collectively attributed to "Negro Students of Atlanta University." Design by Tanvi Sharma. Prototype by Zhou Fang. Implementation by Jay Varner.

American Negro Newspapers and Periodicals.

Image courtesy of the Library of Congress,
Prints & Photographs Division, LC-DIG-ppmsca-33923.

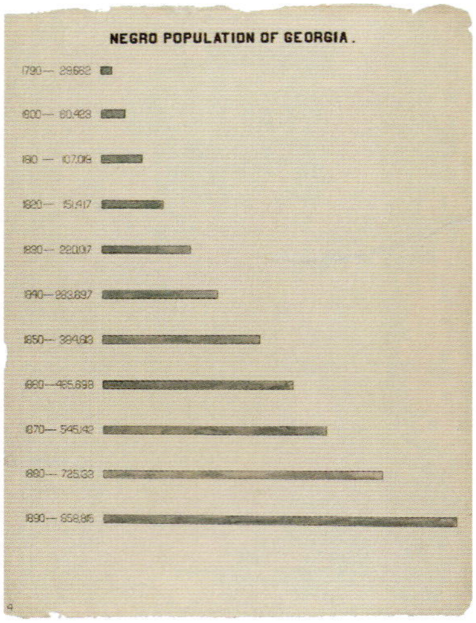

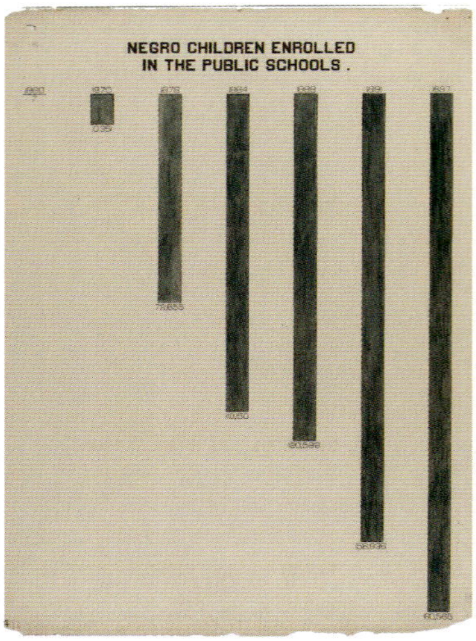

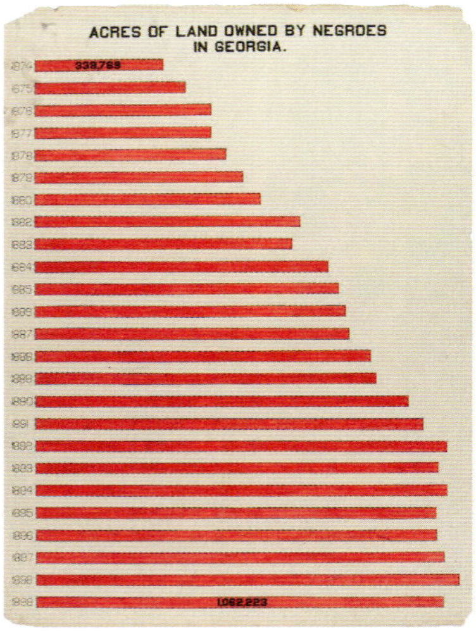

FIGURES 5.6–5.8
LEFT TO RIGHT: The Black population of Georgia; the number of Black children enrolled in Georgia's public schools; the number of acres of land owned by Black Georgians, with an overall form that resembles the shape of the state itself. Images courtesy of the Library of Congress, Prints & Photographs Division, LC-DIG-ppmsca-33866, LC-DIG-ppmsca-33877, and LC-DIG-ppmsca-33881.

whole (FIGURE 5.3). This is both because of its imagery—it depicts a pair of Mercator projections, one of Africa and the other of the Americas, showing the "routes of the African slave trade"—and because of its text. At the bottom of the chart, it reads: "The problem of the 20th century is the problem of the color line." When these words reappeared three years later in *The Souls of Black Folk*, they would become "perhaps Du Bois's most famous indictment of the centrality of race and racism to modern American sociopolitical life," as architectural historian Mabel O. Wilson explains.[24] These words also function as a key to interpreting the actual colored lines of the charts. But I believe that the true key to the charts lies in the contributions of Du Bois's students: Rogers, Lee, Simons, Westmoreland, and Mack. After all, each of them was already living with the effects of the color line every day.

The period beginning in 1877, in the aftermath of Reconstruction and the end of federal support for the laws and policies that would support Black Americans as full and equal citizens, and running through the 1890s, when Du Bois's students were in school, was characterized by state-level laws enforcing racial segregation through a combination of direct and indirect means. The effects of these laws, colloquially known as Jim Crow, were compounded by the rise of the Ku Klux Klan (which had been founded in 1865, on the eve of emancipation) and more generally by the legitimization of racial terror on behalf of Southern whites. The decade leading up to the Paris Exposition had been the most outwardly violent and oppressive since the end of the Civil War.[25]

This was the same context that, just a few years earlier, had prompted the investigative journalist Ida B. Wells to compile the first comprehensive dataset

FIGURE 5.9
"Assessed Value of Household and Kitchen Furniture Owned by Georgia Negroes," which employs the now-iconic Du Boisian spiral, combined with increasingly bright and saturated colors, to emphasize the increasing total value over time. Image courtesy of the Library of Congress, Prints & Photographs Division, LC-DIG-ppmsca-33887.

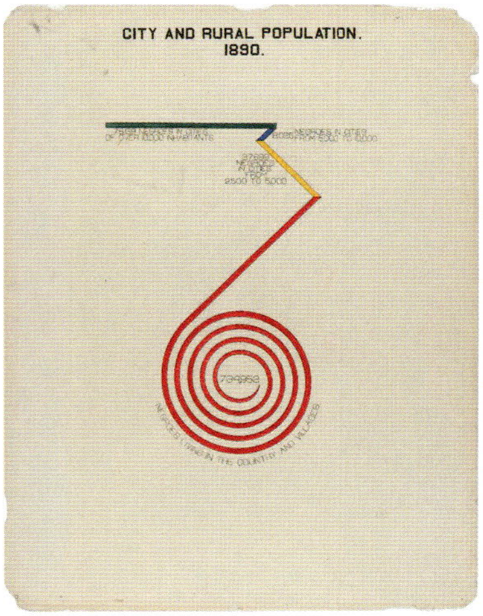

FIGURE 5.10
A zigzag spiral chart of Georgia's city and rural population in 1890. Image courtesy of the Library of Congress, Prints & Photographs Division, LC-DIG-ppmsca-33873.

on lynching, which she published as *A Red Record: Tabulated Statistics and Alleged Causes of Lynchings in the United States* in 1895—a form of counterdata collection more than a century before the term was coined. Wells, like Du Bois at the time, believed that the first step to eradicating this epidemic of racial terror would be to collect data, or "statistics" as they were known, that could document its scope and scale.[26]

As the twentieth century unfolded, data and the (very often spurious) statistical claims that data enabled would become a primary means of racial surveillance, eugenic policy, and social control.[27] But in this particular moment, before the power of data and statistics had been fully consolidated in the hands of the state, the "very newness of these technologies also created loopholes" for Black and white data activists alike, literary scholar Autumn Womack explains.[28] This resulted in "enthusiasm and curiosity" and, in the case of Wells, Du Bois, and his students, the belief that data—when collected, analyzed, and communicated with accuracy and care—could bring the nation closer to the goals that its foundational documents proudly proclaimed.[29]

This combination of enthusiasm, curiosity, and conviction is clearly evident in the Paris Exposition charts. Intent on demonstrating the "progress and prospects" of Black Americans in spite of the constraints on their freedom they faced, Du Bois and his students took active steps to ensure that there would be no misinterpretation of the insights they hoped their charts would convey.[30] They carefully selected specific facets of each dataset, at times also devising novel visualization forms, so as to ensure that this message would be clear. The Black population of Georgia is shown to increase every decade (FIGURE 5.6); the number of Black children enrolled in school is similarly shown to be on the rise (FIGURE 5.7); land ownership is documented as

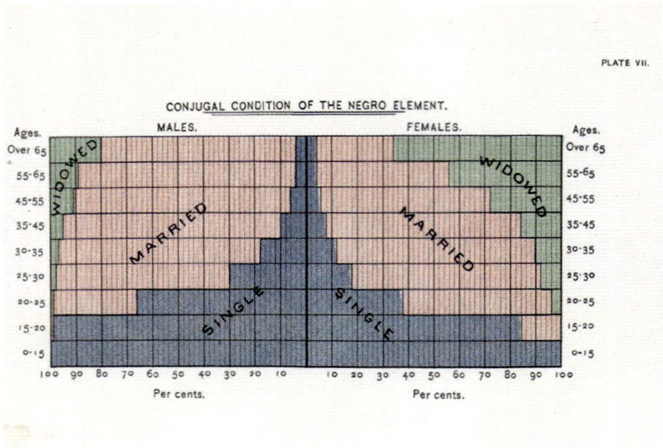

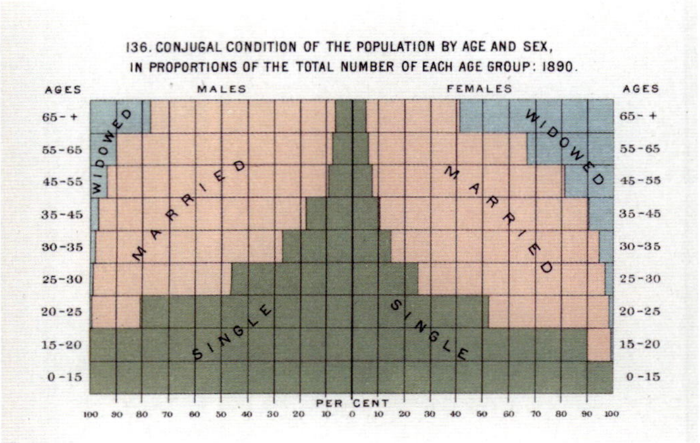

growing nearly every year of the previous 25 (FIGURE 5.8); and the assessed value of Black Georgians' property has exploded so exponentially that it requires a wholly new visual form—a bar chart curved into a spiral—for the most recent values to even fit on the page (FIGURE 5.9). Indeed, the increasingly novel visual forms that characterize the sequencing of the charts also seem to ask viewers to imagine Black progress through the lens of Black creativity: what might come to be if the future of Black America, like the expressive possibilities of data visualization, were also not curtailed by convention and code?

This argument is also supported by the sequence of the charts. Consider how the first several charts in the Georgia series make use of familiar visual strategies—maps, bar charts, and line graphs—in order to introduce the exhibition's international viewership to the state and its significance as an object of study. But the visual style shifts into new terrain once the focus on Georgia's Black residents has been established. The now-iconic "Du Boisian spiral" is introduced, for example, in a comparative representation of the places where Georgia's Black citizens reside, bringing together aspects of the bar chart and the line chart along with this new form of circular graph in order to animate the increasing presence of Black people throughout the state (FIGURE 5.10). While it is impossible to know who, precisely, served as the source of this visual innovation—Rogers is the most likely candidate—it seems significant that the green and blue lines at the top of the chart mirror the path of the northern border of the state of Georgia before zigging and zagging out of known geographical space; the state would be wholly transformed if its Black citizens—its students among them—could fully flourish, this form seems to suggest.[31]

The second set of charts—the set explicitly attributed to "Negro Students of Atlanta University"—continues to communicate this narrative of "progress and prospects." This set is distinguished from the first by how it makes use of comparisons to other populations, as well as to other charts. As it turns out, Du Bois was carrying out his own survey work for the US Bureau of Labor Statistics at the same time as he was instructing his students in how to conduct the

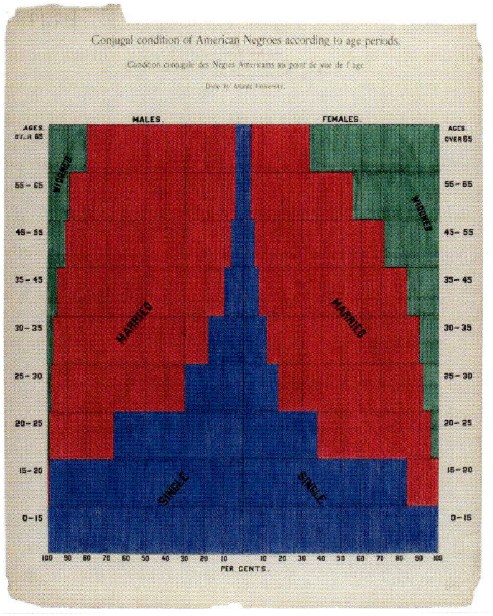

FIGURES 5.11–5.13
LEFT TO RIGHT: The chart of marriage statistics that appeared in Henry Gannett's 1894 *Statistics of the Negroes in the United States*; a visually similar chart of marriage statistics of all races that appeared in the illustrated edition of the 1890 census, also edited by Gannett; and the chart created by Du Bois and his students, visualizing the Black population only, that was created for the Paris Exposition. Images courtesy of the Library of Congress, Rare Book and Special Collections Division; Geography and Map Division; and Prints & Photographs Division, LC-DIG-ppmsca-33915.

"survey of social conditions" that they would fold into the charts.[32] While we do not know whether the results of Du Bois's surveys were ever incorporated into the larger statistical reports that the government released, the design of certain charts from the second set of charts makes clear that Du Bois and his students had the topic of federal data collection—and the information that the government lacked—plainly in their sights.

In 1894, the chief geographer for the US Census, Henry Gannett, had released a pamphlet that compiled and visualized statistics on Black Americans on the basis of one hundred years of US Census data.[33] Among the charts was one labeled "Conjugal Condition of the Negro Element," which Gannett marshaled in the service of a eugenicist argument about the relative weakness of the US Black population in relation to the white one (FIGURE 5.11).[34] (A similar chart of the entire US population appears in the illustrated edition of 1890 US Census, also overseen by Gannett (FIGURE 5.12).)[35] For the Paris Exposition, Du Bois and his students recreated the original Gannett chart as "Conjugal Condition of American Negroes according to Age Periods," the 17th of their 27 (or 28) charts (FIGURE 5.13). Seemingly akin to how, earlier in the century, abolitionists such as David Walker advocated that Black citizens read the racist arguments made by Thomas Jefferson, and other white people of the time, so as to be able to directly refute them, Du Bois enlisted his students in refuting Gannett's arguments through the same form that had been used to advance them in the first place: the chart.[36] This is a clear instance of what we would call "countervisualization" today.

The illustrated edition of the 1890 census in which the Gannett chart also appeared, formally titled *Statistical Atlas of the United States, Based on the Results of the Eleventh Census*, had been published just a year earlier, in 1898, and must have also been a topic of classroom discussion. This is evident in the visual similarity between another chart that appears in the *Atlas*, "Growth of the Elements of the Population: 1790–1890" (FIGURE 5.14), and the graduated area chart, "The Amalgamation of White and Black Elements of the Population in the United States," which immediately follows the "Conjugal Condition" chart in the student set

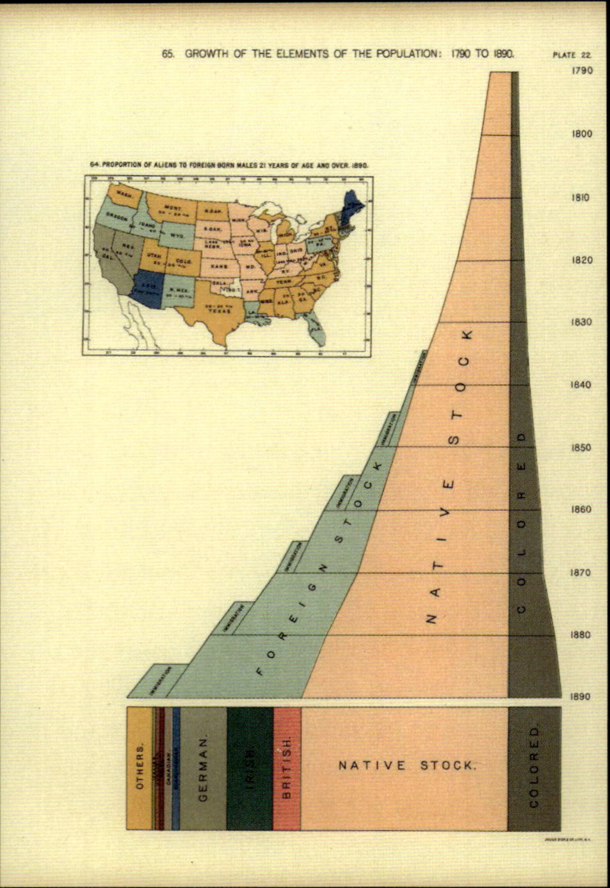

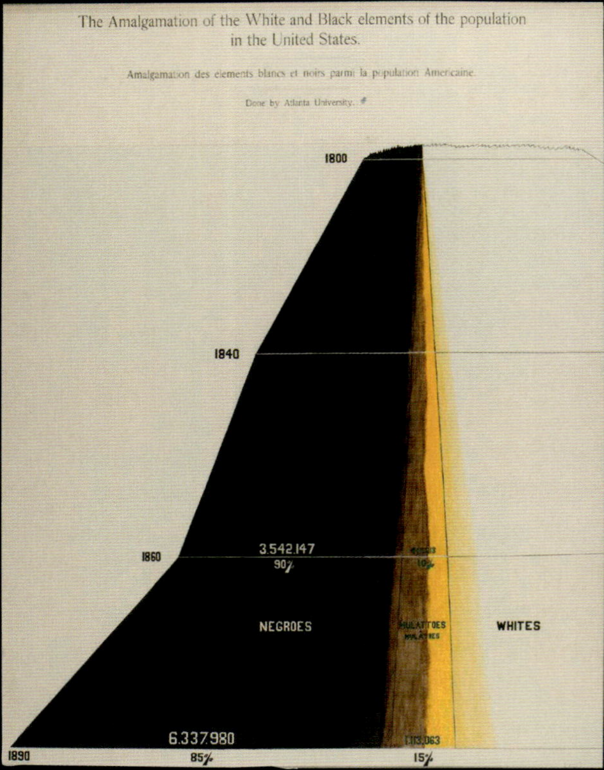

FIGURES 5.14 AND 5.15
LEFT: "Growth of the Elements of the Population, 1790–1890," as it appears in the 1890 *Statistical Atlas of the United States*.
RIGHT: "The Amalgamation of the White and Black Elements of the Population in the United States," as created by Du Bois and his students for the Paris Exposition. Images courtesy of the Library of Congress, Geography and Map Division and the Prints & Photographs Division, LC-DIG-ppmsca-33916.

(and serves as the cover image for this book) (FIGURE 5.15). The notable difference, here, is the focus on the US Black population alone. Whereas in Gannett's version, the white population—titled "native stock"—is in the center of the diagram, Du Bois and his students shift the frame to focus on the Black population, which in this version is also shaded black, so as to command the visual attention it deserves. A thin jagged line at the top, grid lines that extend off the side of the page, and the label (but no associated statistics for) "whites," are all that remain in the students' chart as visual indicators of what has been excluded. But there is no mistaking the reason why the white population has been excluded here: by mirroring the form of the chart in the *Atlas* but focusing on the Black population alone, the charts reinforce the idea Du Bois put into words later that year when recalling the exhibit in Paris: that this "small nation of people" within the larger United States, its triumphs and its challenges, deserved to have a national statistical atlas of its own.[37]

Charts, Photographs, and the Limits of Facts

Today, the charts of the Paris Exposition are often shared as single images, embedded in social media posts, or—as in the case of *W. E. B. Du Bois's Data Portraits*, bound as a dazzling, full-color book. But in their own time, the charts were intended to be viewed in relation to the other artifacts on display. There were the 200 books that had resulted from Murray's attempt to include every Black-authored text that had been published to that point, along with a more complete 1,400-item bibliography. There were physical artifacts, including a bronze statuette of Frederick Douglass and a "remarkable" wood frame carved by a man who was formerly enslaved, among other examples of wood and metalwork from students and faculty at other Black colleges and industrial schools.[38] There was a three-volume compilation that Du Bois titled "Black Codes of Georgia," which contained a catalogue of all of the discriminatory laws enacted in the state since the end of the Civil War—one of the few explicit acknowledgments in the exhibit of the harsh reality of being Black in the post-Reconstruction United States.[39]

And then there were the photographs. The exhibit included over 500 original prints documenting Black Americans at home, at work, and at school—including at Atlanta University, where a group of four women students were photographed on the steps of their classroom building (FIGURE 5.16).[40] Was one of these students Lula Iola Mack? Did any of these students live with Mack in the "plain four story brick building" that served as the women's dorm, on the corner of West Hunter and Vine Streets?[41] Or if none of the women in the photo were associated with Mack, could it be that this group of four women were, instead, in the year below Mack—part of the Class of 1901—who represented the largest cohort of women students at Atlanta University to that point? These questions are tantalizing, but they do not have answers—at least not to date. We do know, however, what Du Bois was trying to accomplish by including photos like this in the exhibit: an overturning of "conventional American ideas" about what the nation's Black citizenry looked like, what social and professional roles they occupied, and, in the case of these four women, what they were poised to achieve.[42]

The art historian Shawn Michelle Smith connects the Paris Exposition photographs to Du Bois's idea of double consciousness, as articulated most famously in *The Souls of Black Folk* as "the sense of always looking at one's self through the eyes of another."[43]

FIGURE 5.16
Four Atlanta University women students posed on the steps of a classroom building, photographed by Thomas Askew for the Paris Exposition. Image courtesy of the Library of Congress Prints and Photographs Division, LC-DIG-ppmsca-08778.

With the photographs, which not only featured Black people but were also created by them—the prominent Black photographer Thomas Askew took the Atlanta University shots—Du Bois sought to present to a predominantly white international audience a view of how Black Americans saw themselves.[44] When we see how the charts were installed alongside the photographs in Paris, it becomes all the more apparent that the charts were also pushing back against this double consciousness, albeit through a different visual form.[45] In direct comparison to the government statistics collected by Gannett, who was white, the charts created by Du Bois and his students can be interpreted as expressing their own assessment of the status of race and racism in America at the time.

At the same time, Du Bois's decision to pair the charts with the photographs also points to his awareness of the limits of what either medium—one quantitative, the other qualitative—could achieve on its own. While the photographs could document the richness of individual lives, they could not document the life of each of the nation's Black citizens. Conversely, while the charts could present powerful evidence of broad trends, they could not expose the individuals behind the data, nor could they express the stories that reflected the warp and weft of each person's life. Considered as a complementary pair, however, the charts and the photographs recall another visual technology of that era, the stereoscope, whose form is suggested by the double-projection layout of the *Georgia Negro*'s introductory chart (FIGURE 5.17).

The stereoscope was a device that spliced together two photographs of the same scene, taken from slightly different viewpoints. When the photos were viewed simultaneously, one by each eye, the stereoscope created the illusion of three-dimensional depth. Similarly, I believe that Du Bois understood the charts and the photographs as two parts of a larger whole. While each was legible on its own, the more complete—and therefore more accurate—picture was gained by viewing the two components together. This might be interpreted as another version of the epistemological pluralism that Elizabeth Palmer Peabody had explored nearly a half-century earlier with her workbooks and variously sized charts—notably, also in work with students. But for Du Bois, Mack, Rogers, and the rest, the stakes were far higher than for Peabody's pupils, whose futures as white citizens were secure. Du Bois's students, and the people for whom they designed their charts, would be required to cultivate not only historical knowledge but also political consciousness: the conviction necessary to pair any insights prompted by the charts with a personal commitment to act.

Du Bois had yet to reveal to his students that he had himself been engaged in a process of

FIGURE 5.17
A stereoscope known as the "Perfectscope," created by the H. C. White Company to commemorate the 1900 Paris Exposition. The Exposition's insignia is stamped on the top of the device.

We have replaced the image card shown in the original photo with one of the Palace of Social Economy and Congresses, where the Negro Exhibit was housed. "Perfectscope" photograph courtesy of the Foticos Museum. Image card of the "Palais de Congrès" from World of Stereo Views, used with permission.

epistemological soul-searching since moving to Atlanta, which by February or March of 1900 had become particularly intense. As he recalls in his 1940 autobiography, *Dusk of Dawn*, he had moved to Atlanta full of optimism and ambition: "I was going to study the facts, any and all facts, concerning the American Negro and his plight, and by measurement and comparison and research, work up to any valid generalization which I could."[46] Here we see the strength of Du Bois's belief in the power of "facts"—the more facts the better—when they could be analyzed and aggregated to point toward larger claims.

But after only a year in Atlanta—and just a few months before he began to assemble the materials for the Paris Exposition—Du Bois experienced nothing less than an epistemological epiphany, and not in any positive sense of the term. It was brought about not by the introduction of any new method but instead by his own firsthand evidence of how white supremacy constrained any work that he might do. In *Dusk of Dawn*, Du Bois narrates this realization in almost metaphysical terms: "At the very time when my studies were most successful, there cut across this plan which I had as a scientist, a red ray which could not be ignored."[47]

This "red ray" was no abstraction. It had a specific and horrific source: the lynching of a man named Sam Hose, which had taken place just outside Atlanta, on April 23, 1899. Ten days before, Hose and his employer had gotten into an argument after Hose requested time off to visit his mother, who was sick. The argument spiraled out of control, resulting in the employer's death; from there, rumors abounded.[48] Du Bois had the idea to write an essay about the incident for the *Atlanta Constitution*, the leading newspaper of the South. He intended to approach it as he would any other inquiry: through a social scientific lens. He recalls how he drafted a "careful and reasoned statement concerning the evident facts."[49] But while walking from the Atlanta University campus to the newspaper office—he planned to deliver his draft in person—Du Bois learned of a gruesome new turn of events: Hose had been lynched, and his knuckles were rumored to be on display in a storefront that Du Bois himself would soon pass on his walk. Du Bois turned around and went home, his essay never to see the light of day.[50]

While Du Bois's desire to publish a statement on the "facts" of the case was extinguished at that point, he continued to contemplate—deeply—the larger function of his research. As he narrates:

> Two considerations thereafter broke in upon my work and eventually disrupted it: first, one could not be a calm, cool, and detached scientist while Negroes were lynched, murdered, and starved; and secondly, there was no such definite demand for scientific work of the sort that I was doing.[51]

From these lines—which Du Bois echoes in an autobiographical audio interview recorded in 1961—it is clear that he emerged from the incident with a greater awareness of the limits of his data-driven approach to advancing knowledge. He goes on to explain how, previously, he had "regarded it as axiomatic that the world wanted to learn the truth and if the truth was sought with even approximate accuracy and pains-taking devotion, the world would gladly support the effort."[52] But the violence brought about by centuries of white supremacy, Du Bois realized at that moment, demanded an epistemological charge greater than what could ever be learned—or communicated—through facts alone. Thus when he turned to the Paris Exposition in the next academic year, it was not facts or data but instead the experience and effects of living within, continually resisting, and, in spite of it all, succeeding within a racist regime that Du Bois must have encouraged his students to give visual form. They did this through the charts they designed that hung on the walls, and through the presence of their classmates pictured in the photographs, pointedly accompanied by the three volumes of Black Codes on the shelves.[53]

Visualization for Liberation

After graduation, Lula Iola Mack moved about two hundred miles south of Atlanta to Albany, Georgia, where she worked as a teacher for several years. In 1902, she married one Mr. F. H. Wilkins, and they spent the next decade together, moving around the south-east.[54] In 1907, they had a child, whom they named Marggette (or possibly Marguerite) Iolamarie.[55]

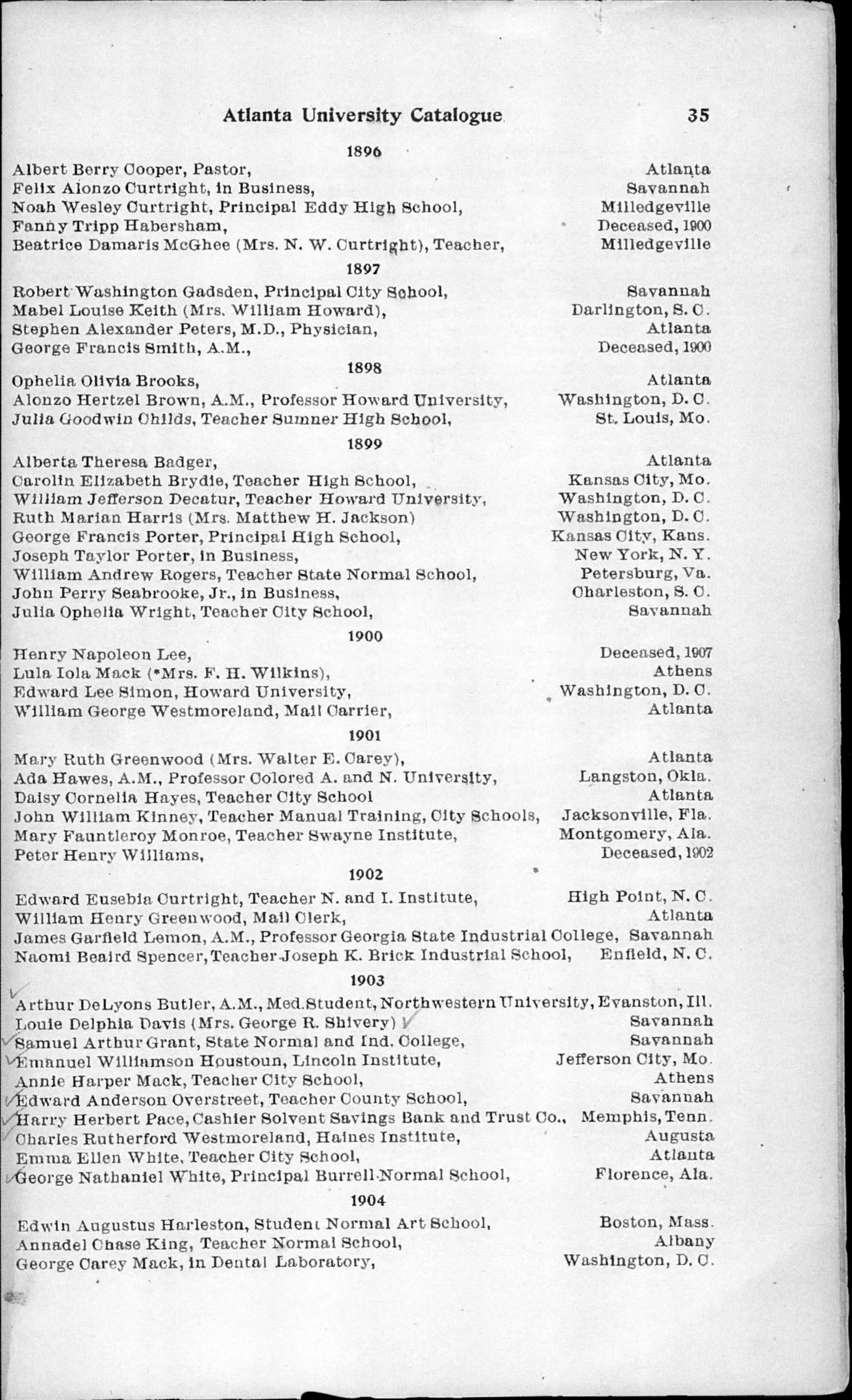

Atlanta University Catalogue 35

1896

Albert Berry Cooper, Pastor, Atlanta
Felix Alonzo Curtright, in Business, Savannah
Noah Wesley Curtright, Principal Eddy High School, Milledgeville
Fanny Tripp Habersham, Deceased, 1900
Beatrice Damaris McGhee (Mrs. N. W. Curtright), Teacher, Milledgeville

1897

Robert Washington Gadsden, Principal City School, Savannah
Mabel Louise Keith (Mrs. William Howard), Darlington, S. C.
Stephen Alexander Peters, M.D., Physician, Atlanta
George Francis Smith, A.M., Deceased, 1900

1898

Ophelia Olivia Brooks, Atlanta
Alonzo Hertzel Brown, A.M., Professor Howard University, Washington, D. C.
Julia Goodwin Childs, Teacher Sumner High School, St. Louis, Mo.

1899

Alberta Theresa Badger, Atlanta
Carolin Elizabeth Brydie, Teacher High School, Kansas City, Mo.
William Jefferson Decatur, Teacher Howard University, Washington, D. C.
Ruth Marian Harris (Mrs. Matthew H. Jackson) Washington, D. C.
George Francis Porter, Principal High School, Kansas City, Kans.
Joseph Taylor Porter, in Business, New York, N. Y.
William Andrew Rogers, Teacher State Normal School, Petersburg, Va.
John Perry Seabrooke, Jr., in Business, Charleston, S. C.
Julia Ophelia Wright, Teacher City School, Savannah

1900

Henry Napoleon Lee, Deceased, 1907
Lula Iola Mack (*Mrs. F. H. Wilkins), Athens
Edward Lee Simon, Howard University, Washington, D. C.
William George Westmoreland, Mail Carrier, Atlanta

1901

Mary Ruth Greenwood (Mrs. Walter E. Carey), Atlanta
Ada Hawes, A.M., Professor Colored A. and N. University, Langston, Okla.
Daisy Cornelia Hayes, Teacher City School Atlanta
John William Kinney, Teacher Manual Training, City Schools, Jacksonville, Fla.
Mary Fauntleroy Monroe, Teacher Swayne Institute, Montgomery, Ala.
Peter Henry Williams, Deceased, 1902

1902

Edward Eusebia Curtright, Teacher N. and I. Institute, High Point, N. C.
William Henry Greenwood, Mail Clerk, Atlanta
James Garfield Lemon, A.M., Professor Georgia State Industrial College, Savannah
Naomi Beaird Spencer, Teacher Joseph K. Brick Industrial School, Enfield, N. C.

1903

Arthur DeLyons Butler, A.M., Med. Student, Northwestern University, Evanston, Ill.
Louie Delphia Davis (Mrs. George R. Shivery) Savannah
Samuel Arthur Grant, State Normal and Ind. College, Savannah
Emanuel Williamson Houstoun, Lincoln Institute, Jefferson City, Mo.
Annie Harper Mack, Teacher City School, Athens
Edward Anderson Overstreet, Teacher County School, Savannah
Harry Herbert Pace, Cashier Solvent Savings Bank and Trust Co., Memphis, Tenn.
Charles Rutherford Westmoreland, Haines Institute, Augusta
Emma Ellen White, Teacher City School, Atlanta
George Nathaniel White, Principal Burrell Normal School, Florence, Ala.

1904

Edwin Augustus Harleston, Student Normal Art School, Boston, Mass.
Annadel Chase King, Teacher Normal School, Albany
George Carey Mack, in Dental Laboratory, Washington, D. C.

FIGURE 5.18
A page from the alumni listings in the *Catalogue of the Officers and Students of Atlanta University, 1909–1910*, which reports Henry Napoleon Lee as deceased. William Andrew Rogers, Class of 1899, is identified as working as a teacher at the State Normal School in Petersburg, Virginia, where he worked until at least 1914, when the catalogue stopped listing individual alumni. Image courtesy of the Archives Research Center, Digital Commons @ Robert W. Woodruff Library, Atlanta University Center.

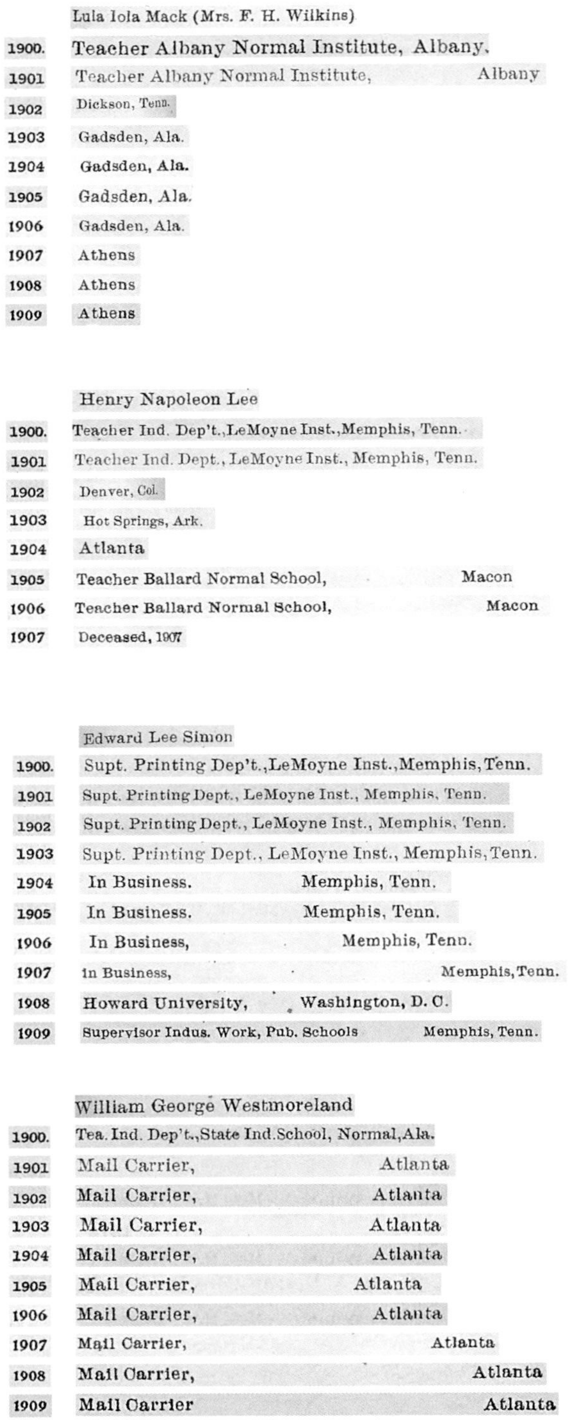
Lula Iola Mack (Mrs. F. H. Wilkins)
1900. Teacher Albany Normal Institute, Albany.
1901 Teacher Albany Normal Institute, Albany
1902 Dickson, Tenn.
1903 Gadsden, Ala.
1904 Gadsden, Ala.
1905 Gadsden, Ala.
1906 Gadsden, Ala.
1907 Athens
1908 Athens
1909 Athens

Henry Napoleon Lee
1900. Teacher Ind. Dep't.,LeMoyne Inst.,Memphis, Tenn.
1901 Teacher Ind. Dept., LeMoyne Inst., Memphis, Tenn.
1902 Denver, Col.
1903 Hot Springs, Ark.
1904 Atlanta
1905 Teacher Ballard Normal School, Macon
1906 Teacher Ballard Normal School, Macon
1907 Deceased, 1907

Edward Lee Simon
1900. Supt. Printing Dep't.,LeMoyne Inst.,Memphis,Tenn.
1901 Supt. Printing Dept., LeMoyne Inst., Memphis, Tenn.
1902 Supt. Printing Dept., LeMoyne Inst., Memphis, Tenn.
1903 Supt. Printing Dept., LeMoyne Inst., Memphis,Tenn.
1904 In Business. Memphis, Tenn.
1905 In Business. Memphis, Tenn.
1906 In Business, Memphis, Tenn.
1907 In Business, Memphis,Tenn.
1908 Howard University, Washington, D. C.
1909 Supervisor Indus. Work, Pub. Schools Memphis, Tenn.

William George Westmoreland
1900. Tea. Ind. Dep't.,State Ind.School, Normal,Ala.
1901 Mail Carrier, Atlanta
1902 Mail Carrier, Atlanta
1903 Mail Carrier, Atlanta
1904 Mail Carrier, Atlanta
1905 Mail Carrier, Atlanta
1906 Mail Carrier, Atlanta
1907 Mail Carrier, Atlanta
1908 Mail Carrier, Atlanta
1909 Mail Carrier Atlanta

FIGURE 5.19
The occupations and places of residence of Lula Iola Mack, Henry Napoleon Lee, Edward Lee Simon, and William George Westmoreland, as documented in the *Catalogues* between 1900–1901 and 1909–1910. Images courtesy of the Archives Research Center, Digital Commons @ Robert W. Woodruff Library, Atlanta University Center. Collage by Tanvi Sharma.

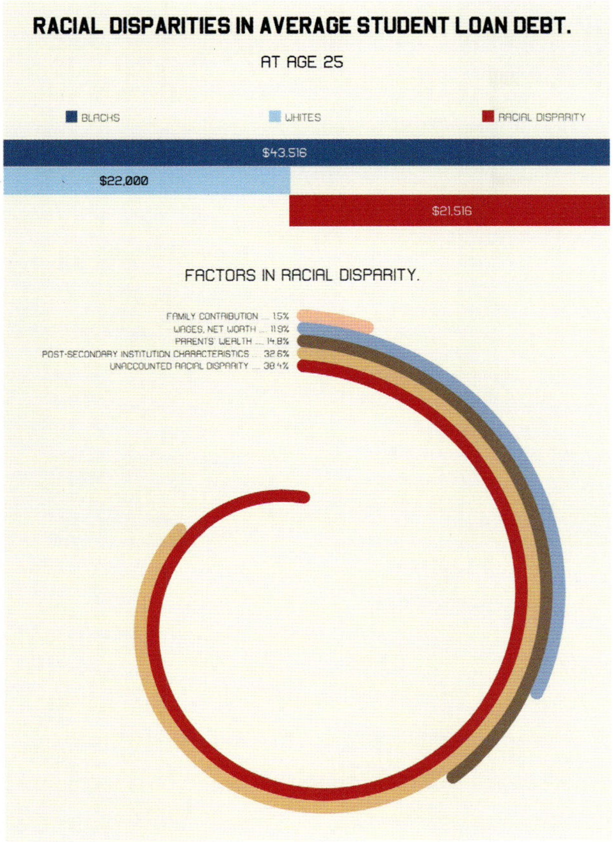

FIGURE 5.22
An animated visualization of disparities in student debt by population, visualized in a Du Boisian spiral. A student loan repayment calculator that allows users to explore how different loan terms may affect their monthly payments appears below. Image courtesy of Jeffrey Himpele, VizE Lab for Ethnographic Data Visualization, Princeton Department of Anthropology, in collaboration with the Dignity and Debt Initiative.

which led her to US Census Bureau data on net worth, resulted in a picture that documented how, "for every dollar a black household in America has in net assets, a white household has 16.5 more."

It is incredibly affecting to see some of the most distinctive of the charts recreated with contemporary data, and, even more, to see how certain trends have so sharply diverged from the narrative of "progress and prospects" that Du Bois and his students sought to convey. From the time that I first encountered Chalabi's recreations and decided to incorporate them in this book, I have asked myself why they are so powerful—far more than even the most impressive of recreations of William Playfair's iconic charts. To be sure, there is a sense of historiographic justice involved in seeing Du Bois and his students elevated to the level of Playfair. There is also some degree of visual novelty; even though the charts of the Paris Exposition have become increasingly well-known, their more unique visual forms, such as the lattice and spiral, are still employed far less often than, say, a line graph or bar chart. But in the end, I think it comes down to the story they tell, both about the continued entrenchment of racial inequality, and about the strength and commitment of Du Bois's and his students' desire to change it. By contrast, we can only speculate as to any change that Playfair sought to bring about (or guard against) with his charts.

For the field of data visualization, the absence of any visually discernable motivation for making the charts—that is to say, their purported objectivity—has

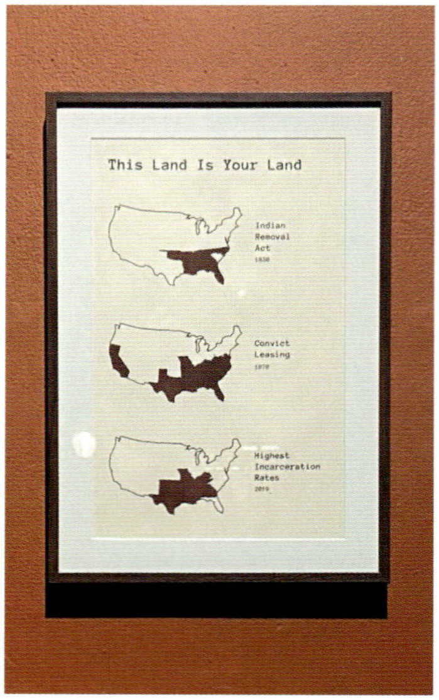

FIGURES 5.23–5.25
Three charts in the "In Absentia" series by Mimi Ọnụọha. Photos by Emile Askey. Courtesy of Mimi Ọnụọha.

long been one of their most celebrated features. But we fail to activate the full insight-producing power of visualization—or, worse, risk prompting inaccurate or otherwise misleading insights in our viewers—when we do not begin our own design process by acknowledging our own goals.

As an example of contemporary work that makes its goals explicit, also inspired by Du Bois and his students, consider the series of interactive visualizations created by the Dignity + Debt Network and the VizE Lab for Ethnographic Data Visualization at Princeton, which were designed to expose racial disparities in student loan debt in the United States (FIGURE 5.22).[66] One visualization tool enables users to enter the amount they originally borrowed and see that amount compared to the average principal for each major racial and ethnic group. The tool makes use of animation to draw a Du Boisian spiral in real time; when the user's loan amount is entered, an additional line in the spiral, in red, is drawn. Whereas in the original charts shown in Paris, the spiral form and the use of red for the largest value focuses the viewer on the increasing nature of the trend, the same color here, combined with the animation, emphasizes the increasing and increasingly constraining effects of the burden of student loan debt. Nevertheless, the new tool shares with the original charts a clear sense of purpose. In this case, the purpose is conveyed in the form of a calculator positioned below the chart, which allows the user to compare their current interest rate to other possible rates, and in the process informs them of how a change in rate (or monthly payment amount) could lead to a different payoff date or total amount owed. On the "About" page, this purpose is made even more explicit: visualizations "are able to reveal what might be either changed or strengthened through policy and individual action."[67]

Of course, those experiencing the negative effects of, for example, student loan debt do not need visualization to "reveal" the nature of the particular problem they face; they live with it every day.[68] This is, in fact, a point made by another contemporary project based on Du Bois's charts, by the artist and educator Mimi Ọnụọha, which will bring us toward this chapter's close. "In Absentia" consists of six charts that reference the sequence and visual forms of the original charts. Their purpose is not to provide additional evidence of what is already known to be true, however. Instead, in Ọnụọha's own words, the charts "form a meditation on interpretability, questioning why such a fact [about racism] should need proving."[69]

Ọnụọha's charts follow a similar progression to those in the *Georgia Negro* series, using typologies that recall the original charts. Maps in small multiples, similar to those that Rogers and Du Bois employed to begin the series, document the spatial relation between three seemingly unrelated events and phenomena: the Indigenous lands that were claimed by the US government through the Indian Removal Act of 1830; the locations in which "convict leasing" was practiced in the 1870s, when prisons required those they imprisoned to work for private companies, which bolstered prison profits but offered no compensation for the imprisoned workers themselves; and the states with, as of 2019, the highest incarceration rates in the country (FIGURE 5.23). (To this we might add the locations of immigrant detention centers that are currently being built.) But rather than continue to provide additional evidence of the unfolding legacy of slavery and dispossession, Ọnụọha employs some of the original series's more experimental forms to offer commentary on the continual need for evidence. A chart that resembles the spiraling bar chart of the value of household possessions, discussed above, here consists of lines labeled "What," "are we," trying to prove" (FIGURE 5.24). A final chart is simply a circle, fully shaded, labeled "A space for truths that cannot be shown" (FIGURE 5.25). The chart is titled "It Could Never Be Large Enough."

In the context of Du Bois's autobiographical writing about the difficulty of connecting facts and data with "the truth," Ọnụọha's use of this term speaks volumes. It reminds us—those of us who seek to create visualizations that bear witness to oppression, those of us who seek to design charts for change, and those of us eager to celebrate the innovative forms that result from both—that we often need no further evidence of what we already know to be true.[70] If our ultimate goal is liberation—that is to say, the realization of a world in which all of us can equally thrive—then we very well may need to look to methods beyond visualization in order to reach it. Some of these methods, like the photographs that accompanied the charts at the Paris Exposition, can complement visualization without much adjustment. But others may require that we recalibrate our own deepest convictions—as did Du Bois in the wake of the lynching of Sam Hose—about how knowledge is best produced and conveyed.

Our Project's Purpose

As our project team considered how we might inhabit Du Bois's approach to visualization, we were keenly aware of the particular challenge posed by this task in relation to *Data by Design* as a whole. As this book has unfolded, we've presented abundant examples of the power of data visualization to change hearts and minds. It's safe to say that we, the members of our project team, believe in this power: it's in large part what has brought (and has kept) us together. But for the visualization component of this chapter, we were required to confront something new: the knowledge that this chapter's central figure—Du Bois himself—had very clearly decided that visualization was not powerful enough to effect lasting change. We came to realize that any new visualization we designed for this chapter would have to confront this contradiction head on. We thus returned to the starting point of this chapter—the story of Lula Iola Mack and her Atlanta University classmates—where we had always found the most kinship as another student-centered project team. We asked ourselves if we might be able to use visualization to prompt additional insights about these students' lives, while simultaneously acknowledging the limits of what we could know.

Over the course of our research involving the Atlanta University *Catalogues*, from which we had surfaced the details of these five students' lives, we had noticed another detail about the Paris Exposition charts that seemed to have gone unremarked upon: the count of the total number of graduates of Atlanta University that is featured (and visualized) on the second series' introductory chart—330—matches exactly with the total number provided in a data table included in the 1898–1899 *Catalogue*, which had been published the year before and would have been available to Du Bois and his students for their work (FIGURE 5.26).[71] The data table in the catalogue also includes summary statistics about the 330 students' occupations. Presumably this table served as the original data source for the occupations pictured in the central pie chart that was displayed on the wall in Paris (FIGURE 5.28).[72]

We were enthralled with the possibility that the students who had collected, analyzed, and visualized the data that appeared in the charts—Mack, Westmoreland, Simon, and Lee—might also be included in the dataset itself. But there was one problem with this line of thinking: the data had been collected during the previous year, so it only included Rogers. It did not include the four seniors, who would graduate and become part of the alumni dataset the following spring. Our team had found its charge.

While the students could not place themselves in the dataset, 125 years later our project team could. Equipped with the tools of archival research and visualization design, we thought we could go further still: we could use visualization to connect their data, and the data of the 330 other Atlanta University graduates visualized in the central pie chart, to the stories suggested by the alumni listings. Illuminating the

Summary of Graduates and their Occupations.

	*COLLEGE.		NORMAL.		†TOTAL.	
	No.	Per Cent.	No.	Per Cent.	No.	Per Cent.
TOTAL	84	100.0	247	100.0	†330	100.0
Male	77	91.7	14	5.7	91	27.6
Female	7	8.3	233	94.3	†239	72.4
Living	74	88.1	215	87.0	†288	87.3
Dead	10	11.9	32	13.0	42	12.7
LIVING	74	100.0	215	100.0	†288	100.0
Male	67	90.5	8	3.7	75	26.0
Female	7	9.5	207	96.3	†213	74.0
OCCUPATIONS.						
Teachers	40	54.0	124	57.7	164	57.0
Religious Work	11	14.9	1	.5	12	4.2
Gov. Service	7	9.5	2	.9	9	3.1
Business	5	6.8	2	.9	7	2.4
Physicians	3	4.0	....	....	3	1.0
Lawyers	2	2.7	....	....	2	.7
Students	3	4.0	1	.5	4	1.4
Dentist	1	1.4	....	....	1	.4
Married Women not otherwise designated	2	2.7	78	36.3	†79	27.4
Undesignated	....	....	7	3.2	7	2.4

*Including three graduates from a theological course.
†One student graduated in two departments.

FIGURE 5.26
The data table included at the end of the 1898–1899 *Atlanta University Catalogue of the Officers and Students*. Image courtesy of the Archives Research Center, Digital Commons @ Robert W. Woodruff Library, Atlanta University Center.

students' stories felt especially important to me in my role as the leader of my own (majority) student project team. It also felt important to me in my own role as a student of the archive of slavery and the many, like Du Bois and his students, who lived in its wake. What I have learned from scholars such as Saidiya Hartman is that the lives of those who have not been "endowed with the gravity and authority of historical actor" are precisely those who should demand our attention today.[73] This is, after all, why I have placed so much emphasis on Lula Iola Mack in this chapter. It is because of my own deep-seated belief that it is in the lives of those most marginalized in the archive, as in the history of data visualization, that the most powerful stories can be found.

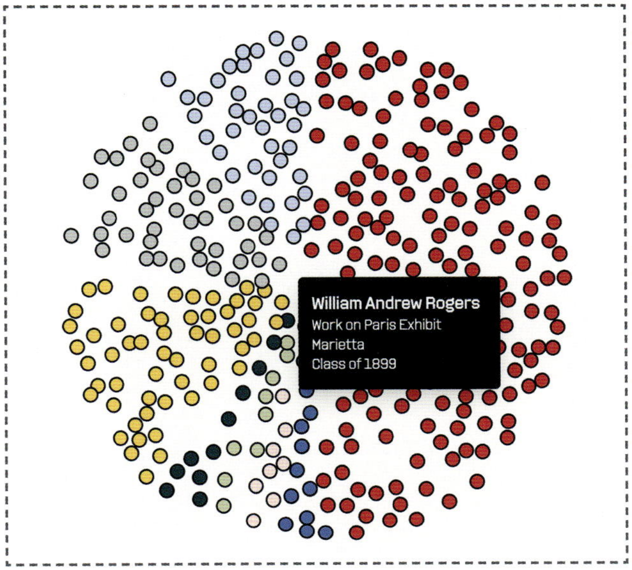

FIGURE 5.27
A detail view of the pie chart, displaying information about William Andrew Rogers. Hovering on any single dot reveals the information about that graduate from their alumni listing. Design and implementation by Tanvi Sharma.

The task of bringing back Mack and her classmates to the center of the story was one we approached as a full project team. Tanvi first transcribed all of the names and associated information from the 1898–1899 *Catalogue* into a spreadsheet. Then she designed an expanded version of the original pie chart with interactive datapoints representing each of the students in the original dataset. The members of the class of 1900 were placed in the chart according to the occupations they would assume upon graduation. In a meaningful echo of the student-led process of creating the original charts, Anna and Nicholas implemented Tanvi's design, which Jay then finessed into its final form (FIGURE 5.29).

Hovering on each dot in our expanded visualization reveals the person's name, self-reported occupation, and place of residence (FIGURE 5.27). Additional categories represent the 35 alumni with no occupation provided and the 42 alumni who were recorded in the catalogue as "Deceased." These details were our way to give more texture to the lives behind the data in the original chart—to the people whose education and accomplishments mattered so much to Du Bois and his students that they devoted the most prominent visualization in the entire set to them. (It should be noted that this chart was also used to raise money for the college; a call for funding appears below the image, ensuring similar opportunities for students ahead.)

But even as this visualization provided a more detailed picture of the students in the dataset—more "facts," if you will—we knew from our research that this was not the only way in which these students contributed to Du Bois's larger intellectual project. Du Bois was recruited to his professorial position in large part to assume direction of the Atlanta University Studies. These annual reports, on topics of Black American life ranging from business to religion to public health to crime, were presented each spring at a large public

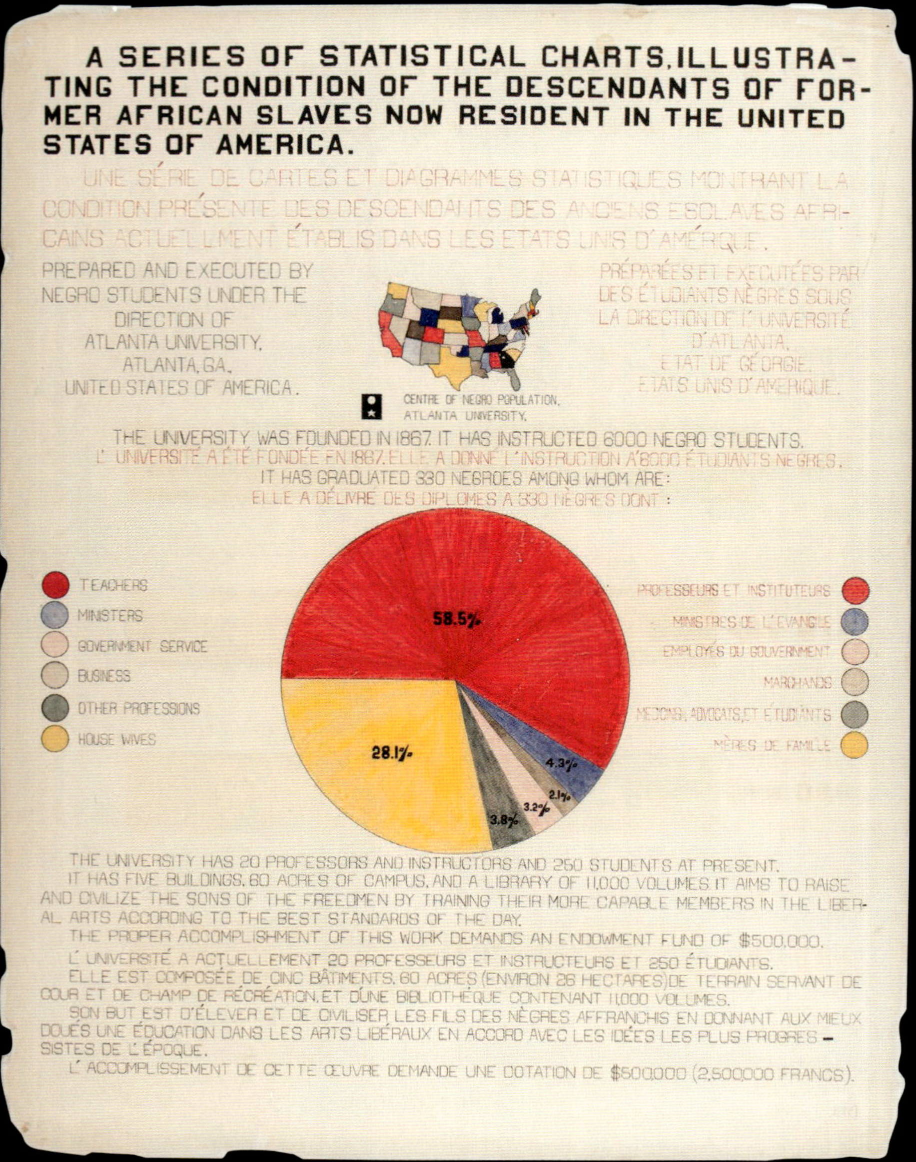

FIGURE 5.28
The title image of the second set of charts prepared for the Paris Exposition, *A Series of Statistical Charts Illustrating the Condition of the Descendants of Former African Slaves Now Resident in the United States of America*. Image courtesy of the Library of Congress, Prints & Photographs Division, LC-DIG-ppmsca-08994.

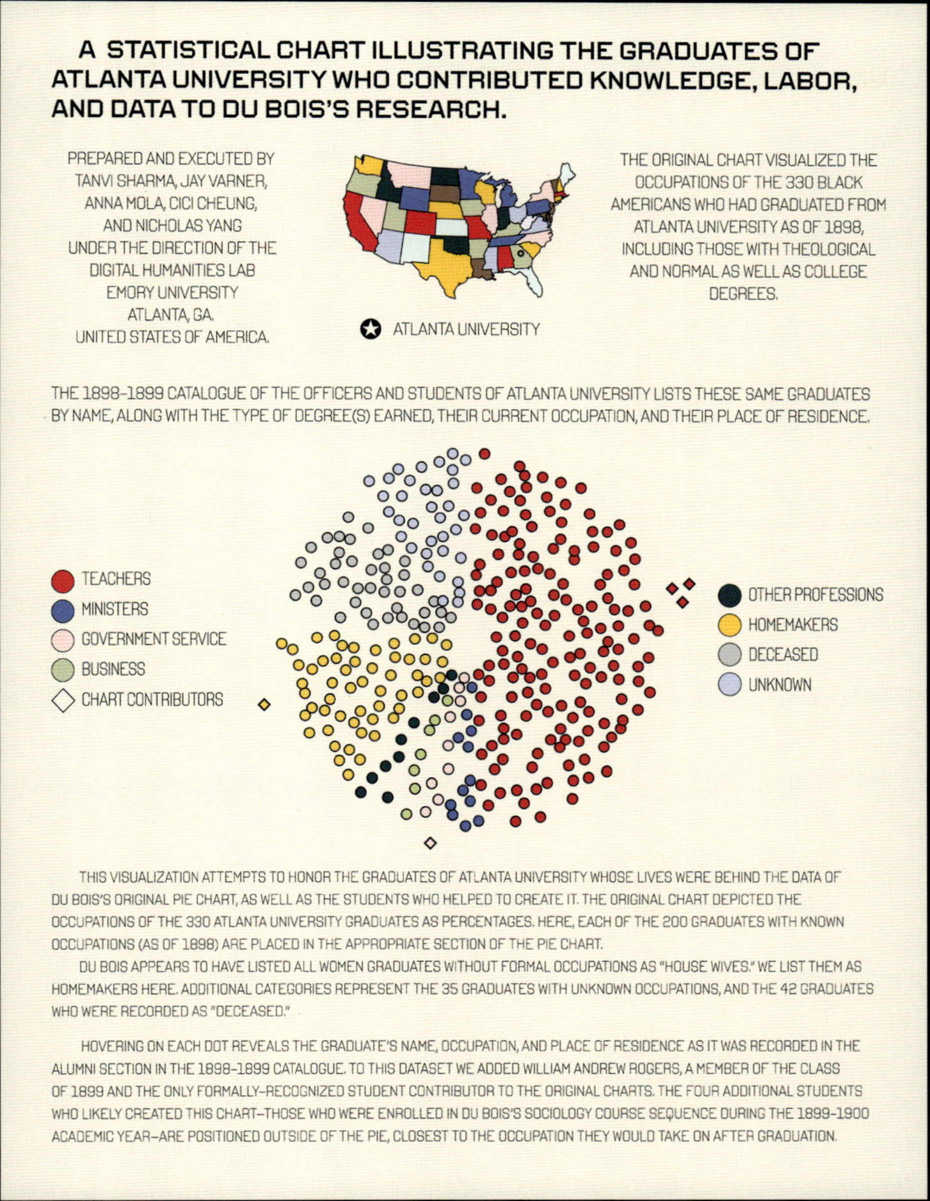

FIGURE 5.29
An expanded visualization of the 330 graduates of Atlanta University through 1899, with the four members of the class of 1900 added to the chart. An interactive version of this visualization appears at dataxdesign.io. Design by Tanvi Sharma. Prototype by Anna Mola and Nicholas Yang. Implementation by Jay Varner.

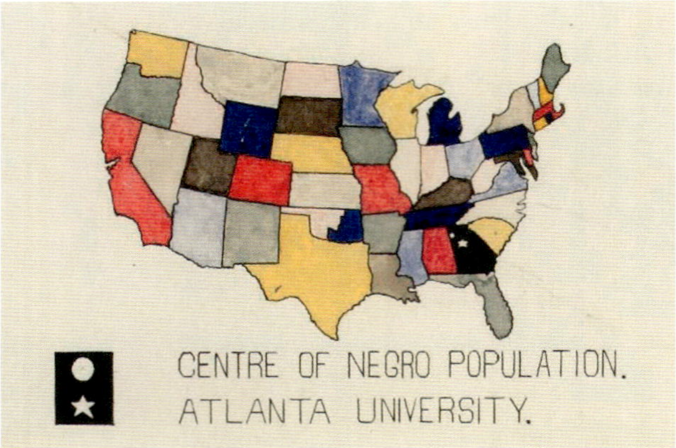

FIGURE 5.30
A detail view of the original map that appears in *A Series of Statistical Charts Illustrating the Condition of the Descendants of Former African Slaves Now Resident in the United States of America.* Image courtesy of the Library of Congress, Prints & Photographs Division, LC-DIG-ppmsca-08994.

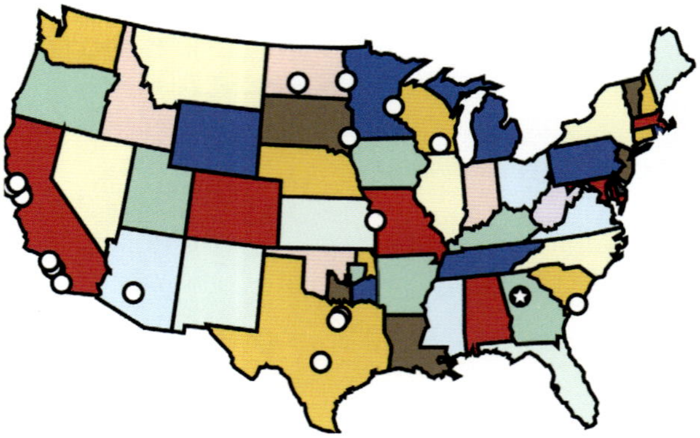

FIGURE 5.31
An updated map showing the locations of the additional 33 Black colleges included in the 1910 Atlanta University Study. Design by Tanvi Sharma. Prototype by Anna Mola and Nicholas Yang. Implementation by Jay Varner.

conference and, several weeks or months later, were published by Atlanta University's in-house printing press for broader circulation.[74] The first had been published only two years before Du Bois's arrival, but they had already become one of the school's defining intellectual contributions. Not incidentally, some of the data collected for the studies can be found in visual form in the Paris Exposition charts.[75]

In order to collect this data, as sociologist Aldon Morris has described, Du Bois enlisted not only his current students but all those who had ever taken his sociology course sequence. Together, the students and alumni constituted "two tiers of volunteer researchers," Morris explains.[76] While the current class of seniors would prepare the surveys and, when the surveys were returned, help to analyze the results, it was members of the Atlanta University alumni network who, each year, became unofficial field site supervisors. Du Bois asked this ever-expanding group of graduates to further distribute the surveys in their own communities, and to ensure that they were completed and returned. This collaborative work became the basis for what Morris has retrospectively honored as the "Du Bois-Atlanta School of Sociology," a means of formally recognizing the first data-driven sociology program in the United States.[77]

Was there a way for us to honor this network of alumni for the crucial role they played in contributing to Du Bois's efforts at Black data collection? This was the next challenge that our project team took on. We returned to the form of the introductory chart. This time, however, we did not look to the Atlanta University *Catalogues* as our source of data; we looked to the Atlanta University Studies themselves. We had known that the study that Mack and her classmates produced after completing the charts for the Paris Exposition was, very appropriately, on the subject of college education.[78] (The spring term must have been intense!) We also knew that, a decade later, Du Bois had returned to the topic, conducting a follow-up study entitled

The College-Bred Negro American, based on a new survey of all known living Black college graduates across the country. Per the published study, 800 responses were received.[79]

Among the respondents were, presumably, many of the ten years of graduates of Atlanta University whom Du Bois had personally taught by that point—perhaps even Mack, who had since returned to her hometown of Athens, Georgia with her three-year-old daughter.[80] Few individual responses to the survey have been preserved, so we cannot know with any certainty whether Mack or any of her classmates chose to respond. But the study does contain a data table of all Atlanta University graduates to that point; Mack, Lee, Simon, Westmoreland, and Rogers were definitely counted in number if not credited by name.

We thus created an updated version of our initial chart with the additional data from the 1910 Atlanta University Study. In the original chart, a map at the top of the page informed viewers of the location of Atlanta University (FIGURE 5.30). In our version, we added the locations of the 33 additional Black colleges also included in the study (FIGURE 5.31).[81] The location of Atlanta University (which is also, notably, the location of the majority of our own research team) is still marked with a star.

We also expanded the pie chart (FIGURE 5.32). We documented all 3,856 of the Black college graduates described by the study, living or dead, whether they returned the survey or not. The 163 graduates of the "college course" at Atlanta University, whom Du Bois describes in the study in depth, are named in our chart. We once again used the alumni section of that year's *Catalogue* to group the graduates by occupation and position them in the appropriate area of the pie chart. Mack is positioned as a "homemaker," as she had stopped teaching for a time. Westmoreland is positioned in "government service," as he was still working as a mail carrier. Simon and Rogers are positioned as "teachers." As in our own previous chart, we've added categories to represent those with unknown occupations, as well as those—at that point also including Lee—who were recorded as "Deceased."

An additional 3,693 gray dots represent the graduates of the other 140 colleges included in the study, whose names and exact occupations we do not know.[82] (In addition to the 34 Black colleges, Du Bois also surveyed the Black graduates of 107 predominantly white institutions.)[83] We used the counts of the occupations reported by the 800 survey respondents to calculate the approximate proportion of dots to place in each category, which we see as holding visual space for each graduate's contributions to the study, even if we do not know their exact occupation or name.

We do know something more about the students who returned the survey, however: we know a little about how they thought. In addition to the quantitative data that was collected and analyzed for the published study, there was also data that was qualitative in form. The survey that Du Bois and the seniors in that year's sociology course created and sent out across the United States—and that perhaps Mack in Athens, Simon in Memphis, Westmoreland in Atlanta, and Rogers in Petersburg, Virginia, helped to further distribute—did not just solicit information about the lives and accomplishments of graduates that could be aggregated as descriptive statistics; it also included four questions that encouraged long-form response.

These included questions about the respondents' reflections on their "early life and training," their plans to educate their own children, the "chief hindrances" they had faced since graduation, and their "present

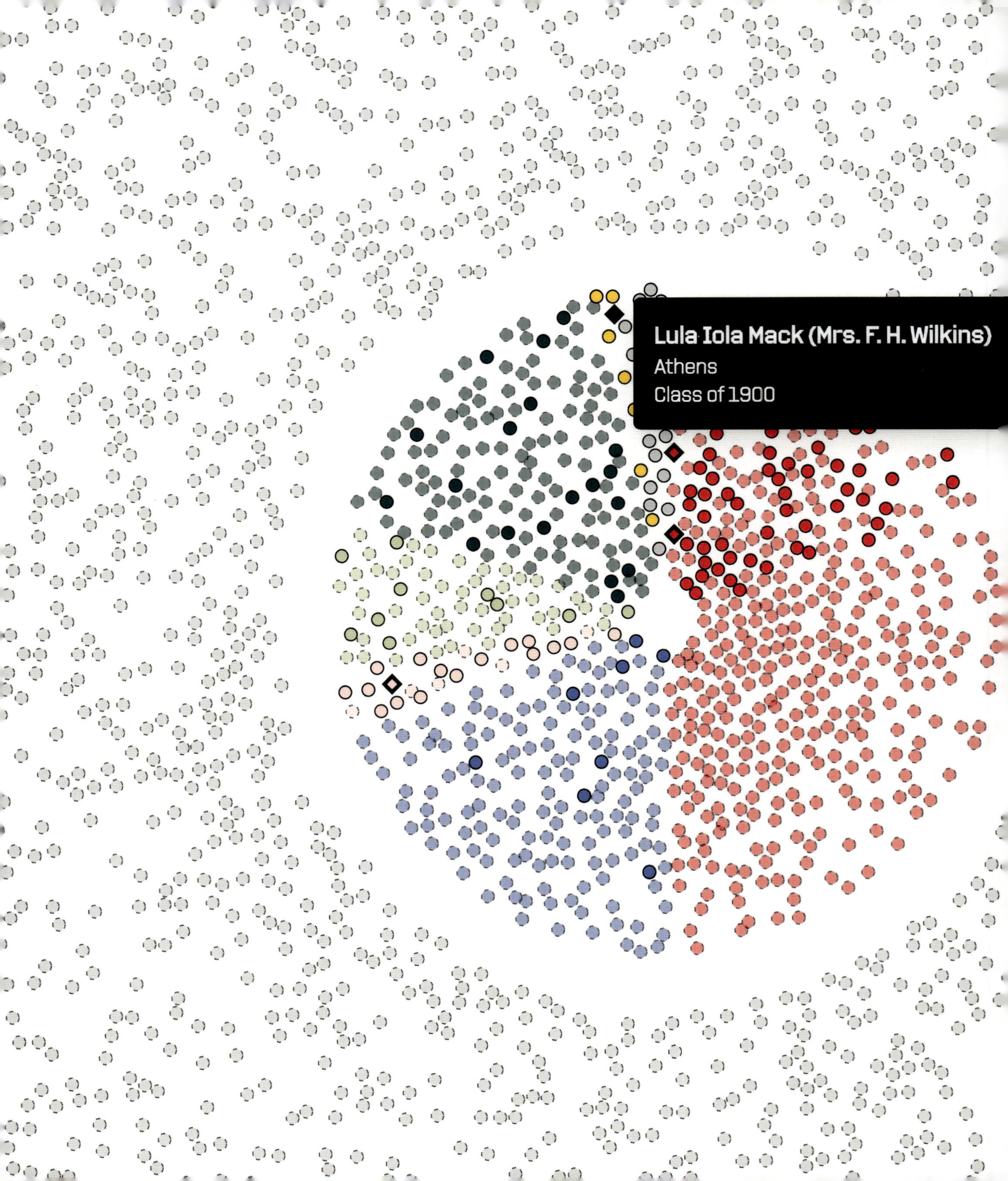
Lula Iola Mack (Mrs. F. H. Wilkins)
Athens
Class of 1900

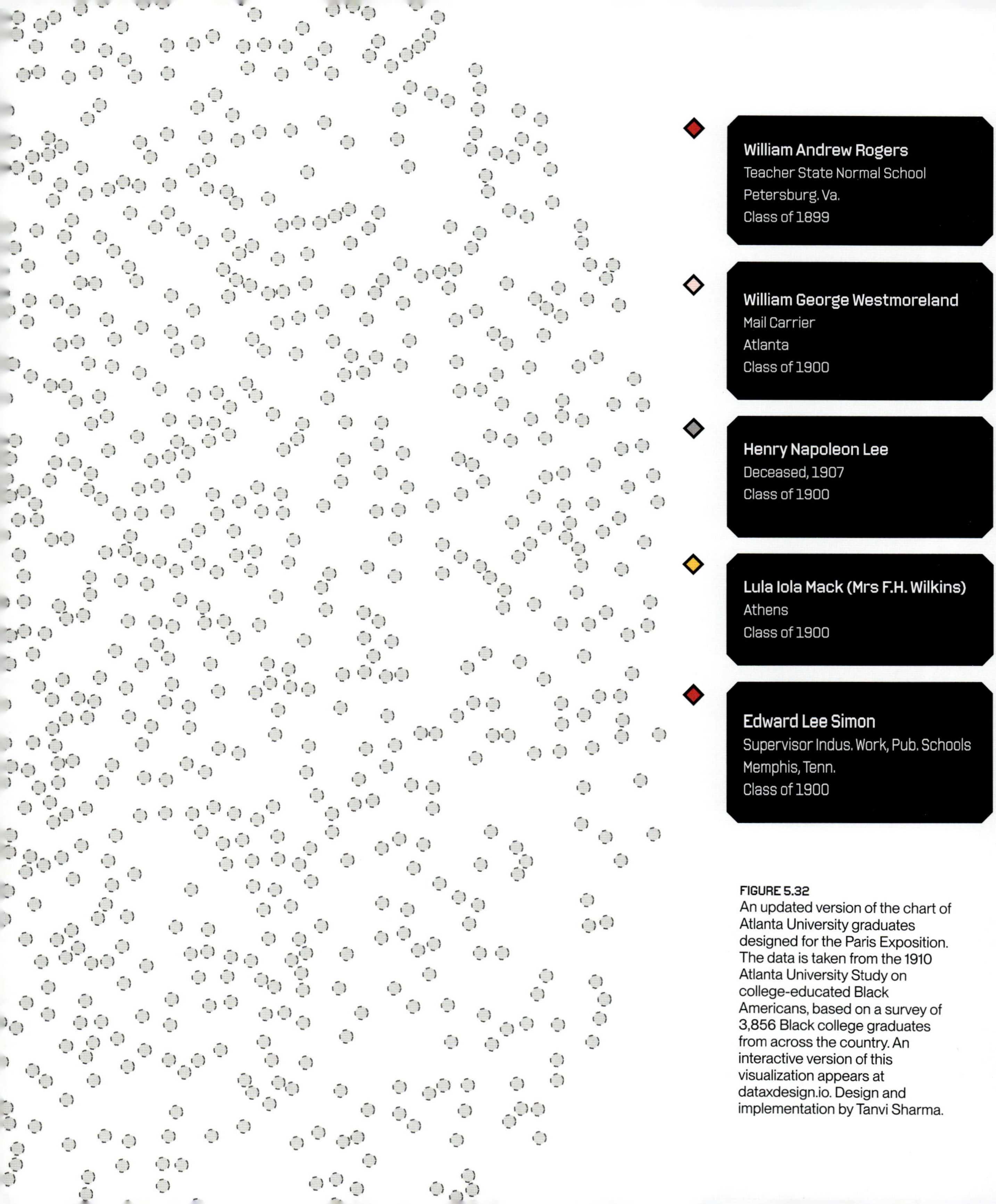

FIGURE 5.32
An updated version of the chart of Atlanta University graduates designed for the Paris Exposition. The data is taken from the 1910 Atlanta University Study on college-educated Black Americans, based on a survey of 3,856 Black college graduates from across the country. An interactive version of this visualization appears at dataxdesign.io. Design and implementation by Tanvi Sharma.

Track View | Text View | View All Responses

Reports of early life and training

How shall you educate your children? (i.e., to a trade, in College, in a profession, etc.)

What have been your chief hindrances?

Present practical philosophy in regard to the Negro race in America

38 responses available

← 1 of 38 →

Reports of early life and training

How shall you educate your children? (i.e., to a trade, in College, in a profession, etc.)

What have been your chief hindrances?

Present practical philosophy in regard to the Negro race in America

38 responses available

Back to Single View

1. It is my present intention to give my boys a full university training in order that they may be equipped to take high rank in whatever calling or profession they may choose.
2. I believe in educating the child to make the best citizen; a college education to those who will take it.
3. In obedience to their inclinations and gifts and without prejudice for or against any particular training. Technical, agricultural, mercantile, professional training are of equal importance if preparation and research are sufficient, there being urgent need for real high grade leadership in...
4. First public school, secondly college, third university, then if possible provide for her to study abroad. I shall teach the importance of attainment of the highest possible type of culture and refinement and the importance of possessing something that people who have money want.
5. I have but one daughter. I plan to give her a college education in southern, eastern and European institutions.
6. I am striving to be in shape to give my children a thoro, practical education which will best fit them for the daily pursuits of life.
7. I shall endeavor to give that training which in my judgment will be of the highest good to him.

practical philosophy in regard to the Negro race in America," which was abbreviated in the published study as "philosophy of life." These responses seemed, to our team, a reflection of Du Bois's own desire to arrive at "the truth" through a combination of quantitative and qualitative means.

To visualize this textual data, we were required to revise our chart yet again. Lauren and Tanvi used optical character recognition (OCR) to create a machine-readable dataset of the responses, and then hand-corrected any errors. Tanvi then designed a new chart of her own: in place of the pie chart that occupied the center of the original chart, she plotted dots representing each of these responses, color-coded by the question at hand. Adapting a method of interaction developed by visualization designer Tommaso Elli and his team, also involving long-form textual data, Jay designed a mechanism such that clicking each dot displays a key phrase from the response it represents (FIGURE 5.33, TOP).[84] Clicking again on the phrase reveals the full response from which it is drawn (FIGURE 5.33, BOTTOM).

The result is a visualization that at once draws directly from the original Paris Exposition charts and diverges sharply from them. We do not present all of the information it contains "at a glance," to return to Du Bois's original words. Instead, the user must interact with the visualization at length, and with sustained attention, in order to derive insight from the data it contains. The data is also, quite intentionally, not quantitative. Instead, in a callback to the photographs presented alongside the charts in Paris, it prioritizes qualitative data instead. We see this choice as honoring Du Bois's evolving view of the utility of visualization, as our visualization enables meaning-making from data of a variety of forms.

In the end, we do not know which of these survey responses, if any, can be traced to Mack, Simon, Westmoreland, or Rogers, although there are some suggestions as to the stories they might have told. Like Mack, several respondents report having daughters. On the subject of education, one writes: "I plan to give her a college education in southern, eastern and European institutions." "I shall teach the importance of attainment of the highest possible type of culture and refinement and the importance of possessing something that people who have money want," another proclaims. "I will try to educate my children according to their inclinations. I am not partial," writes a third. Thus, even if we cannot trace these words back to specific people, we can still use this visualization to expand our knowledge about the "small nation of people" that, like the charts created for the Paris Exposition, it seeks to represent.[85]

We can also use this visualization to expand our own understanding of what visualization, itself, can do. Thinking back to chapter 1 and Thomas Clarkson's "Description of a Slave Ship," we can now see how his

FIGURE 5.33
OPPOSITE: An interactive visualization of responses to the survey of Black college graduates conducted during the 1909–1910 academic year. Clicking on any dot reveals an excerpt of one student's response, and clicking again reveals the full response. The full visualization is viewable at dataxdesign.io. Design and implementation by Tanvi Sharma.

"immediate impression of horror" achieved some but not all of its goals. While it contributed to the halting of the actual traffic of human souls—a necessary step in the abolition of slavery—it did not halt the injustice that Du Bois and his students confronted a full century later, and an ocean away.[86] The survey responses that they compiled testify to that fact. In visualizing the survey responses—and thus, the unfinished nature of the struggle for racial justice—over one hundred years after that, we are reminded of the task that we now all must share. This task is to illuminate knowledge through all available methods, in the service of contributing to a more liberatory world.

Just as Du Bois sought to show with his charts and photographs at the Paris Exposition, the visualizations presented in this book, all together, illustrate a deeper truth: that data visualization, while tremendously powerful and highly persuasive, still cannot stand on its own. Visualization must always be accompanied by additional methods of making knowledge, by an acknowledgment of the broader context of the work, and by a commitment to recognizing when we must look beyond our own expertise. This is how visualization leads to change: by serving not as sole arbiter of "the truth," but as one vital method—among many—in bringing that truth to light.

Takeaways

FOR VIEWERS

Combine visualization with other sources of insight

Recognize when knowledge demands change

Know that change requires a commitment to act

Always aspire to more complete knowledge

FOR DESIGNERS

Connect your visualization with the world it represents

Consider whether change should be your goal

Explore how to combine visualization with other methods

Teach others to continue the work

CHAPTER 6

LABOR

Data by Design and What We Learn from Visualization Work

By Margy Adams and Tanvi Sharma,
with Shiyao Li, Jay Varner, and Lauren Klein

*Any visualization project is the work of many hands.
How can we surface the meaning of this labor?*

At precisely 2:18 pm on Wednesday, August 13, 2025, Tanvi Sharma, the lead designer of this project and one of the authors of this chapter, posted a link to the *Data by Design* team Slack. This communication platform, used by so many of us "knowledge workers," is often the source of an impossible-to-manage number of messages and notifications. But this notification was a welcome one. Margy Adams, a PhD candidate in English and the lead accessibility researcher on the project, as well as the other author here, was the first to notice the link. She clicked.

The link opened to an entry on *Folklore*, a website dedicated to compiling and preserving anecdotes by the people who created the original Apple Macintosh computer.[1] The entry was written by Andy Hertzfeld, a software engineer who worked at Apple at the time, and was submitted to the archive in 1982. Hertzfeld recalls how Bill Atkinson, one of his fellow Apple software engineers, was frustrated by a new management policy that required that he report the number of new lines of code he'd written each week. The intent was to track the productivity of each worker, but as a software engineer, Atkinson knew that lines of code was a flawed measure. After a week spent optimizing a core graphics library called QuickDraw, the code was much more efficient. But the number of lines had decreased by 2,000. In frustration, Atkinson typed "-2000" into the reporting form. As Hertzfeld recalls, the reporting requirement persisted for only a few more weeks.

The story made Margy chuckle, but it also struck a chord—not only in light of the similar reporting requirements we saw implemented in the US federal government under the auspices of DOGE, with serious repercussions that are still felt today, but also in light of conversations that our project team has had over the past several years. These conversations have been about how we work with numbers and data, and about the limits of each in capturing a full picture of what it is that we do. We especially liked that the Apple story gives center stage to work that is *undone*. In the absence of any new lines of code, what is left in that space is the person who did the undoing.[2] Over the course of the five years that we—the authors of this chapter—have been involved in creating *Data by Design*, we've each done and undone significant amounts of work. No numbers or data or even the most well-designed visualization can capture the full extent of that labor ... or can it?

This is the central question this chapter takes up. But before we begin to answer it, we must acknowledge the reality of the present moment: data-veillance is everywhere. Data is used to determine what counts as "productive" and to measure worth, overriding our basic humanity in the process. Exposés of workplace surveillance in areas as far ranging as package delivery and keyboard monitoring have called attention to how the data we generate through our work, when placed in the hands of bosses and others in positions of power, leads down a very dark path.[3] Even working on this project, we've long been aware of how putting our code on GitHub (an online collaborative code management system) creates a data trail of each contributor, from which (often faulty) conclusions about the significance of that person's role in the project can be drawn.

But as we have learned through the process of working on *Data by Design*, this same data can still tell us something—just not everything. The question of this chapter, then, restated more directly, is this: how far can we push the data associated with our work to tell us something meaningful? Something more than just about the number of lines of code written or removed? Can we use what we've learned about visualization—both how it helps to shape knowledge and what forms of knowledge it can help shape—to gain additional insight into the labor of this project and the people behind it? Can we use what we've learned about the importance of context to surface additional insight about our own processes, both for our edification and for yours? And finally, and most importantly, can we surface these insights with intention and care, and with an awareness of power, so that we can celebrate the wide range of work that went into this book, the emphatically human skills and strengths that work required, and the joy (and, at times, the frustration) that we found along the way?

By answering these questions through writing and through additional visualizations by our project team, our goal is not merely to resist technocapitalist definitions of what productivity entails and how it should be measured. Rather, as we have argued and illustrated throughout this book, we hope to reorient ourselves—and you, our readers—away from the narrow and ultimately limiting goals of increased efficiency, immediate insight, and optimizing measurable tasks (even though, in some contexts, these are still germane) and toward a more capacious goal of working toward more complete knowledge. Across centuries and contexts, from pens and pencils to computers and code, this is the goal that has animated some of the most powerful—indeed, the most effective—examples of data visualization. By considering which aspects of human experience are and are not captured by data, as we saw with SEAST's "Plan of an African Ship's Lower Deck" and its variants; by considering by and for whom each chart is designed, as we saw in that chapter and those that followed; and by considering which forms of knowledge are enabled by existing visualization approaches and which require additional approaches—including but not limited to visual ones—we can continue to push forward the bounds of what we know. In so doing, we can contribute to creating a more informed and more liberatory world.

Toppling the Stack

We will be up-front about the argument of this chapter: that too much focus on technical questions about how visualizations are made is detrimental to understanding the full scope and significance of the labor involved in making them. Because when most people ask technical questions, they generally only receive technical answers: information about the software libraries or other visualization tools used to implement

WRITING

google docs
microsoft word
notes
xcode
vs code
intelliJ

ARCHIVE

zotero
google sheets
google docs

DESIGN

sketch → figma
adobe photoshop
glyphs
procreate

COMMUNICATION

slack
google calender
zoom
IRL

FRONT END

p5.js
d3.js
tailwind
vue → react
node.js
remix
typescript
html
css

VERSION CTRL

github

OS

mac
windows

BROWSER

google chrome
apple safari
mozilla firefox

DNS

gandi

HOSTING

reclaim
aws
netlify → ecds

DATA

adobe acrobat
anthropic claude
google sheets
python

FIGURE 6.1
The tech stack of *Data by Design*.
Compiled and designed by Tanvi Sharma.

it, and details about its data sources and how they were processed and analyzed. By the same token, when a visualization is integrated into a web platform, as are ours, people often ask additional questions about the "stack": what libraries and frameworks were used to make the website, what platforms are used to store, optimize, or otherwise serve up the content, and what service provider is responsible for ensuring that the project remains consistently online. The answers to these questions can be illuminating, especially for people who want to create similar projects of their own, and before we get too deep into our critique of how the stack is structured, we are very glad to answer these questions here. Also, we are very big nerds and love talking about how we made our project all (mostly) work.

As the list above makes clear, we relied upon a range of libraries, servers, databases, browsers, and hosting providers to bring this project to fruition (FIGURE 6.1). More specifically, the web platform for this project was created with HTML, JavaScript, and CSS, using the React.js and Tailwind frameworks. We used a combination of D3, p5, and raw SVG to create the visualizations that appear on the site (and in this book). The project is hosted on a server maintained by Emory's Center for Digital Scholarship, using content delivered by AWS. You can find the entire project codebase on GitHub.

But this is not the entire stack that we relied upon to bring *Data by Design* to completion. Before the final implementation came the much longer and more iterative design phase, which we (and largely Tanvi) conducted using Figma. When designing the visualization of the Voyages Database in the first chapter, for example, our weekly meetings often included a segment when Tanvi or our visualization prototyper, Shiyao Li, or our lead developer Jay Varner, would walk the team through their new design ideas. It was during these feedback sessions that we would discuss heady questions about, for example, the relationship between time and linearity as it was experienced by the enslaved; or whether it was possible to employ a horizontal mapping of our data without it being interpreted spatially by the viewer; or the ethics of employing water as a visual metaphor given its associations with the Atlantic Ocean, the site of the Middle Passage.[4] In each of these cases, Figma was a necessary tool, enabling Tanvi to create quick mockups in real time that focused our discussion around possible designs and their conceptual implications. But Figma is not usually viewed as a core component of a project's tech stack.

Neither does Google Docs, which we used for our meeting minutes, or Zotero, which Lauren used to compile in one place the notes, images, and metadata from her visits to archives. There was iCal to schedule the meetings, and Zoom for the meetings themselves. These platforms are also part of our tech stack, as they are for many collaborative project teams. But they tend not to be included in responses to questions about how visualizations and web projects get made because they are not directly connected to the finished product—that is to say, the work that you can see.

We could just expand the list of items in our tech stack to include these other platforms and tools, as we did in figure 6.1. But the expanded list does not address the underlying reason why these tools and platforms are rarely viewed as part of the stack: the emphasis placed on web development (or "engineering") as the sole process that is required to bring a visualization or website from idea to implementation. Our team has reflected on this throughout our work on *Data by Design*, particularly around the example of GitHub—both because of how central it has become in today's engineering culture, and because of the visualizations that are built into the site itself (FIGURES 6.2 AND 6.3). The explicit function of these bar charts and heat maps and the like is to give the otherwise invisible process of writing code visual form.

On the one hand, we love them! The work of writing code takes so much time and thinking, but in the end, all the user sees is what is rendered in their web browser. GitHub's "Code Frequency" layout, by contrast, shows exactly how many lines of code were added to and (per the anecdote that begins this chapter) subtracted from the codebase over time. Lest individuals not get credit for their work, the "Contributor" layout visualizes these contributions in small bar charts associated with every contributor to the project.

But on the other hand, the GitHub visualizations fall far short. Key contributors to our project (such as Margy) are not included in these charts, since they do not use GitHub for their work, while others are not represented in proportion to their actual contributions, since GitHub is not the primary platform they employ. If we do not think to question the source of the data that is being visualized, and—per chapter 2, on William Playfair—what might be hidden from sight, we can easily fall prey to the belief that we are looking at a visualization of the entire project, and the entire project team.

The problem with emphasizing engineering alone has increased with the advent of the "vibe coding" approach to web development, motivated by the belief that it's possible for anyone to make anything using generative AI. It is not lost on us that there is often another visual metaphor at play when vibe coding is discussed: a crystal ball that generates (seeming) clarity from a fog of obscurity, as if there is no additional thinking required—no questioning about why something needs to exist and, if it does, how to build it. Here

FIGURE 6.2
The default visualizations of contributors (top) and code frequency (bottom) made available on GitHub.com.

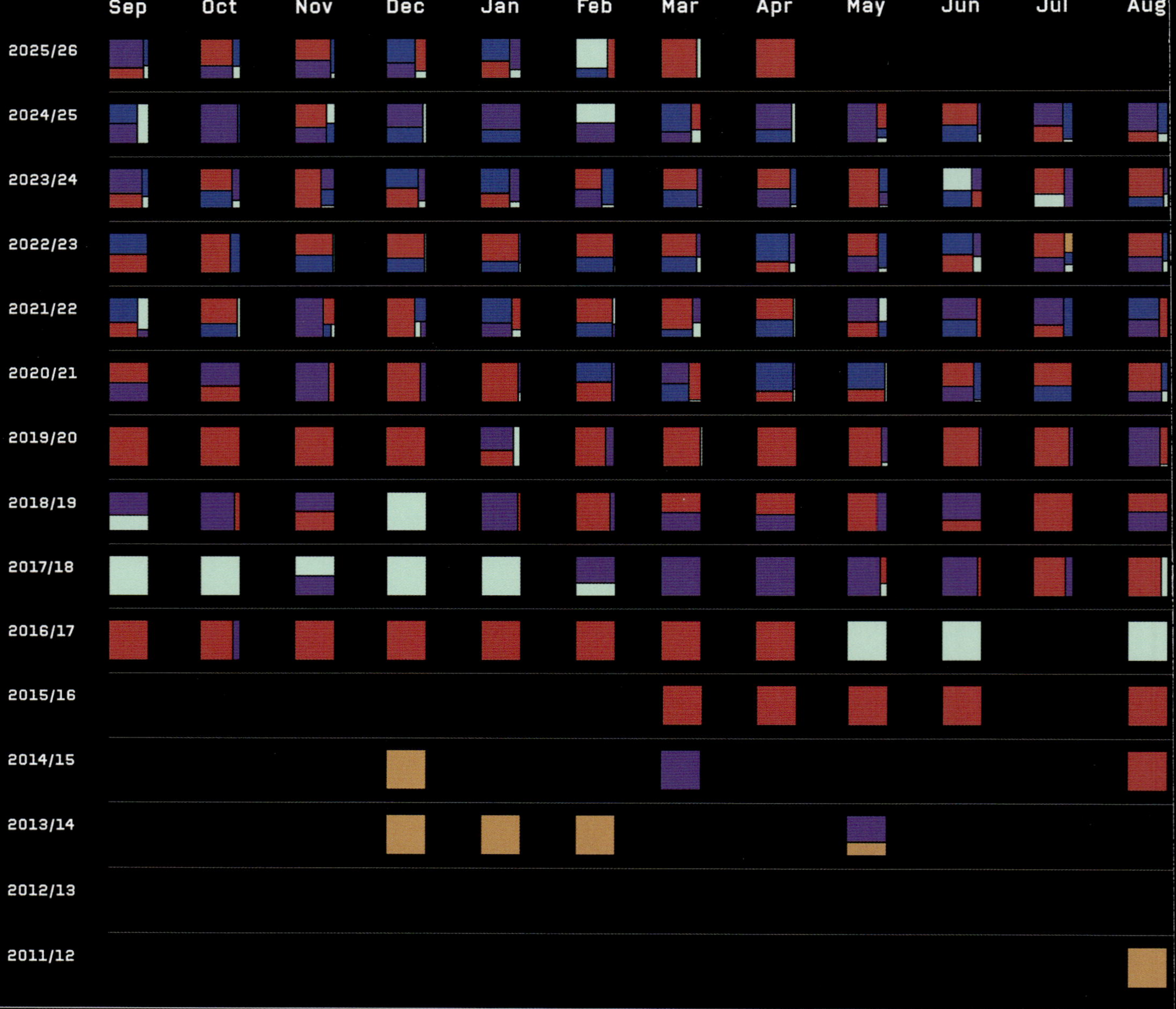

Contribution Across Time

The Making of DxD, file by file

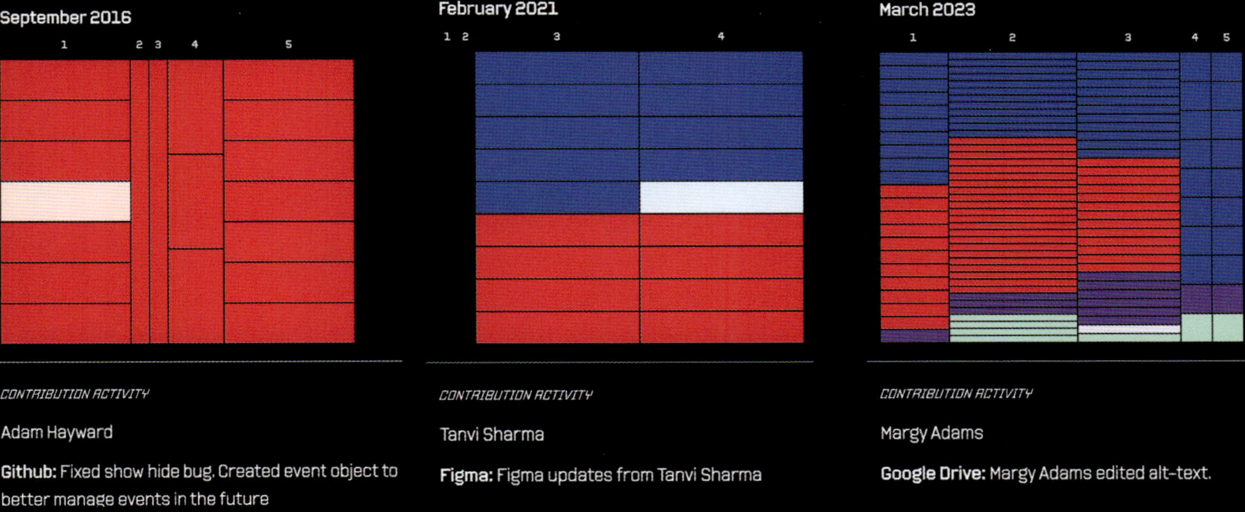

IGURE 6.4
etail views of our quilt chart documenting
EFT TO RIGHT: an early contribution by Adam Hayward, who created he first project prototype; the arrival of Tanvi (and Figma) on the roject team; and the arrival of Margy on the project team.

we return to the problem with the stack: a presumption that what matters most are the tools themselves, and not the process—or people—involved in using them.

So we decided to topple the stack. Instead of visualizing the platforms we employed, we decided to visualize our process. We borrowed the idea from GitHub of visualizing contributions over time, but we included the data we generated from all of our platforms: GitHub, of course, but also Google Docs, iCal, Zotero, and Figma. After a significant ideation phase, involving everything from Playfair-esque time series charts to a data-driven music video, we arrived at a visualization strategy that could capture what we wanted to convey: treemaps, arranged in small multiples over time, with contributions from each platform encoded in colors reminiscent of the Gee's Bend quilts discussed in chapter 4 (FIGURE 6.3).

We liked the idea that our core visual form was the treemap, which originated with Ben Shneiderman: his argument—that visualization enhances human cognition—had set us (and the field of visualization) on our path. We also liked that our choices of color, and—without even realizing it—the rhythms of the entire project lifecycle, aligned with the "polyrhythmic, nonsymmetrical, and nonlinear" patterns that Elsa Barkley Brown had identified in the Gee's Bend quilts and abstracted into her classroom pedagogy. After all, our project was pedagogical as well, and in the polyrhythms of our quilt chart we see our patterns of learning at multiple scales: the early phases of development, involving Georgia Tech students we'd never met (FIGURE 6.5); our own arrival on the project (2020 for Tanvi and 2023 for Margy) and with Tanvi, the introduction of Figma into our project workflow, which resulted in a new pattern of design followed by implementation, which would continue to the project's end (FIGURE 6.4). In contrast to the tech stack, this visualization gives a better sense of the range of processes that are involved in web development projects—processes that far exceed lines of code.

Process Is People

Of course, a large part of how this project came together is still missing from this visualization: the people involved in these processes, who employed the tools that produced the data, guided by their own (and our collective) decisions and their own (and our collective) expertise. In nearly every team meeting—that ever-present mint-green slice in the visualization above—someone uttered the name of an absent presence: Adam, who had first prototyped the Peabody chapter; Qing and Morgan, who had envisioned the first site design; Dan and Jianing, who had carried the project from wireframe to its first full instantiation, allowing the rest of us to glimpse what this project could eventually be; Anna and Zhou, the first student workers from Emory, who prototyped elements of the chapters on Playfair and Du Bois; Nicholas, who joined from NYU; Yang, who joined from Emory's Center for Digital Scholarship; Silas, who joined from Polymode. For us—Tanvi, Margy, Shiyao, and Jay, the team members who joined Lauren over the years and stayed until the final release—these ongoing utterances functioned like an ephemeral citation, a repeated calling-back-to. Some names were almost like beloved ancestors, others were familiar colleagues, while others were by that point the closest of friends. Though we did not explicitly consider the prospect in our earliest design processes, the echo of those who were not present drove us to reflect on the ways we could acknowledge how this project grew out of a multitude of voices, hands, disciplines, technologies, and communities. We wanted a way to account for all of those who touched the project—as well as how

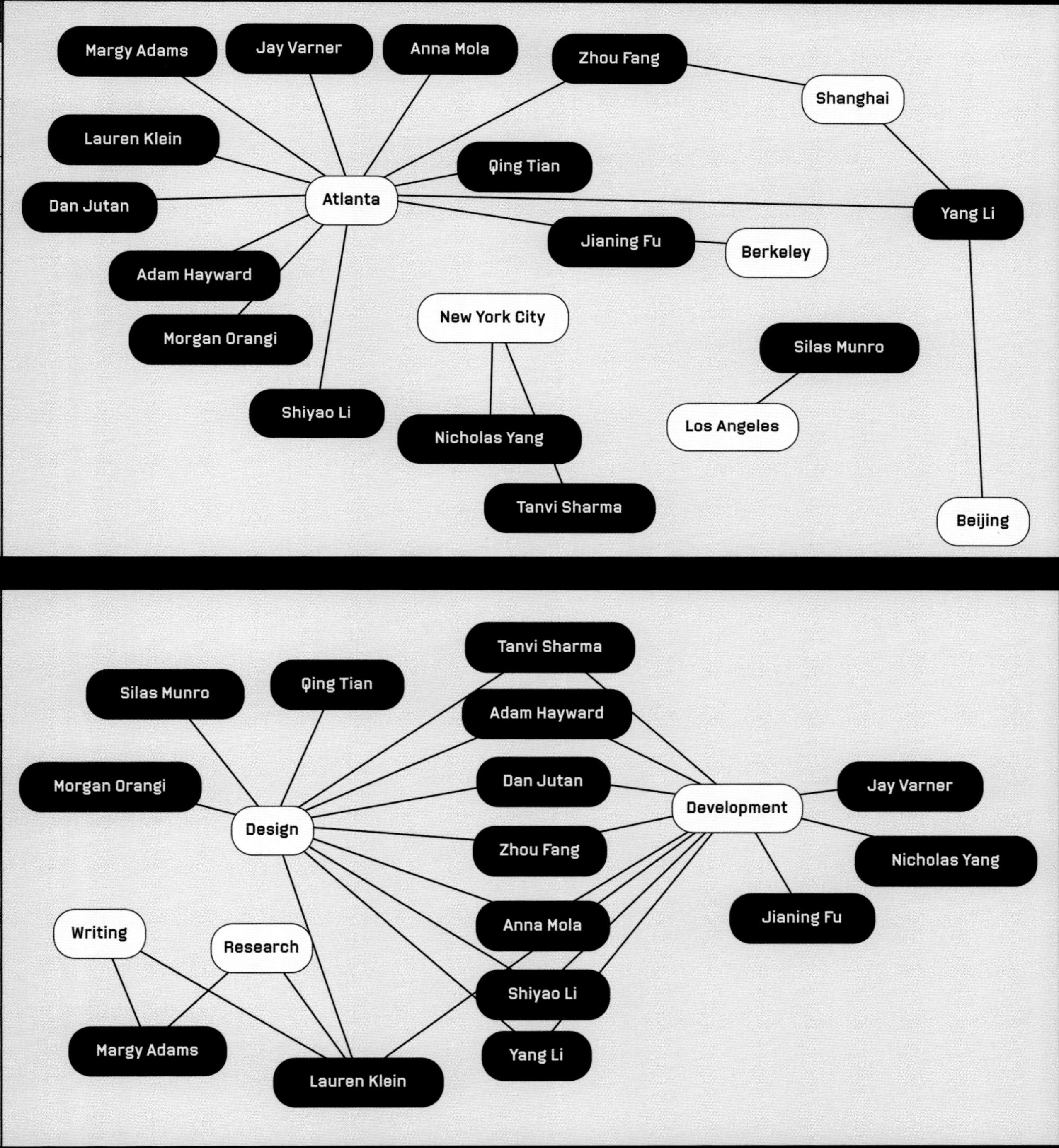

FIGURE 6.5
The people involved in the creation of *Data by Design*, configured as nodes in a network connected by locations and roles. Design by Tanvi Sharma. Implementation by Jay Varner.

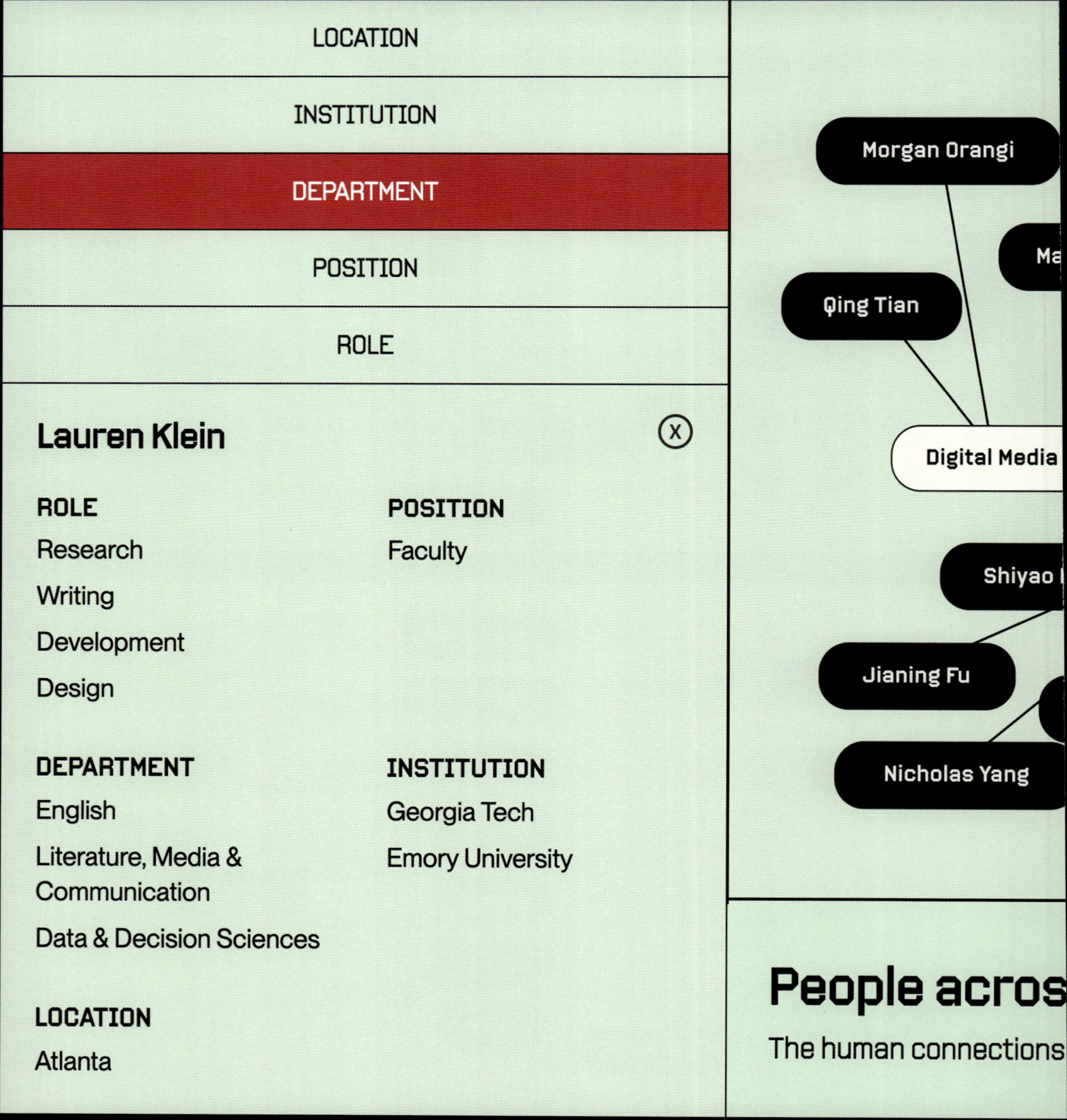

FIGURE 6.6
Our interactive network diagram of the many connections that connect the contributors to this project across versions, locations, institutions, roles, and disciplines, as it appears in the web version of *Data by Design*. Design by Tanvi Sharma. Implementation by Jay Varner.

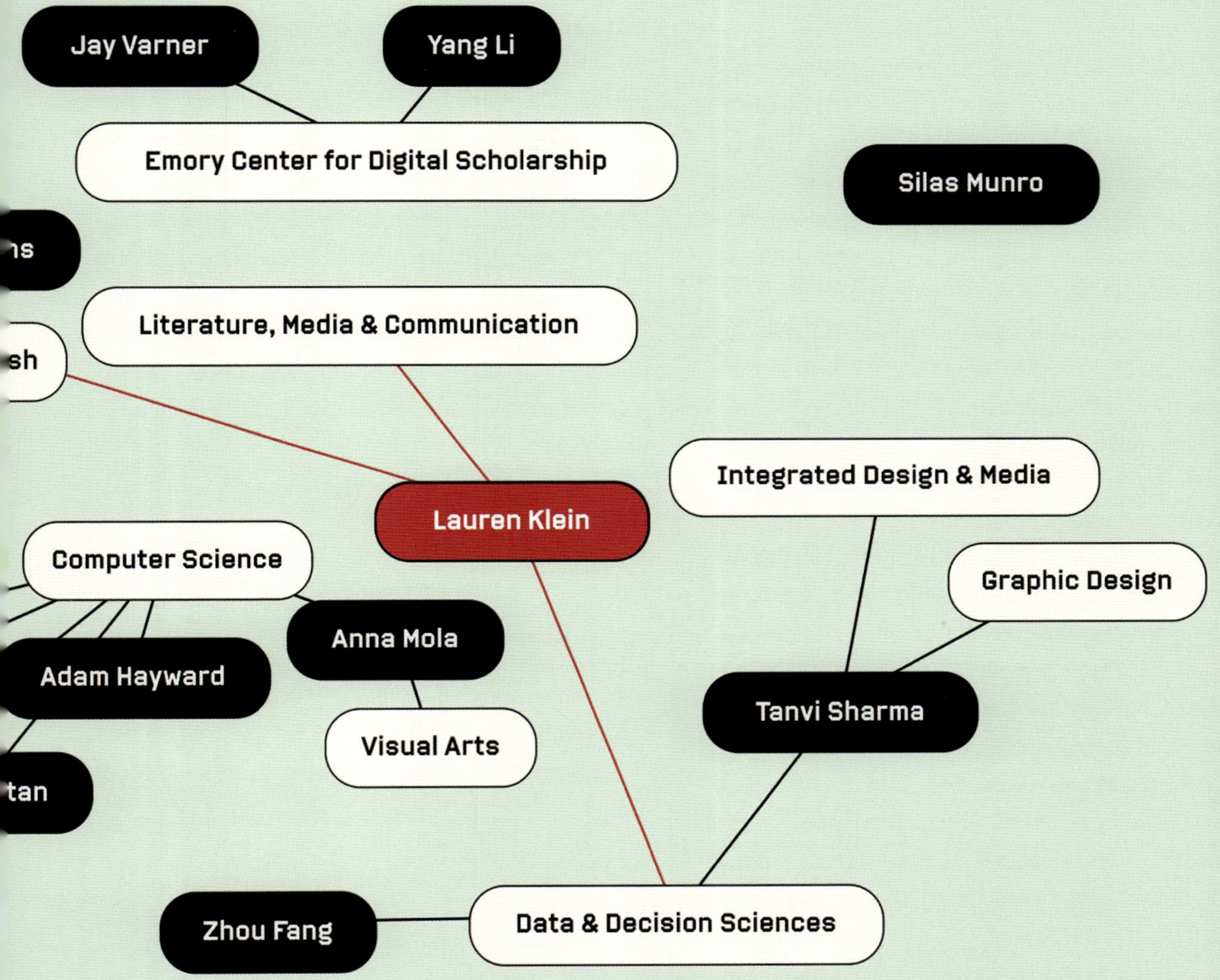

ersions

tructure *Data by Design*

V.1 V.2 V.3

they did so—a way of visualizing the project's expanse (FIGURE 6.6).

This visualization is structured by the three major versions of the project and the people who connect them. In the web-based version, clicking on "location" shows how our team members were connected across geography (FIGURE 6.5) by "roles," the positions we played on the project team (for example, writing and research in the case of Margy, design in the case of Tanvi); by "institutions" or academic affiliations (for example, Emory or Georgia Tech); and by "department," a way of representing our various academic fields.

While the points and paths of connection change depending on the view that is selected, the people remain constant. A stack isn't the right visual metaphor for these relations, since there's no bottom or top. At every meeting we were discussing and designing and presenting new implementation work; it was never a linear path. Different specialists moved in and out of the project over time—because of their graduation, or because of changing institutional roles. There were also people such as Laura Wasowicz, Curator of Children's Literature at the American Antiquarian Society, who first showed Lauren a Peabody chart; she is not recorded in this network diagram, although she made a crucial contribution to the project. There were also scholars or artists whom none of us had ever met who nevertheless left an impression on the project, like art historian Tina Campt, whose work Margy discusses in depth in the pages that follow. We cannot name every individual who is connected to this project, for that constellation is too vast to visualize, or likely too vast to even imagine. But we hope that our visualization, combined with this narrative, can give you an idea of its expanse. Now you, too, are part of our constellation—as is every reader of this book. One of the pure delights of knowing that your eyes are reading these particular words is also knowing that you will expand this visualization further still.

Reflections on our "Every Labor"

Every visualization has a point of view, as we've argued throughout this book. As we've also tried to surface, so too does every team member. While the visualizations that appear in this book reflect the composite perspective of the project team, the reason why this perspective is so generative is because of the individual vantage points it brings together. Here, the final cohort of team members each describe their own vantage point on the project.

Tanvi Sharma, lead designer (FIGURE 6.7)

Tanvi Sharma ⓧ

ROLE	POSITION
Design	Undergraduate Student
Development	MS Student
	Research Staff

DEPARTMENT	INSTITUTION
Graphic Design	New York University
Integrated Design & Media	Emory University
Data & Decision Sciences	

LOCATION
New York City

FIGURE 6.7 The project team metadata "card" for Tanvi Sharma.

I was asked by Silas Munro in 2020 to join this project. I had just graduated from my BFA program at MICA (Maryland Institute College of Art). It was the early days of the COVID pandemic, and the job market felt uncertain and fragile. At the time, I was working with Silas on *BIPOC Design History: Black Design in America*, and he told me that *Data by Design* would be "right up my alley." I trusted his instinct. I was excited to spend more time learning from him and to explore what a design practice grounded in poetic research—a method of drawing inspiration from the sensory, the sacred, and the set-aside—could look like as a career. What I didn't know then was how formative this invitation would be—not just for my design practice, but for my sense of what sustainable, care-driven collaboration could feel like.

Formally, I'm the lead designer on the project, but my role has always been fluid. Lauren welcomed me to participate in whatever way I wanted to: as a researcher, a designer, an engineer, and a student. I was hungry to take on any task—from data entry to researching interaction patterns to chasing down every book and resource she shared in our Zoom meetings. That openness gave me the freedom to experiment and grow without restraint. It felt like entering into a reciprocal relationship with the project itself: it trusted that I would grow alongside it, and I trusted that it would hold space for me to evolve. That mutual care is at the heart of how I think about *Data by Design*.

In the "Change" chapter, I transcribed the names and occupations of the Atlanta University graduates from successive editions of the annual course catalogue. These are the same students who are visualized in Du Bois's "data portraits," and their names became the data that I used for my own design work. I relied upon them (and my own transcription) when considering what a follow-up visualization would look like for students who graduated in the subsequent years. I did this in p5.js, which took me away from Figma, my primary design tool, but allowed me to explore more interactive approaches, pushing my ideas further toward a final prototype. When it came time to design the final visualizations that appear in the book, I'd not only participated in a set of steps very similar to those of Du Bois's actual students—from creating data to designing visualizations to implementing them—but I felt an intimate connection to the students themselves. I'd tracked their lives through each year's catalogue, then into a dataset, and finally into visual form. I saw them through marriages, name changes, change in jobs and relocating to different parts of the country. I'd come to care for their place in the project just as the project had, in a way, cared for me: expanding to allow my role to evolve.

What's incredible about a project like this, as opposed to the work that I've done for major tech companies and design firms, is that the needs of the people involved are prioritized before the needs of the project. Without the pressure of delivering within a specific quarter, or the often-toxic demands of design work in the service of capitalism, it shows what becomes possible when care is prioritized: creativity alongside curiosity and community. Through five cities, a breakup, two degrees, a marriage, and many jobs, *Data by Design* has always been a constant in my design career. I know I'll feel a little lost when the project sunsets. The project has shaped me by the values it embodies, and is a testament that design can be sustained by care rather than extraction, and can create space for all of us to thrive.

Jay Varner, lead software engineer (FIGURE 6.8)

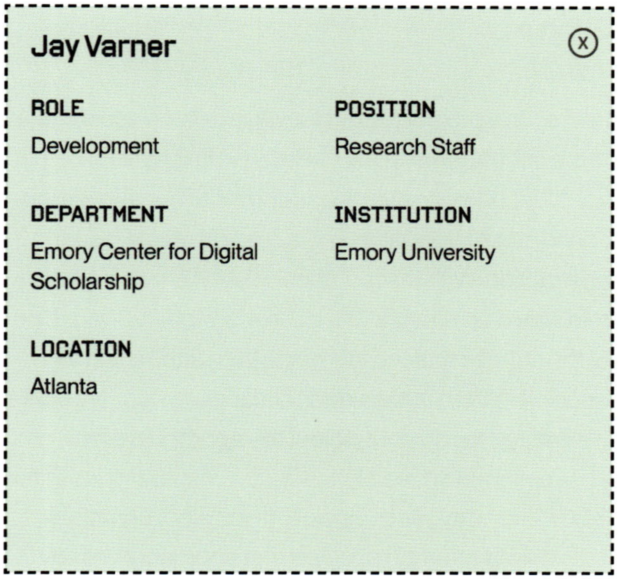

FIGURE 6.8 The project team metadata “card” for Jay Varner.

As a software engineer for the Emory Center for Digital Scholarship (ECDS), I lead the development work for the center’s bespoke projects and platforms. My role is to take the ideas from the Emory community and turn them into code. To put it another way: I develop new ways for people to interact with old shit. People think of this as a tech role, but creativity and problem-solving are at its core. My projects also need to work in the real world; for my master’s degree, I studied user experience, which helps make that happen.

I joined the project in fall 2022 when Lauren reached out to ECDS for support. I made my first commit to the project’s GitHub repository in December of that year. (A “commit” is a way of contributing new or revised code to a project.) The thought was I would spend a few months helping get the site over the finish line. That’s not how it went. I think we were very fortunate to be able to take the time we did.

In software development, process is paramount. There are steps we always go through and established ways to come up with an efficient result. In a for-profit setting, the timeline would be more strict and finite. In this case, because I’m on the staff of ECDS, we were able to stretch the timeline and truly experiment and iterate, adjusting to feedback and internal deliberation along the way. This was the necessary process. The time and collaboration created the confidence that we got it right, or as right as possible for this project in this moment.

Challenges and hard choices were the norm. The re-creations of Elizabeth Palmer Peabody’s charts that appear in the “Knowledge” chapter were particularly tedious. There were 218 unique events across the four centuries that Dan Jutan, a student from Georgia Tech who worked on the first version of the website, had compiled from the text. When you read an explanation of Peabody’s process, it seems pretty straightforward to turn into code. But at times Peabody took some liberties and deviated from her own system. Improv is harder to translate into code. We had to strike a balance between writing efficient code that could be reused for all four centuries and three different visualizations, and code that was flexible enough to handle human mistakes. To stay true to Peabody’s original images, we would have to intentionally introduce an error into the underlying data. The other alternative was to create a visual discrepancy. We went with the latter.

It’s unusual for me, as a developer, to see my name listed as an author. In the past, I probably wouldn’t have thought twice about being left off. It aligns with the spirit of the project to see it there now.

Shiyao Li, visualization researcher (FIGURE 6.9)

I joined this project as a junior visualization researcher in the second year of my PhD, carrying the mindset that visualization leverages the strengths of human perception to make information more discoverable through abstraction. With that perspective, I initially assumed my role would be straightforward: review the 293 variables of the dataset that the team was then attempting to visualize—from the Voyages Database—identify those we wanted to highlight, select the appropriate form of abstraction, produce clear, effective visualizations of the 36,079 voyages, and consider the project complete.

My first attempt reflected that approach. I initially designed a stream chart to show the changes in the trend in the number of voyages from 1565 to 1858, expecting this historical trajectory to be the central focus. But that assumption was quickly challenged as the team was less interested in broad trends and more concerned with representing the lived realities behind each individual voyage. What followed was nearly two months of collaborative exploration, during which we experimented with different ways of representing each voyage and the resistance of the people on board, before arriving at the design that ultimately appears in the "Data" chapter.

This process reshaped my understanding of visualization design. My own research focuses on confirmation bias in data visualization—how people interpret visualizations in ways that align with their prior beliefs. Much of my academic writing has emphasized abstraction: by reducing data points to patterns, visualizations make insights easier to see, and in turn make those that fit prior beliefs even more accessible. Working on this project, however, revealed a different side of abstraction—how it can also flatten and even erase the human realities behind each data point. In digital humanities projects, visualizations are not only about revealing patterns but also about inviting reflection and contemplation. Each data point carries weight and meaning, and abstraction can easily obscure those nuances. The most important lesson I took away from this collaboration is that sometimes the power of visualization lies not in simplifying complexity, but in holding on to it—encouraging audiences to reflect on both what is revealed and what is lost when we visually represent human lives.

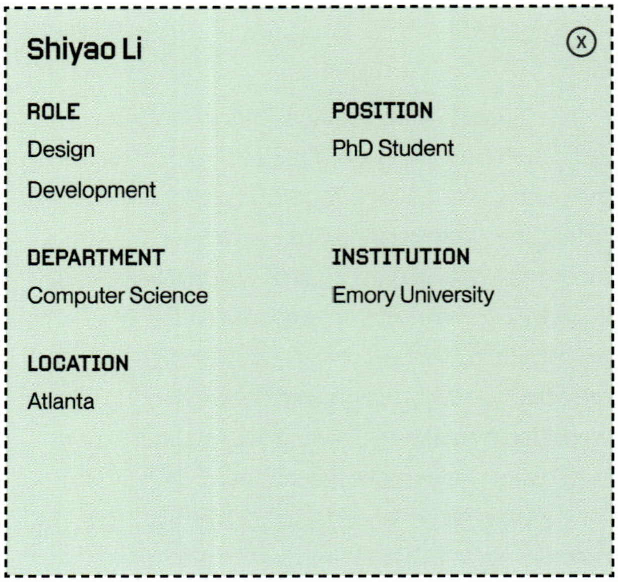

FIGURE 6.9 The project team metadata "card" for Shiyao Li.

Margy Adams, lead accessibility and humanities researcher (FIGURE 6.10)

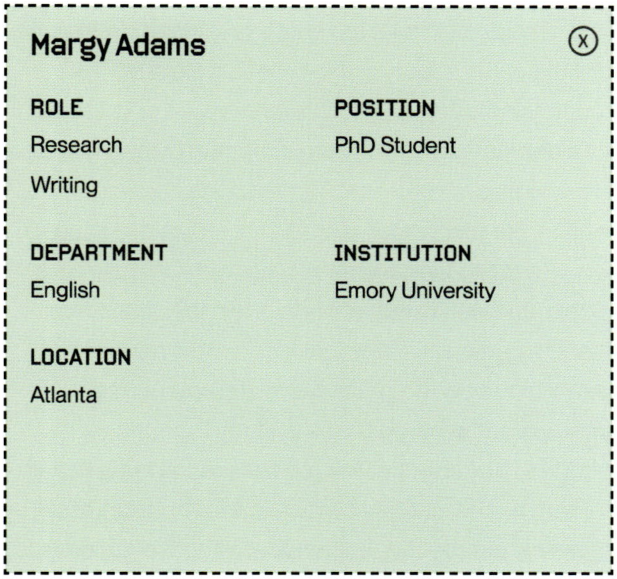

FIGURE 6.10 The project team metadata "card" for Margy Adams.

I've been part of this project team for its final three years, and as such, I've been involved in its numerous pushes toward the finish line. Throughout this process, I've been intrigued, again and again, by the idea that visualizations are not simply something we see. I study Black sound. For years, Tina Campt's *Listening to Images* has served as a methodological and even spiritual guide for me, and it especially came into focus as I began working on this project. Operating from a Black feminist standpoint, Campt sets up a practice of "listening to images" as a multisensory encounter that emphasizes "looking beyond what we see and attuning our senses to the other affective frequencies ... of images and how they move, touch, and connect us to the event of the photo."[5] While we need to be careful of how we apply this framework to images beyond Campt's primary corpus (Black vernacular photography), the ethic of looking "beyond" is crucial to disrupting the alleged objectivity of visualizations of data, especially those that pertain to marginalized populations and the production of histories. Listening to images allows us to understand visualizations not as neutral presentations of data, but instead as *re*presentations of a particular dataset, as well as how and why it was collected. To reiterate the foremost point of this project: data visualizations inevitably fall short of presenting a totalizing picture of the visualized content because that task in itself—completely *capturing* data—is vacuous.

However, instead of viewing this as a problem to solve, as I've learned through my work on this project, we find other possibilities for visualization emerge when we focus on *why* we know that prospect is impossible: we are able to *read* data visualizations in multiple, sometimes contradictory, even infinite ways. Reading—or, as Campt and I prefer, *listening*—to the visual, a mode assumed to be beyond linguistic and sonic capture, drives how we help make the visual *produce meaning*. The practice of listening, in turn, informed my primary contribution to this project, the composition of the alternative text (or "alt text") that is displayed in place of each image in the project for those using screen readers (or other adaptive technologies).

Instead of following a set of universal standards for writing image descriptions, which can be appropriate for some projects, I composed the alt text with listening in mind. The visual content of *Data by Design* is not as straightforward as, for instance, a photograph of a bowl of fruit, which might be described as "a ceramic bowl on a wooden table that contains a banana, an apple, and a bunch of grapes." Each of our chapters includes historical charts, vast datasets, and intricate symbolism that already challenge the process of meaning-making. But that is the beauty of this project: it elevates the responsibility we carry as interpreters—that is to say, as individual (and collective) producers of meaning. So I tasked myself with listening: with considering how my writing, like one track in a studio recording of a song, would accompany and harmonize with the main body of text; and with using words to retain the impression that the particular image left on me. I saw my task as describing how the image resonated with someone (me) whose primary job was to study the image and attune to it deeply. The impressions recorded in the alt text are necessarily my own, so it was important that I make my perspective known, too, without overriding the most important visual information.

This hierarchy of visual significance changed with every chapter, and with every set of images, just as every visualization designer's purpose (and our argument about it) changed too. The "Change" chapter, for example, which centers on the charts created by W. E. B. Du Bois and his students for the 1900 Paris Exposition, showcases their skillful use of color. Standards for alt text often suggest omitting details about color. But not mentioning these charts' vibrant reds, yellows, blacks, and greens seemed to diminish the effects of what Du Bois and his students sought to convey through their color choices. So I decided to note the colors of the charts and my sense of their connotations, as well as the larger shapes that the charts, to me, seemed to form.

With the "Data" chapter, however, the abstraction in the images posed a very different challenge. The text of the chapter discussed how the process of abstracting bodies into numbers contributed directly to the dehumanization of slavery. Simply describing the abstracted bodies in the images seemed like it would only reproduce that original violence. So for "Plan of an African Ship's Lower Deck," for instance, I decided to explain not only that the ship's hold held people but also *how* it did so, both on the actual vessel and on the page. My alt text could work alongside the argument of the chapter, exposing abstraction as one of the insidious mechanisms that allowed slavery to persist.

Listening proved to be such a powerful technique for writing the alt text for this project because it required me to interpret and translate each image's form and content, as well as how that content was constructed. What I came to realize through my process of listening to images, especially as I listened to "Plan of an African Ship's Lower Deck," was what Campt meant when she described her own practice as, foundationally, *grammatical*.[6] Following literary scholar Hortense Spillers's landmark work on the linguistic enslavement of Black diasporic peoples since the Middle Passage, Campt's discussion of grammar invites her readers to consider how our looking practices—our ways of engaging with visuals—have also been structured by a set of rules, by a system of *visual* syntax. She asks us to consider what has resulted from this regimented framework: reductive, abstract narratives about Black subjecthood. The grammatical practice that she is after here, then, is not so much about "correctness" or exactitude, but about contending with how these rules *govern us,* their users.[7]

In Campt's own words, her grammatical reframing is one that "grappl[es] with precarity, while maintaining an active commitment to the every labor of creating an alternative future. Indeed," she continues, "it is this grammatical practice of futurity that constitutes my definition of *freedom*."[8] To Campt, listening to images is concerned with the future real conditional tense—the grammatical "what will have had to happen," the enactment of a future that has not yet occurred but *must*, if we are to ever really be free.[9] It seems to me that this chapter, maybe this entire project, is an attempt to attend to the precarity of the "every labor"—in other words, the quotidian yet necessary work that is required to achieve liberation. This "every labor" is precarious because it is very often taken for granted, or "quiet," to invoke Kevin Quashie, whose book *The Sovereignty of Quiet* we discussed in chapter 1. It also reminds me of Toni Morrison's famous words delivered at Barnard in 1979: "the function of freedom is to free somebody else."[10] The very argument of this project, that data visualizations have points of view, seems itself to create the possibility for an "alternative future," to return to Campt's words—one where visualization can act as a language that propels us toward freedom.

To propose that visualization itself can function grammatically is to insist that meaning does not end once the image is created; in fact, it keeps living *through* our interpretations of and interactions *with* it. Campt's project "question[s] the grammar of the camera (as both an event of photography and a photographed event) as well as the haptic temporalities of photographic capture as pernicious instruments of knowledge production."[11] Put more simply, she does not only consider what the photograph is of, but also the contexts in which the photograph was taken, and of the camera itself—the "pernicious instrument"—in order to glean what the very act of the recording itself might tell us. Likewise, with data visualization we must be thoughtful of the contexts in which the data was recorded and the "pernicious instruments" employed to visualize it—the strategies (or, one might say, *grammars*) and traditions of the field—to ultimately reflect on the systems that have allowed the subject to document and the other to be (un)documented. How does a visualization speak? What is it saying, what is it hiding? And why might it be hiding certain things?

Thinking of visualization as a grammar in the sense that Spillers and Campt understand it enables creators of data visualizations to convey rather than overdetermine, to gesture toward rather than eclipse. It also allows witnesses to actively build up worlds of meaning with creators rather than receive information through a hierarchical, passive structure. Reconstruing visualization as a grammar means that the unknown becomes central. But instead of fearing it, we might instead expose the contours around it that made it so, and in turn see what we can do to produce knowledge anew.

Listening, Labor, and Liberation

Like some of the best visualizations, which offer up new insights when we return to the original image (or repeated interactions with the interface) with additional questions, the best humanities scholarship—like Campt's and Spillers's—also rewards repeated engagement. As I see it (and it is now Lauren writing these final lines), there is a close kinship between the aim of humanities scholarship and the work that visualization seeks to do. When Margy brought *Listening to Images* to the project team, she explained (as she does in her reflection) how Campt allows us to see the "grammar" of visualization, like the construction of language itself, as an active practice rather than a rulebook to follow. This interpretation of visualization's core components—its grammar in the sense that Leland Wilkinson used the term—is empowering, or should be, because it shifts the emphasis from the components themselves to the *people* who assemble and derive meaning from them.[12]

Now, returning to *Listening to Images* as this project reaches its final phases, I see how Campt's work can push us further still. Campt's phrasing underscores how the "grammatical practice" of listening to images (or in our case, to visualizations) has "existential" stakes. The future we all hope to inhabit is, by definition, precarious; it is conditional—both grammatically and experientially—until we take active steps to make it real. Margy helped our team to see how the very argument of *Data by Design*, that visualizations have points of view, itself creates the possibility of alternative futures. I also now see our work in similarly existential terms: the power of visualization designers to shape what we see is the same power that can be harnessed by those doing the seeing—that is to say, by all of us. As viewers, we all have the ability to use the insights we gain in order to imagine alternate worlds. This, in the end, is how visualization can propel us toward freedom. But to do so, it requires an attention to the "every labor" (Campt's words) of that imaginative work.

This chapter represents a literal attempt to attend to our "every labor." By insisting that our process is as important as our product, we have exposed deeper layers of the decision-making that has culminated in the book you are reading now. By amplifying each of our individual perspectives, we have exposed the seams that bind these perspectives into a composite whole. By uplifting our collective contributions, we have shown—we hope—how multiple skillsets and multiple minds open up new possibilities for understanding visualization, both its history and its future.

When we listen to labor, we can gain insight into the small decisions—and the larger ones—that result in the visualizations that we would otherwise perceive as objective or self-contained. Indeed, these decisions, the people who made them, and the perspectives they inhabit, are what help us to understand visualizations more fully: as representations of the process by which data is given new form, whether visual or interactive. In the end, this process is what determines the nature of the insights that we as designers are able to prompt in our viewers; and our viewers' understanding of that process, in turn, enriches the knowledge they ultimately gain. We hope that this chapter has prompted new insights for you—both about the specific visualizations that you have encountered throughout this book, and about the possibilities of visualization as a practice, one that can directly contribute to a more informed, more equitable, and more liberatory world.

Takeaways

FOR VIEWERS

Reorient practical questions from tools to process

Probe the meaning behind all design decisions

Ask about the people who designed each chart

Place yourself in the meaning-making process

FOR DESIGNERS

Name and celebrate all contributors to your project

Remember that process matters as much as product

Recognize the many forms of labor behind each chart

Approach visualization as work toward freedom

EPILOGUE

The story of visualization that we have told in these pages is intended to be a hopeful one. While we have learned how colonialism, capitalism, and the institution of slavery all intertwine with the roots of modern data visualization, an acknowledgment of these roots does not foreclose future possibility. On the contrary, when we are honest about these origins, we are able to better identify—and more fully celebrate—the contributions of those who, in spite of the oppression or outright violence they faced, or to which they bore witness, continued to believe that a better world was on the way. Their contributions, in turn, affirm the role that visualization has and can continue to play in charting a course to a more liberatory future.

But I am under no illusions: to believe in this future now, in the current technopolitical moment, presents a challenge. We are living through a time in which it increasingly feels like the future has been ceded to Big Tech. Already decisions about the development, distribution, and use of new technologies are made by a handful of absurdly rich Silicon Valley CEOs. These so-called "broligarchs" seem to want nothing more than to consolidate power and maximize profit. They are actively undermining the value of education and the dignity of creative work. Even as journalists (another of their targets) continue to surface incredibly damning direct quotations and incriminating internal reports, it remains truly difficult to know whether they are making their decisions because they have never personally experienced the full richness of the human experience, or because the five or so of them in fact see very clearly how much they stand to benefit—both financially and epistemologically—from a society of "users" who can no longer tell what is real.

What is certain is that, with every new data center constructed or chatbot released, and every safeguard against potential (or actual) harm rejected, this dystopian tech future is getting closer. In truth, it might already be here, with these men and those who enable them now fully invested in the fascist regime being advanced by the Trump administration. This regime is one that places the interests of the corporate over that of the community, that chooses to persecute rather than provide support, and that actively seeks to destroy institutions that serve the public—including but not limited to universities like the ones that made this book possible.

The increasing coziness between Big Tech and MAGA, while deeply distressing, is also clarifying. We can now plainly see why universities and their students were among the first targets of this hate-fueled regime:

our collective knowledge—about how we have arrived at this moment, about what we stand (and in fact have already begun) to lose, and about why we must continue to fight—is one of our most powerful tools of resistance. Because what we learn in universities, what we *teach* in universities, and *especially* what we teach in humanities departments, equip us to imagine alternative worlds.

When I set out on this project over a decade ago, I knew that I wanted to model the value of history and context, of complexity and critique. But I did not anticipate just how much, in the end, I would want this project to stand for the value of the humanities writ large. This value manifests in the deep thinking born of disciplinary training and domain expertise; the fresh ideas that are sparked when disciplines and domains come together; and the time and labor—*so much time and labor*—that are required for the cultivation of any truly transformative idea. Now, looking back, I can very clearly see how this project only exists because of my PhD in English—because of what I would learn about visualization, its uses, and its possibilities, only *after* leaving the tech job that paid me (in part) to design actual charts.

The possibilities that emerge from this type of humanities research are precisely those that the current administration has sought to foreclose. One of the very first acts of Trump 2.0—after freezing over $400 billion in active research grants and arresting student activists simply for speaking (or writing) their mind—was to repurpose the National Endowment for the Humanities (NEH) as an ideological enterprise and to dismantle its Office of Digital Humanities (ODH) altogether. Not coincidentally, the NEH ODH funded this project not once but twice. *Data by Design* may in fact be one of the last projects—at least for the moment—to reap the benefits of what ODH grants enabled: dedicated time for experimentation and iteration, increased visibility and potential for impact, and cold hard cash (OK, direct deposit) to provide students with real research experience. In the case of my Digital Humanities Lab students, they would take this experience on to graduate school and professorships, to nonprofit and community work, and to jobs at Microsoft, Meta, and the NSA, to name just some of the places where my formerly NEH-funded students are now employed.

In this context, I've been thinking back to W. E. B. Du Bois and the students he taught at Atlanta University, which is located just a few miles from Emory, where I teach today. Du Bois also sent his students into an actively disillusioning world—one in which they nevertheless went on to pursue graduate study, careers as teachers at all levels, roles in government and as entrepreneurs, as well as homemakers, members of the clergy, and more. As our project team learned from reading these students' words, what they achieved—in the face of deeply entrenched racism and hostility—was in large part due to the strength of their kinship networks, their communities, and their convictions. But a small part of the reason for their flourishing, or so I like to think, was that in college, they had been immersed in a sustained collaborative research project. They had been guided through difficult questions about data collection. They had been pushed to consider what the data could and could not represent. Most crucially, they had each personally experienced the enhanced perspective that always comes with accumulated knowledge and additional research.

To be clear: Du Bois and his students faced a very different set of conditions than my students and I do today. Speaking now just for myself, as a person with tenure, white skin, and—in the context of the present moment—a birth certificate from New York State, I am infinitely more protected. But the frequencies of Du Bois's project that resonate with the one that I have envisioned and sustained over the past twelve years are real, and they are affirming. They attest to the value of the work that takes place in universities: work that is slow, work that students must learn from their professors how to do, and work that we as their professors hope they will carry with them wherever they go.

In today's world, this work matters precisely because it is not fast or "trending," because it is not profitable or "scalable," because it is not attractive to fascist governments. Why this work matters is because it *is* so full of possibility, it *is* so full of community, and at times (my personal writing process notwithstanding) so full of joy. This is the kind of work we need more of right now. It's the kind of work we need our institutions to actively support. Indeed, it's the kind of work that we need more people—more students, more tech workers, and more students who will become tech workers—to learn to recognize, to value, and perhaps even themselves to learn how to do.

***

A funny thing happened on our team as we attempted to finish, once and for all, the final visualization for this book. It was the "quilt chart" that appears in chapter 6, which aggregates all of the data produced by all of the platforms we used to make *Data by Design*—our attempt to visualize all of the work on the project, over all of its years, by all of the people involved. We needed a complete dataset of this work in order to create the final version of the visualization, and in order to do that, we had to decide on a specific date to stop our data collection. But we realized that, in the act of creating the final visualization, we would do more work, and thus produce more data. It was a classic case of chicken-or-egg: if we wanted to document all of our work, we would be caught in an endless cycle of ensuring that the chart reflected the most up-to-date data of our work, and that the data reflected the most up-to-date work on our chart.

We were stymied by the question of the stop date for at least several weeks. But then I realized that I was looking at the problem in the wrong way. Its recursiveness was a metaphor. We are of course always working, just as we are of course always learning. The more important question to ask—and the more important question to answer—is what all of this work, and all of this learning, is actually *for*. Eventually, we figured out the answer, which in the end is the gambit of this book: we do this work in order to design a better future, to design the future we need.

***

So how do we actually do this? How do we design, together, for this future we need? For the future we need now, more than ever?

For our quilt chart, the answer was easy. We just chose a date and stopped collecting the data. Time is an arbitrary construction anyway, we told ourselves. We then turned our energies to acknowledging the enormity and expanse of the efforts that were recorded (and unrecorded) on the quilt, as chapter 6 explains in detail.

For our world as a whole, the answer is far more difficult to fully realize. But in the end, it is just as easy to state. We must begin by recognizing the goals we all share: of working toward greater knowledge, and from there enlisting this knowledge in the service of creating a more liberatory world. Across disciplines and institutions, both within academia and beyond, we are partners in these goals. We will achieve more together.

At the same time, we must recognize that our partnerships are rarely founded on equal ground. Speaking from the place of my own work—higher education—we might observe that technical researchers are only now facing the prospect of vanishing funding streams, reduced student cohorts, detrimental government restrictions, and questions about the utility of their work. But these have been the baseline conditions of humanities research for decades. I say this not to draw battle lines. On the contrary, it is to show how our struggle as seekers and creators of knowledge has become a common one, and how we must insist on solidarity—intellectual as much as political—even when (and in fact precisely because) our work has become more precarious and our resources more scarce. Or, to put it in feminist terms, our differences of discipline and method, of (perceived) relevance and prestige, can no longer divide us; we must bring the perspectives born of these differences together to forge a composite whole. This is how (in addition to organizing) we build collective strength.

If there is a vision in this book that extends beyond the ivory tower, one that can help us design not merely images or interactions but the very future we seek to inhabit, it is this: in spite of everything, even still, we hold the power to shape the stories we tell. And by actively shaping the story about how we have arrived to this point, we can also shape our sense of the possibilities for the future.

From the first lines of this book, I have sought to emphasize what can be gained from greater attention to context, and from pursuing answers to questions about who is doing the work, for whose purposes, and to what ends. To do so, I have selected a set of figures that each, through their charts and the context that gave rise to them, surface aspects of the history of visualization that have not yet been considered in standard accounts of the field. I have labored (in the full meaning of that term) to show how these historical figures and their stories—sometimes inspiring and sometimes deeply depressing—push us in new and unexpected directions. This is the transformative force of new knowledge, which we now must all defend. Across academia and industry, in our roles as researchers and designers and people in the world, we must continue to seek out stories that offer new possibilities for the future. We must resist the future that is being forced upon us, and we must envision—and then work to enact—the future we need.

ACKNOWLEDGMENTS

It is humbling to recognize how this project, which began in fall 2012 with an idea prompted by a fleeting mention of an obscure historical figure, has now spanned nearly the full length of my professional career. It is not an understatement to say that a debt of gratitude is owed to every single person who has supported me on my academic journey—which is to say, so many of you, named and unnamed, who are at long last reading this note.

Thanks first go to the *Data by Design* project team: Tanvi Sharma, who was the one to teach me what design can truly do; Jay Varner, whose patience, experience, and skill are much of why this project was able to reach its final form; Margy Adams and Shiyao Li, whose ideas are embedded in this project and who stayed to see it through; Silas Munro, whose intentional design practice remains a beacon for my own collaborative work; and Dan Jutan, Jianing Fu, and Adam Hayward, my Digital Humanities Lab students at Georgia Tech, whose early contributions to this project first helped me see what it could become. Also at Georgia Tech, Qing Tian and Morgan Orangi provided crucial early design assistance. At Emory, Anna Mola and Zhou Fang became my first DH Lab researchers and taught me about the place that would become my new academic home. Nicholas Yang joined the project team from NYU and led a crucial implementation phase. At Emory's Center for Digital Scholarship (ECDS), Yang Li also helped with parts of this project's implementation at a transitional time. Thanks also go to Chase Lovellette and Allan Tullos, the co-directors of ECDS, for supporting Jay and Yang in their work on this project, and for first assigning Margy to work on this project. Thanks to Emily Wall, in the Department of Computer Science, for first enabling Shiyao to do a rotation in my lab.

David Lobenstine read multiple drafts of this book and offered expert editorial guidance, essential feedback on content and tone, and the crucial encouragement I needed to see this project through. You have my deepest gratitude. The contributors to my book workshop, Catherine D'Ignazio, Alberto Cairo, Miriah Meyer, Derya Akbaba, Britt Rusert, Alex Gil, Jessica Marie Johnson, Laura Harjo, and Matt Jones, offered exactly the tough and targeted feedback needed to push this project forward at a crucial phase. Catherine, Miriah, and Derya—I have learned so much from writing with all of you. I hope you see how your thoughts have shaped these pages. Alberto, thank you for supporting this project from its start.

When I decided to run an open community review for this project, I had no idea that the visualization community would join my digital humanities friends by honoring this work with such deep engagement. Thanks to RJ Andrews, Arvind Satyanarayan, Derya Akbaba (again), Yanna Yannakakis, Roopika Risam, Michael Correll, Alfie Abdul-Rahman, Martin Wattenberg, Yanni Loukissas, Crystal Lee, Geoffrey Rockwell, Michael Friendly, Bill Rankin, Jer Thorp, Mandy Suhr-Sytsma, Stephanie Evergreen, Georgia Panagiotidou, Daniel Rosenberg, Allan Hillery, Jason Forrest, Rahul Barghava, Jonathan Zong, Melanie Walsh, Alex Sherman, Ben Schmidt, Attila Batorfy, Clayton Hayes, Frank Elavsky, and everyone else whose usernames I could not match to real names. Please know that my team and I read each and every comment, which shaped this project's final form.

Talks at Outlier 2025, the University of Miami, Stanford University, Yale University, the Turing Institute, Sony Research, Maastricht University, the University of Tennessee-Knoxville, the University of Washington, the University of North Carolina-Wilmington, Birkbeck College, the University of London, Clermont Graduate University, Brigham Young University, Penn State University, Columbia University, the University of Hamburg, Davidson College, Louisiana State University, Northeastern University, Umea University, Bucknell University, the CUNY Graduate Center, Bard College, the University of Pennsylvania, and the University of Alabama allowed me to refine the ideas presented in this book, and the audiences for whom I was writing. Thanks to Alberto Cairo (again!), Giovanna Ceserani, Susan Schreibman, Nick Holliman, Daniela Rosner, Joshua Ortiz-Baco, Hannah Alpert-Abrams, Morgan Scheuerman, Kayla Shipp, Brian Croxall, Laura Wexler, Alex Gil (again!), Matt Jones (again!), Johanna Drucker, Geoffrey Rockwell, Mark Sample, Lauren Coats, Patrick Svensson, Elizabeth Maddock Dillon, David Smith, Diane Jakacki, Maria Cecire, Matt Gold, Jim English, and everyone else who honored me with these invitations. The smart, engaged, and varied audiences whom you took the time and care to assemble became the readers I imagined while writing this book.

At the American Antiquarian Society, I thank Paul Erickson, Lauren Wasowicz, Lauren Hewes, Elizabeth Pope, Molly Hardy, Vincent Golden, and Nan Wolverton for recognizing what this project could become when the ideas it contains were so new, and offering my first sustained exposure to archival research. At the Library Company of Philadelphia, I thank Jim Green, Erica Piola, Connie King, and the Rights and Reproductions Team, for additional crucial early research support, and for continuing to support this project as it developed. At the New York Public Library, I thank Thomas Lannon for his guidance and expertise. At The Rooms, I thank Lori Temple and Mark Ferguson for opening up their archives to me, and for being so generous in their feedback and fact-checking. The community of partnership, knowledge, and trust that you have built around your holdings remains an inspiration. I also thank Granville Ganter, Stephen Stigler, Ian Spence, RJ Andrews, Jason Forrest, and everyone else who humored me with responses to my email queries about obscure charts and requests for confirmatory cell phone pics. I hope you see your efforts rewarded in these pages.

An NEH-Mellon Fellowship for Digital Publication (FEL-257658-18) provided me with the time away from teaching that allowed me to draft the early chapters of this book. An additional grant from the NEH Office of Digital Humanities (HAA-281011-21) allowed me to fund my student team during a crucial time. It pains me to know that, as a result of the devastating budget cuts and narrowness of vision of the Trump administration, future scholars will not have access to these same necessary support structures. I hope this book, the many students it involved, and the varied and impactful careers across industry and academia that they've gone on to pursue can become part of the evidence of the value of these funding streams if and when the political will to support them returns.

At Emory, a Chronos Fellowship from the Katz Foundation enabled me to push this project to completion. A Furthermore Grant in Publishing from the Kaplan Foundation provided crucial financial support for Polymode to design the print version of this book. A subvention from Emory's Digital Scholarly Publishing Initiative enabled the book to be offered in open access form. Research fellowships from the American Antiquarian Society, the Library Company of Philadelphia, and the New York Public Library during the 2013–2014 academic year enabled me to conduct the archival work that would become the foundation of this book. A more recent grant from the Mellon Foundation (G-2211-14240) to found the Atlanta Interdisciplinary AI Network, supporting students in research at the intersection of AI and the humanities (among other goals), has allowed me to carry the work of this project forward into my future work.

At Emory, I found unparalleled support for this project. Thanks go to Carla Freeman, Deboleena Roy, Cliff Carrubba, Ben Reiss, Deepika Bahri, Sarah McKee, Mae Velloso-Lyons, Kaelyn McAdams, Lora McDonald, Sadie Hannans, Corian Hill, Ria Buford, Anita Corbett, and Ravi Bellamkonda (now at Ohio State) for all that you have done to support me and my work at Emory. The trust and value that you have placed in my scholarship mean so much. To my Emory students (and former students), especially those who have worked with me in the DH Lab and on AIAI projects, including Ololade Faniyi, Dani Roytburg, Debbie Olorunisola, Diana Duarte Salinas, Dez Miller, and Odelia Larbi-Amoah: thank you for reminding me why I do this work. Thanks also to Luz Mary Salinas for helping me to secure image permissions in Bogotá. Moms are always the answer.

At Georgia Tech, Richard Utz enabled me to take my first year of fellowship leave to begin this book. Through the Women in Science and Technology program, Carol Colatrella funded my first student researchers. My Georgia Tech students Courtney Allen, Noah Sutter, Gwendolyn McGuire, and Sarah Schoemann inspired me with their technical expertise and collaborative know-how. Erica Pramer and Catherine Roshelli were also essential contributors to early research that led to this project.

At the MIT Press, Gita Manaktala offered expert guidance and unwavering support in more areas that I can count: brainstorming about book titles, strategizing about design, offering feedback on chapter drafts, and simply more time. Thank you for placing your trust in this project. I'm sorry that it took so long that you could not see it all the way through. Also at MITP, Suraiya Jetha provided essential knowledge and guidance throughout the process. Victoria Hindley took over the project in its final months, lending additional expertise. Matthew Abbate guided the project through the production process, along with Jay Martsi. Mandy Keifetz produced the excellent index. Thanks, also, to Friederike Sundaram at Stanford University Press, who supported this project in its early phases. Thanks, also, to my anonymous (and formerly anonymous) peer reviewers. I hope you see how your thoughtful feedback shaped the project into what it became.

There are far too many friends to name who have supported me on this long journey. While this is not a comprehensive list, in Atlanta I thank Nihad Farooq, Todd Michney, Kate Diedrick, Yanni Loukissas, Carl DiSalvo, Betsy DiSalvo, André Brock, Yanna Yannakakis, Maggie Penn, John Patty, Kevin Quinn, Andi Quinn, Sarah Salter, Susan Gagliardi, Cassandra Belton, Troy Belton, Maggie Fan, Joseph Monti, Jo Guldi, Dan Sinykin, Masha Vlasova, Ben Miller, Heather Cristle, Ben Reiss, Cliff Carrubba, my colleagues in the departments of Data & Decision Sciences and English, my climbing mom crew, and so many others who have listened to me and offered encouragement. In New York (and now also Philly), my forever writing group Sarah Blackwood, Kyla Schuller, and Karen Weingarten remain in my mind as my kindest and toughest critics. To my additional collaborators and co-conspirators Matt Gold, Sari Altschuler, Meredith Martin, Miriam Posner, Shannon Mattern, Lauren Tilton, Nicole Guidotti-Hernández, Natalia Cecire, Maria Cecire, and although they have been thanked already in other capacities, Catherine D'Ignazio, Miriah Meyer, Derya Akbaba, and Alberto Cairo: you've offered not only feedback but also sincere friendship. It means so much. Ian Loew, Jen Liu, Luca Marinelli, Nora Morrison, Eamon Johnson, Aileen Brophy, Sarah Madigan, Loren Hough, Lauren Wilcox, Shawn Ramirez, Jacob Eisenstein, Jenny Davidson, Leora Bersohn, Toby Moore, Mary Beth Kennedy, Frances Wall Jha, Saurabh Jha, and Ben Schmidt: thanks for caring about my work while also reminding me that the pleasures of life most often begin when the laptop is closed. Thanks also to Mary Coffey, Jon Zinman, and Dick Zinman, who are probably very excited for a new topic of conversation next year Up North.

A very key note of thanks goes to Cam Tran, who cared for our children while I finished this book.

Finally, thanks go to my family: my husband Greg, my children Loie and Aurora, my parents Diane and Francis, and my sister Amy and her partner Tomás. To Loie, who was there in the Baby Bjorn on my very first day in my very first archive; to Rora, who arrived on the scene to make our family complete; and to Greg, who is always the first and last person to read my writing: this book is for you.

NOTES

Preface

1 A "facsimile edition" is the term for a book that's been scanned and reprinted, its original formatting preserved. By contrast, a "modern edition" (which we encounter far more often) is the term for a book that's been transcribed and reformatted so that its contents are presented in a more contemporary style.

2 Howard Wainer and Ian Spence, "Introduction," in William Playfair, *The Commercial and Political Atlas* and *Statistical Breviary*, ed. Howard Wainer and Ian Spence (Cambridge UP, 2023 [1801]), 3.

3 Thomas Jefferson, *The Writings of Thomas Jefferson*, vol. 1, *Memoir, Correspondence, and Miscellanies from the Papers of Thomas Jefferson*, ed. Thomas Jefferson Randolph (Gray and Bowen, 1830), 2.

4 Lauren Klein, "The Image of Absence: Archival Silence, Data Visualization, and James Hemings," *American Literature* 85.4 (2013): 661–88.

5 Notably, Adam Smith wrote against the practice of slavery in *The Theory of Moral Sentiments* (1759) and touches on its economic inefficiency in *The Wealth of Nations* (1776). He stops short of challenging the racial hierarchies proposed by his contemporaries, including Hume and Kames (and Jefferson).

6 "Modern data visualization" is not a phrase that should go unexamined. See pp. 25–29 for this discussion.

7 The first, *An Archive of Taste: Race and Eating in the Early United States* (U of Minnesota Press, 2020), has a significant focus on Thomas Jefferson. The second, *Data Feminism*, coauthored with Catherine D'Ignazio (MIT Press, 2020), is characterized by a different blend of humanistic theory and technical expertise.

8 The 2010s were not the first time that humanities scholars had thought to use computers for their research, of course—that tradition dates back to the 1940s, if not before. But the early 2010s were the years when more and more scholars from the humanities—students, professors, and library and research staff alike—were turning to the web as a platform for creating "born digital" work. On the long history of the digital humanities, see Rachel Sagner Buurma and Laura Heffernan, "Search and Replace: Josephine Miles and the Origins of Distant Reading," *Modernism/modernity* 3.1 (2018). For an overview of born digital scholarship, see Matthew Kirschenbaum, "The .txtual Condition: Digital Humanities, Born-Digital Archives, and the Future Literary," *Digital Humanities Quarterly* 7.1 (2013).

9 Lauren Klein, Tanvi Sharma, Jay Varner, Margy Adams, Shiyao Li, Dan Jutan, Jianing Fu, Anna Mola, Nicholas Yang, Yang Li, Zhou Fang, and Silas Munro, *Data by Design: An Interactive History of Visualization and Power, 1789–1900*. Viewable at: dataxdesign.io.

Introduction: A History of Visualization and Power

1 I came across this phrase through a Google Image search, and while I am somewhat mortified that the very first citation of this introduction is not to a scholarly text but instead to a tourism website, I have not been able to come up with a better descriptor for Oaxaca's terrain on my own. Thus, Ivan Phillipsen, "The Natural Wonders of Oaxaca, Mexico," *Wild Latitudes*, https://wildlatitudes.com/natural-wonders-of-oaxaca-mexico/, accessed May 16, 2025.

2 The Ñudzahui, like the Mexica (Aztec), used both solar and sacred calendar systems, but did not count years in the same way as the Gregorian (European) calendar system. For more on Mesoamerican calendars and codices, see Alfonso Caso, *Los calendarios prehispánicos* (Universidad Nacional Autónoma de México, 1967), and more recently Maarten E. R. G. N. Jansen and Gabina Aurora Pérez Jiménez, "Chronological Correlations in Aztec and Mixtec History," in *Mesoamerican Manuscripts: New Scientific Approaches and Interpretations*, edited by Maarten E. R. G. N. Jansen, Virginia Lladò-Buisán, and Ludo Snijders (Brill, 2019), 341–379.

3 The cosmographer was a man by the name of Juan López de Velasco, who had been appointed Chief Cosmographer-Chronicler of the Indies a decade earlier, in 1571. In this role, López de Velasco compiled several substantial texts documenting colonial knowledge of the Americas as well as charts and maps of his own. See Felipe E. Ruan, "Prudent Deferment: Cosmographer-Chronicler Juan López de Velasco and the Historiography of the Indies," *Americas* 74.1 (2017): 27–55.

4 Visualization researcher Ben Shneiderman is credited with the oft-quoted phrase, "The purpose of visualization is insight, not pictures." See, for example, Jessica Hullman, "The Purpose of Visualization Is Insight, Not Pictures: An Interview with Ben

Shneiderman," *ACM Interactions*, August 19, 2019, https://interactions.acm.org/blog/view/the-purpose-of-visualization-is-insight-not-pictures-an-interview-with-ben, accessed May 16, 2025. For a more extended discussion of insight, including its roots in Stuart Card, Jock Mackinlay, and Shneiderman's "visualization reference model," see Derya Akbaba, Lauren Klein, and Miriah Meyer, "Entanglements for Visualization: Changing Research Outcomes through Feminist Theory," *IEEE Transactions on Visualization and Computer Graphics* 31.1 (2025): 1279–1289.

5 This statement is inspired by the definition of "local data" as articulated by Yanni Loukissas in *All Data Are Local: Thinking Critically in a Data-Driven Society* (MIT Press, 2019). There he writes: "Data are cultural artifacts created by people, and their dutiful machines, at a time, in a place, and with the instruments at hand for audiences that are conditioned to receive them" (2).

6 Thanks to Jihan Sherman for alerting me to this method of attribution, developed by the curators of "Hear Me Now: The Black Potters of Old Edgefield, South Carolina" at the High Museum in Atlanta, which ran from February 16 through May 12, 2024. It is intended to acknowledge the social, historical, and political conditions that result in the names of certain artists being preserved for posterity and others not. See https://high.org/exhibition/hear-me-now/, accessed May 16, 2025. On the mapmaker's training as a painter, see Alex Hidalgo, *Trail of Footprints: A History of Indigenous Maps from Viceregal Mexico* (U of Texas Press, 2019), 40.

7 A later section of this introduction problematizes the framing of modernity. For now, I employ scare quotes with all that they imply to indicate that this designation requires significant contextualization, and significant critique.

8 Howard F. Cline, "The Relaciones Geográficas of the Spanish Indies, 1577–1586," *Hispanic American Historical Review* 44.3 (1964): 341–374.

9 Henry Kamen, *Philip of Spain* (Yale UP, 1997).

10 Thanks to Yanna Yannakakis for her detailed feedback on an earlier version of this introduction, and in particular her suggestion that I introduce the concept of *yuhuiayu* so as to place Ñudzahui and Spanish worldviews alongside each other. This reflets how Teozacoalco itself was a place that was viewed through multiple lenses by multiple groups, often at the same time. For more on the concept of *yuhuiayu*, see Yanna Yannakakis, *Since Time Immemorial: Native Custom and Law in Colonial Mexico* (Duke UP, 2023), and Kevin Terraciano, *The Mixtecs of Colonial Oaxaca: Ñudzahui History, Sixteenth through Eighteenth Centuries* (Stanford UP, 2002).

11 David E. Shoemaker, "El Mapa de Teozacoalco" (Foundation for the Advancement of Mesoamerican Studies, 1999), http://www.famsi.org/reports/98032/index.html.

13 Both Barbara Mundy and Yanna Yannakakis provide detailed readings of these diagrams. See Mundy, *The Mapping of New Spain: Indigenous Cartography and the Maps of the Relaciones Geográficas* (U of Chicago Press, 2000), 114, and Yannakakis, *Since Time Immemorial*, 99.

14 In addition to Mundy, Hidalgo, and Yannakakis, see also Alfonso Caso, "El Mapa de Teozacoalco," *Cuadernos Americanos* 47.5 (1949): 145–181; René Acuña, ed., *Relaciones geográficas del siglo XVI* (UNAM, 1984); and Francisco de Solano, ed., *Cuestionarios para la formación de las relaciones geográficas de Indias: Siglos XVI/XIX* (Consejo Superior de Investigaciones Científicas, 1988).

15 Yannakakis, *Since Time Immemorial*, 106.

16 Sixty-three of these charts have been preserved in the Library of Congress, but visualization designer Jason Forrest has provided compelling evidence that there were originally 64. See Forrest, "Discovering an Unknown Chart from W. E. B. Du Bois's 'The Exhibition of American Negroes,'" *Nightingale: The Journal of the Data Visualization Society*, January 1, 2019, https://medium.com/nightingale/discovering-a-new-chart-from-w-e-b-du-boiss-the-exhibition-of-american-negros-part-6-66e7b340b90c, accessed August 18, 2025.

17 For an elaboration of this point, see Yannakakis, *Since Time Immemorial*, 106.

18 Orit Halpern, *Beautiful Data: A History of Vision and Reason since 1945* (Duke UP, 2015), 21.

19 Chart, n.," *The Oxford English Dictionary*, September 2025, https://doi.org/10.1093/OED/4890659491. For an in-depth

exploration of the history of the term, see the multivolume *History of Cartography* project, edited by J. B. Harley and David Woodward and published by the University of Chicago Press.

20 Daniel Rosenberg and Anthony Grafton, *Cartographies of Time: A History of the Timeline* (Princeton Architectural Press, 2012), 96–149.

21 Numerous large-scale visualizations of US Census data can be found in the catalogues of the American Antiquarian Society and the Library Company of Philadelphia, dating as far back as the 1840s; this was a trend I observed during my archival research. Beginning with the 1870 census, the Census Bureau itself published dazzling illustrated statistical atlases, discussed in depth in chapter 5.

22 RJ Andrews, "Don't Say 'Data Visualizations,'" *Chartograph*, January 28, 2025,1yh6 https://www.chartography.net/p/dont-say-data-visualizations, accessed May 16, 2025.

23 In 2022, Georges Hattab and RJ Andrews published a modern English translation, *The Graphic Method* (Visionary Press).

24 Howard Grey Funkhouser, "Historical Development of the Graphical Representation of Statistical Data," *Osiris* 3 (1936): 269–404; Jacques Bertin, *Sémiologie graphique. Les diagrammes, les réseaux, les cartes* (Gauthier-Villars, 1967); Edward R. Tufte, *The Visual Display of Quantitative Information* (Graphics Press, 1983). Another prominent work of this era is John Tukey's *Exploratory Data Analysis* (Addison-Wesley, 1977), which advances lightweight visualizations (e.g., scatter plots and bar charts) as tools to support early, exploratory work with data.

25 Halpern, *Beautiful Data*, 21.

26 Stuart K. Card, Jock D. Mackinlay, and Ben Shneiderman, *Readings in Information Visualization: Using Vision to Think* (Morgan Kaufman, 1999), 6.

27 See, for example, Michael Friendly and Daniel J. Denis, "Varieties of Data Visualization," which accompanies their monumental website, *Milestones in the History of Thematic Cartography, Statistical Graphics, and Data Visualization*, https://www.datavis.ca/milestones/index.php?page=varieties+of+data+visualization, accessed May 16, 2025.

28 Tamara Munzner, *Visualization Analysis and Design* (CRC Press, 2014), 1. One could also point to the 2021 merging of the flagship IEEE VIS conferences on information visualization, scientific visualization, and visual analytics into a single conference now called IEEE VIS: Visualization & Visual Analytics. See Christoph Garth, Min Chen, Alex Endert, Petra Isenberg, Alexander Lex, Shixia Liu, and Anders Ynnerman, "Things Are Changing in 2021: The New VIS Conference," *IEEE VIS* (blog), October 7, 2020, https://ieeevis.org/year/2024/blog/things-are-changing-2021, accessed January 15, 2026.

29 On data physicalization, see Pierre Dragicevic, Yvonne Jansen, and Andrew Vande Moere, "Data Physicalization," in the *Springer Handbook of Human Computer Interaction*, ed. Jean Vanderdonckt, Philippe Palanque, and Marco Winckler (Springer, 2021). On "data hunches," see Haihan Lin, Derya Akbaba, Miriah Meyer, and Alexander Lex, "Data Hunches: Incorporating Personal Knowledge into Visualizations," *IEEE Transactions on Visualization and Computer Graphics (VIS)* 29.1 (2022): 504–514. On "visualization vibes," see Michelle Morgenstern, Amy Fox, Graham M. Jones, and Arvind Satyanarayan, "Visualization Vibes: The Socio-Indexical Function of Visualization Design," *IEEE Transactions on Visualization and Computer Graphics (VIS)* 32.1 (2026): 1284–1294.

30 We should all now be familiar with the history of women "computers," as documented by historians of technology such as Nathan Ensmenger and Mar Hicks, in which the introduction of formal technical credentialing requirements pushed out women who had taught themselves. But the phenomenon goes back further still. In *Witches, Midwives, and Nurses: A History of Women Healers*, Barbara Ehrenreich and Deirdre English detail the history of obstetrics in the United States, in which experience-directed women midwives were replaced by book-smart men obstetricians after the establishment of formal medical schools. The same pattern can be found in the kitchen, with women performing most home cooking without formal training (or recognition), while men attend culinary school to become celebrity chefs. See Ensmenger, *The Computer Boys Take Over: Computers, Programmers, and the Politics of Technical Expertise* (MIT Press, 2010); Hicks, *Programmed Inequality: How Britain Discarded Women Technologists and Lost Its Edge in Computing* (MIT Press, 2017); and Ehrenreich and English, *Witches, Midwives, and Nurses: A History of Women Healers* (New York: Feminist Press, [1973] 2010).

31 On the "continuum" between data visualization and data communication, see Alberto Cairo, *The Functional Art: An Introduction to Information Graphics and Visualization* (New Riders, 2012), 15.

32 On the significance of visualizations that occur naturally in the world, see Dietmar Offenhuber, *Autographic Design: The Matter of Data in a Self-Inscribing World* (MIT Press, 2023).

33 This argument largely parallels the argument about data science that I make, alongside Catherine D'Ignazio, in *Data Feminism* (MIT Press, 2020), 21. It also bears mention that I am not alone in rejecting binary distinctions between visualization forms. In describing the relationship between visualizations and infographics, for example, theorist-practitioner Alberto Cairo explains, "every infographic and every visualization has a *presentation* and an *exploration* component: they present, but they also facilitate the analysis of what they show, to different degrees." Cairo, *The Functional Art*, xvi.

34 On cochineal, see Elena Phipps, *Cochineal Red: The Art History of a Color* (Metropolitan Museum of Art, 2010), and, for a probing creative engagement with cochineal and its darker valences (as well as the Met), see Robyn Schiff, *Information Desk: An Epic* (Penguin, 2023).

35 Another difference to register may be the distinction between a visualization and a map. Visualization researchers, including Munzner as well as Shneiderman, Card, and Mackinlay, tend to posit geospatial visualization as a subcategory of visualization. Some in the field of cartography, however, insist quite strongly that specialized skills are required to represent geospatial information, e.g., Alan M. MacEachren and D. R. Fraser Taylor, eds., *Visualization in Modern Cartography* (Pergamon, 1994); and more recently, Menno-Jan Kraak and Ferjan Ormeling, *Cartography: Visualization of Spatial Data*, 4th ed. (Routledge, 2023).

36 While this anecdote has never been formally verified, it remains a part of Galileo lore, viz. Steven Strogatz, "From a Swinging Chandelier to Global Positioning Systems," *American Scientist* 109.2 (2021): 106.

37 See Funkhouser, "Historical Development," Tufte, *Visual Display*, and more recently Sandra Rendgen, *History of Information Graphics* (Taschen, 2019), and Michael Friendly and Howard Wainer, *A History of Data Visualization and Graphic Communication* (Harvard UP, 2021). While many of these describe how the generally exploratory air of the Renaissance set these developments in motion, especially with the example of illuminated manuscripts, it was the Scientific Revolution's turn to empirical observation and measurement that provided the data which modern visualization would require.

38 This is, of course, an exceedingly general gloss. In an Islamic epistemology, for example, empirical evidence has been considered a part of divine knowledge from its very inception. Even within Enlightenment Europe, empiricism was not the only governing theory. The theory of rationalism, as exemplified by figures such as René Descartes and Baruch Spinoza, offered a wholly alternative explanation to how knowledge was produced, one that held that certain core ideas, such as the notion of language or the shape of a curve, came to be understood not by experiment but instead through the inner workings of the mind. These fundamental concepts were not observable but innate, the rationalists posited, and were transformed into knowledge by a process of deduction: by moving from the abstract to the particular. This process was the opposite of the inductive approach advanced by the empiricists, who believed in moving from the particular to the abstract.

39 The idea of a hierarchy of the senses, in which sight is at the top, dates to the ancient Greeks. In *De anima* (ca. 350 BCE), for example, Aristotle posits that the sense of sight is the highest sense because it can act alone. In *Sense and Sensibilia*, he further articulates the five senses and maintains that sight is at the top because it retains distance from its objects of perception. For more of this history, see Carolyn Korsmeyer, "The Hierarchy of the Senses," in *Making Sense of Taste: Food and Philosophy* (Cornell UP, 2015), 20–53.

40 His air pump experiments notwithstanding, Boyle's own visual impulse would seem to have been toward precision and detail rather than abstraction, as evinced by the detailed illustrations that accompanied his *New Experiments*, as literary scholar Alex Sherman observes. Comment on open review, January 13, 2025, https://hyp.is/_GCQpNH6Ee-wcGtkgLC9FQ/dataxdesign.io/chapters/intro.

41 If you are distressed to see the entirety of the Enlightenment summarized in a single sentence, rest assured that this is not the last time we will discuss the contributions associated with this pivotal time.

42 Friendly and Wainer, *A History of Data Visualization*, 17.

43 On the role of the barometer, see Friendly and Wainer, *A History of Data Visualization*, 20.

44 Friendly and Wainer, in the subtitle of their chapter on Playfair, call him "the father of modern graphics" (*A History of Data Visualization*, 95). Friendly and Daniel J. Denis's website, *Milestones in the History of Thematic Cartography, Statistical Graphics, and Data Visualization*, which, before the publication of Friendly and Wainer's book, served as the field's most comprehensive history, is the source of the "golden age" terminology, defined as the period between 1840 and 1910. Also see Friendly with Matthew Sigal and Derek Harnanansingh, "The Milestones Project: A Database for the History of Data Visualization," January 11, 2013, https://datavis.ca/papers/MilestonesProject.pdf, accessed January 25, 2026.

45 Tufte, *Visual Display*, 24. On the mythology of John Snow and his visual and epidemiological antecedents, see Howard Brody et al., "Map-Making and Myth-Making in Broad Street: The London Cholera Epidemic, 1854," *The Lancet* 356.9223 (2000): 64–68.

46 The sole woman (sometimes) included in this chronology, Florence Nightingale, has been anointed a "master" but, significantly, never (at least to my knowledge) a "founder" or "pioneer" (or even a "mother"); e.g., Noel-Ann Bradshaw, "Florence Nightingale (1820–1910): An Unexpected Master of Data," *Patterns* 8.1/2 (2020).

47 Deborah Shapley, "The Da Vinci of Data," *New York Times*, March 30, 1998; Tufte, *Visual Display*, 40. In *The Minard System: The Complete Statistical Graphics of Charles-Joseph Minard* (Princeton Architectural Press, 2018), Sandra Rendgen describes Minard as a "forefather of modern information visualization." Friendly, in his early work on Minard, also calls him a "true pioneer." See Friendly, "Visions and Re-Visions of Charles Joseph Minard," *Journal of Educational and Behavioral Statistics* 27.1 (Spring 2002): 31.

48 Tufte, *Visual Display*, 22.

49 Halley would go on to captain the British naval vessel *Paramour* on a mission to the South Atlantic that has been claimed as the first "purely scientific voyage" to take place on a royal ship (Susan Borowski, "The Multi-layered Wisdom of Edmond Halley," American Association for the Advancement of Science, October 12, 2012). Forced to return due to personal conflicts with his crew, Halley returned to the South Atlantic two more times in the next three years, his work there culminating in a 1701 map showing the isogonic lines of the entire Atlantic Ocean. This map was, in turn, used to ease navigation across the Atlantic, not only for future scientific research but also for the trade of goods—goods which, at times, took human form.

50 I am not the first to have made this argument. See, for example, Sylvia Wynter, "1492: A New World View," in *Race, Discourse, and the Origin of the Americas: A New World View*, ed. Vera Lawrence Hyatt and Rex Nettleford (Smithsonian Institution Press, 1994), 5–57.

51 The impact of this devastation continues into the present, as contemporary Indigenous scholars such as Mishauna Goeman (Seneca) and Max Liboiron (Métis/Mechif) remind us, profoundly disordering social structures and relationships (Goeman), as well as relations with land and resources (Liboiron). As Liboiron, citing Glen Sean Coulthard (Yellowknives Dene) explains, "colonialism is a way to describe relationships characterized by conquest and genocide that grant colonialists and settlers ongoing state access to land and resources that contradictorily provide the material and spiritual sustenance of Indigenous societies on the one hand, and the foundation of colonial state-formation, settlement, and capitalist development on the other." Max Liboiron, *Pollution Is Colonialism* (Duke UP, 2021), 9.

52 Russell Thornton, "Native American Demographic and Tribal Survival into the Twenty-first Century," *American Studies* 46.3/4 (2005): 23–38.

53 Cline, "The Relaciones Geográficas of the Spanish Indies," 350.

54 In *Horizons: The Globalization of Modern Science* (Mariner, 2022) James Poskett documents how, in fact, the Spanish learned many methods of data collection from the Indigenous societies over which they sought to rule..

55 Ian Hacking, *The Taming of Chance* (Cambridge UP, 1990), 18. For a more recent account of the rise of modern statistics, see Chris Wiggins and Matthew L. Jones, *How Data Happened: A History from the Age of Reason to the Age of Algorithms* (Norton, 2023).

56 Admittedly not in as much depth or with as much urgency as is deserved. As a supplement, see Banu Subramaniam, *Ghost Stories for Darwin: The Science of Variation and the Politics of Diversity* (U of Illinois Press, 2014), and Anita Chan, *Predatory Data: Eugenics in Big Tech and Our Fight for an Independent Future* (U of California Press, 2025).

57 This is the enduring contribution of feminist science and technology studies scholarship of the 1980s and 1990s, viz. Donna Haraway's "Situated Knowledges: The Science Question in Feminism and the Privilege of Partial Perspective," *Feminist Studies* 14.3 (1988): 575–599. It has since been renewed and expanded by scholars such as Katherine McKittrick, whose *Dear Science and Other Stories* (Duke UP, 2020) exposes the limits of even Haraway's partial perspective; this work is discussed in chapter 4.

58 On Tawantinsuyu, or "the land of four parts," see *The Inka Empire: A Multidisciplinary Approach*, ed. Izumi Shimada (U of Texas Press, 2015).

59 Xiyao Mei, Yu Zhang, Chaofan Yang, Rui Shi, and Xiaoru Yuan, "ZuantuSet: A Collection of Historical Chinese Visualizations and Illustrations," arXiv preprint, February 26, 2025, https://arxiv.org/pdf/2502.19093.

60 On the shogunal map and early modern Japanese mapping practices, see Mark Ravina, "Algorithmic Maps and the Political Geography of Early-Modern Japan," *Cultural Analytics* 8.3, https://doi.org/10.22148/001c.84860. On the map of Jambudvīpa, see Katherine Anne Paul, Richard A. Pegg, and Katie E. Osborne, "A Jain Cosmographic Map of the Rose-Apple Continent (Jambudvipa)," Norman B. Leventhal Map & Education Center, Boston Public Library, https://www.leventhalmap.org/projects/jain-map-chat/, accessed March 17, 2026.

61 Some, such as Bruno Latour, have argued that Europe has never been modern at all, and that all claims to modernity that are founded on the distinctions between science and culture that the Enlightenment would introduce are in fact false distinctions, and therefore untrue. Latour, *We Have Never Been Modern* (Harvard UP, 1993).

62 The prominent Scottish philosopher David Hume, for example, claimed that there was a racial hierarchy of "four or five different kinds" of humans, with "all other species ... naturally inferior to the whites." "On National Characters," in *Political Essays*, ed. Knud Haakonssen (Cambridge UP, 2012), 78–92.

63 Some readers may recognize this as a gloss of Michel Foucault's theorization of counterhistory, which he develops in relation to the historiographic theory of Friedrich Nietzsche. For more on Foucault and counter-history, see Foucault, "Nietzsche, Genealogy, History," in *Language, Counter-Memory, Practice: Selected Essays and Interviews*, ed. D. F. Bouchard (Cornell UP, 1977), 139–164.

64 Ruha Benjamin, *Imagination: A Manifesto* (Norton, 2024), 23.

65 E.g., Michael Friendly, "The Golden Age of Statistical Graphics," *Statistical Science* 23.4 (2008): 502–535.

66 On racial capitalism, see Cedric Robinson, *Black Marxism: The Making of the Black Radical Tradition* (U of North Carolina Press, 1993), or just read chapter 1.

67 On positivism as the dominant epistemology of visualization research, see Akbaba, Klein, and Meyer, "Entanglements for Visualization."

1 **Data:** The Society for Effecting the Abolition of the Slave Trade and Responsible Design

1 Recent scholarship has raised questions about Equiano's account of his early years in Benin. Literary scholar Vincent Carretta, most notably, has pointed to several later records that list Equiano's birthplace as South Carolina. As Carretta himself makes clear, this evidence is "not absolutely conclusive" (xvi). Because of this lack of certainty, I have chosen to prioritize Equiano's own account of his homeland and early life experiences. See Carretta, *Equiano, the African: Biography of a Self-Made Man* (U of Georgia Press, 2005).

2 Olaudah Equiano, *The Interesting Narrative of the Life of Olaudah Equiano, or Gustavus Vassa, the African, Written by Himself*, ed. Werner Sollors (Norton, 2001), 39, 41.

3 Say it out loud. Contemporary computer science project acronyms have nothing on nineteenth-century abolitionists! More seriously, SEAST was a major (but of course not the only)

force that led to the abolition of the slave trade in England. It coalesced in 1787, after a decade of increasing momentum around the movement to end slavery in England, and was run mostly by Quakers (although Clarkson himself was not one). As indicated by the distinction between the Plymouth Committee and the London Committee noted above, its structure was geographically distributed. Other committees existed in Manchester and Birmingham, among other British locales. The committees communicated with each other via letters, many of which have been preserved in local Quaker archives, as well as in the British National Archives. Clarkson himself, radicalized through his senior thesis work at St. John's College, Cambridge, also traveled from committee to committee, circulating pamphlets and other documents and giving speeches of his own. SEAST was also a reference point for the abolitionist movement in the United States. In fact, that first "Plan of an African Ship's Lower Deck" was copied and reengraved by the famed Philadelphia printer Matthew Carey, who circulated it throughout the young United States. As we know, however, it would take more than 50 years for full legal emancipation to take place in the United States, with its struggle for racial justice continuing into the present. For more on SEAST, see John Oldfield, *Popular Politics and British Anti-Slavery: The Mobilization of Public Opinion against the Slave Trade, 1787–1807* (Frank Cass, 1998) and Adam Hochschild, *Bury the Chains: Prophets and Rebels in the Fight to Free an Empire's Slaves* (Houghton Mifflin, 2005). For more on Clarkson's role in particular, see the texts cited throughout this chapter.

4 The meeting between Equiano and Clarkson can be confirmed by a public letter authored by Equiano and published in a London newspaper, *The Public Advertiser*, on February 14, 1789. In the letter, Equiano registers "having seen" the engraving sent to Clarkson, whom he describes as a "worthy friend of mine." See "Letter to the Committee for the Abolition of the Slave Trade," in Equiano, *The Interesting Narrative*, 205. While Equiano offers scant additional details about their exchange, the art historian Cheryl Finley, in her landmark study *Committed to Memory: The Art of the Slave Ship Icon* (Princeton UP, 2018), powerfully imagines the impact of this request from Equiano's perspective (23). We will return to this letter, and the significance of the exchange between Equiano and Clarkson, in this chapter's final section.

5 Equiano, *The Interesting Narrative*, 41. Scholars have made much of Cugoano's assessment, in his 1787 treatise, *Thoughts and Sentiments on the Evils of Slavery*, that his own experience "may be easier conceived than described" (15). As literary scholar Joanne Chassot explains, "His silence over the details of the experience might therefore be explained by a form of censorship, which could be imposed by white editors for reasons of decorum, or self-imposed either out of a personal or cultural aversion at relating such intimate matters or out of the psychological refusal to recall events that are too painful." Chassot, *Ghosts of the African Diaspora: Re-visioning History, Memory, and Identity* (Dartmouth College Press, 2018), 41. In the context of our present discussion, we might draw generative connections between Cugoano's narrative refusal and contemporary arguments for data refusal. See, for example, Marika Cifor, Patricia Garcia, TL Cowan, Jasmine Rault, Tonya Sutherland, Anita Say Chan, Jennifer Rode, Anna Lauren Hoffmann, Niloufar Salehi, and Lisa Nakamura, *Feminist Data Manifest-No* (2019), https://www.manifestno.com/, accessed May 20, 2025.

6 Thomas Clarkson, *The History of the Rise, Progress, and Accomplishment of the Abolition of the African Slave-Trade by the British Parliament*, vol. 2 (London, 1808), 111.

7 The use of the word "impression" is the tell. For Enlightenment thinkers such as John Locke and David Hume, the impression that objects in the world make on sensory organs signals the start of the knowledge-making process. This epistemological theory would go on to influence British literary sentimentalism, as well as the sentimentalist undercurrent of white-led abolitionist movements, of which Clarkson was certainly a part. On literary sentimentalism and its relationship to empiricism, see Jessica Riskin, *Science in the Age of Sensibility: The Sentimental Empiricists of the French Enlightenment* (U of Chicago Press, 2002); on the colonial implications of sentimentalism, see Lynn Festa, *Sentimental Figures of Empire in Eighteenth-Century Britain and France* (Johns Hopkins UP, 2006); on the imbrications of racism and sentimentalism, and Clarkson's in particular, see Ian Baucom, *Specters of the Atlantic: Finance Capital, Slavery, and the Philosophy of History* (Duke UP, 2005); on sentimentalism's journey from the British empiricists to the United States, and its transformation into nineteenth-century race science, see Kyla Schuller, *The Biopolitics of Feeling: Race, Sex, and Science in the Nineteenth Century* (Duke UP, 2018).

8 While the Parrey report would become a central piece of evidence for the British abolitionist movement, one important fact that failed to enter the historical record was Parrey's first name. See Finley, *Committed to Memory*, 34.

9 While not directly relevant to this chapter's main argument, two points are worth noting here: first, that while the origin story of Clarkson's "Description" usually begins with Elford's "Plan," in fact it has a longer and more diffuse history; and second, that while the story would seem to be one of the visual and moral acuity of the elite, the idea for the "Plan," if not its application, originates with working-class sailors.

10 Finley, *Committed to Memory*, 34.

11 These harms may be retrospective, in the influential words of literary scholar Saidiya Hartman, the result of "the uncertain line between witness and spectator" that scholars of slavery often must walk. They may also carry into the future, in the form of the desensitization brought about by ever more depictions of the "routinized violence of slavery," as Hartman further explains. For viewers whose own ancestors were enslaved, such depictions can also exacerbate existing intergenerational trauma, as historian Jennifer Morgan describes. For this reason, we gave online viewers the choice to "opt in" to view these images by clicking a toggle at the top left of the screen. If they did not opt in, they would proceed with the default: to read short and intentionally generalized descriptions of the images discussed. (We also provided more detailed descriptions of the images, toggled when opting in, for users who rely on alt text as their primary means of encountering visual material.) Saidiya Hartman, *Scenes of Subjection: Terror, Slavery, and Self-Making in Nineteenth-Century America* (Oxford UP, 1997), 4; and Jennifer Morgan, *Reckoning with Slavery: Gender, Kinship, and Capitalism in the Early Black Atlantic* (Duke UP, 2021).

12 Hortense Spillers, "Mama's Baby, Papa's Maybe: An American Grammar Book," *Diacritics* 17.2 (Summer 1987): 72.

13 As is true of the history of visualization, the history of data tables is also long, with examples drawn from all over the world. The earliest known data tables can be traced to the Neolithic cities of Sumerian Mesopotamia ca. 3500 BCE. For a short and accessible history, see "A Brief History of Data," in Rahul Bhargava, *Community Data: Creative Approaches to Empowering People with Information* (Oxford UP, 2024), 16–50. For a visualization-oriented assessment of data tables as knowledge-making devices, see Lyn Bartram, Michael Correll, and Melanie Tory, "Untidy Data: The Unreasonable Effectiveness of Tables," *IEEE Transactions on Visualization and Computer Graphics* 28.1 (2002): 686–696.

14 Jessica Marie Johnson, "Markup Bodies: Black [Life] Studies and Slavery [Death] Studies at the Digital Crossroads," *Social Text* 36.4 (2017): 70.

15 Spillers, "Mama's Baby," 72.

16 There exist several commemorative engravings of the *Marie-Séraphique*. Finley makes a similar point with respect to a different image of this ship, which dates to 1772–1773, in *Committed to Memory*, 38–42.

17 Wood's use of the word "flesh" is also likely a reference to Spillers's distinction between bodies and flesh, the latter resulting from the dehumanizing process of the Middle Passage. Marcus Wood, *Blind Memory: Visual Representations of Slavery, 1780–1865* (Manchester UP, 2000), 27.

18 Celeste-Marie Bernier also lends her support to this interpretation: "Working not to humanize but to propagandize black subjects bought and sold into slavery, the iconographic emphasis across these broadsides betrays a white British commitment to exposing slavery's atrocities by imagining black women, men, and children not as individualized subjects but as unindividualized objects defined solely by their incarceration on board a slave ship." Bernier, "The Slave Ship Imprint: Representing the Body, Memory, and History in Contemporary African American and Black British Painting," *Callaloo* 37.4 (2014): 998. We might trace the visual contrast forward to the work of contemporary artist Glenn Ligon, whose 1992 screenprint *Untitled (I Feel Most Colored When I Am Thrown against a Sharp White Background)*, itself a quote from author Zora Neale Hurston, emphasizes the racialized dimensions of black-on-white design. This echoes a larger critique of modern art and design for its glorification of the color white, and therefore whiteness, alongside its other signal features of clean lines and other forms of visual minimalism. While not engaging the racialized dimension of modernism, Helen Kennedy, Rosemary Lucy Hill, Giorgia Aiello, and William Allen analyze how specific modernist visual conventions, such as clean lines and clearly demarcated boundaries, are strongly associated with the authoritativeness of visualization. Kennedy et al., "The Work that

Visualization Conventions Do," *Information, Communication and Society* 19.6 (2016): 715–735.

19 In fact, Otto and Marie Neurath, the inventors of the Isotype icon system, incorporated their Isotype icons into a reproduction of Clarkson's "Description" in their pamphlet on the history of global trade, *Wereldverkeer* (1935). They also held a ca. 1800 version of the "Description," created by the German printmaker Jacob Xaver Schmunzer, in their personal archive, as Otto Neurath himself documents in *From Hieroglyphics to Isotype: A Visual Autobiography* (Hyphen Press, 2010). It is possible to make an argument that their icons, like those on the slave ship, were similarly intended to serve as abstract representations of "observed particulars"—in their case, of the population of the world. What's more, they similarly sought to enlist their icons in a political cause, though their icons of human figures fell prey to the deeply offensive cultural stereotypes of non-European cultures in circulation at the time. For more on Isotype and the Neuraths in a visualization context, see Christopher Burke and Gunther Sandner, *History and Legacy of Isotype* (Bloomsbury, 2024).

20 Edward R. Tufte, *The Visual Display of Quantitative Information* (Graphics Press, [1983] 2001), 13.

21 Catherine D'Ignazio and Lauren F. Klein, *Data Feminism* (MIT Press, 2020), 92.

22 Notably, Clarkson also published the transcripts of his interviews with these men at roughly the same time as he released his revised version of the Plymouth Committee's "Plan." For more on the interviews, see Marcus Rediker, *The Slave Ship: A Human History* (Penguin, 2007), 319–324. On the concept of counterdata, see D'Ignazio and Klein, *Data Feminism*, as well as Seyi Olojo, "Counterdata," in *Keywords of the Datafied State*, ed. Jenna Burrell, Ranjit Singh, and Patrick Davison (Data & Society, 2024), 170–181. On the history of data being enlisted in the service of abolition, see Ellen Gruber Garvey, "'facts and FACTS': Abolitionists' Database Innovations," in *"Raw Data" Is an Oxymoron*, ed. Lisa Gitelman (MIT Press, 2013), 89–102.

23 These lines come from the first paragraph of text included in the chart itself. Finley provides a full transcription in *Committed to Memory*, 62–67.

24 A notable recent exception is Peter A. Hall and Patricio Dávila's *Critical Visualization: Rethinking the Representation of Data* (Bloomsbury, 2022). There, they herald the diagram as "one of the most impactful visualizations in the history of the genre" (55). For their full discussion, see 55–59.

25 Michael Friendly and Howard Wainer, *A History of Data Visualization and Graphic Communication* (Harvard UP, 2021), 27.

26 Wood, *Blind Memory*, 28.

27 Wood, *Blind Memory*, 19.

28 Clarkson's motivation for these design choices is difficult to discern. Why leave the women unbound when evidence points to the fact that they were also "known to organize, lead, and assist with rebellions aboard slave ships," Finley wonders (*Committed to Memory*, 59). Is the exaggerated sexuality of the women figures a comment, on the part of Clarkson, about the "objectified status of the black female body and the history of rape and sexual abuse to which black female slaves were subjected by their white male captors," or because of his own racist perceptions about the sexuality of Black women, Finley further enquires (59)? We may never be able to answer these questions with any degree of certainty, but their possible answers expand our understanding of the chart nonetheless. More specifically, they underscore the perspective inhabited by the chart, and point to the additional perspectives we must consider if we are to achieve a more complete picture of the actual subject at hand.

29 Clarkson, para 22, reproduced in Finley, *Committed to Memory*, 67.

30 Wilberforce, quoted in Baucom, *Specters of the Atlantic*, 218. Across Europe, other abolitionists shared Wilberforce's sentiment. Upon seeing Clarkson's diagram, the French Comte de Mirabeau went one step further, commissioning a wooden replica of the ship depicted in the chart about "one yard long," with "little men and women" painted black, which he kept on his dining room mantel to show to potential converts to the cause (Finley, *Committed to Memory*, 78). Clarkson himself saw the replica, and when he told Wilberforce about it, Wilberforce had his own replica commissioned, which he brough to Parliament and "passed from hand to hand" (Finley, 81). In *Community Data*, Bhargava discusses this replica in the context of an argument about the value of data physicalization (125–127).

31 Donna Haraway, "Situated Knowledges: The Science Question in Feminism and the Privilege of Partial Perspective," *Feminist Studies* 14.3 (1988): 589.

32 Simone Browne, *Dark Matters: On the Surveillance of Blackness* (Duke UP, 2015) 49.

33 Haraway, "Situated Knowledges," 589. It is also worth noting that this observation applies not only to the perspectives we impose when we design static visualizations, but also to those we leave unexamined when we design for "viewer empowerment." In these cases, we must similarly probe our default assumptions about who our viewers are, as well as what their goals might be.

34 Quoted in Baucom, *Specters of the Atlantic*, 267. Here, Baucom argues, Clarkson is mobilizing the central tenet of the Scottish strain of Enlightenment empiricism, which held that meaning was created first by verifiable evidence and then through a responsibly used imagination. In this case, Clarkson enlists the "facts" of the slave trade in the interest of engaging the "otherwise disinterested spectator in the sufferings of an actually unseen other" (268). But for this transformation to take place, the viewer would also be required to enlist their own imagination in imagining the suffering of others—the point at which theories of sentimentalism, as previously discussed, also enter in. There are many flaws in this theory. Note 18 discusses its implicit racism. Among the others is the presumption that all viewers, even limited to all white viewers, would respond in the same way.

35 Browne, *Dark Matters*, 49.

36 Browne, *Dark Matters*, 50.

37 "Trans-Atlantic Slave Trade—Estimates," *Slave Voyages*, https://www.slavevoyages.org/assessment/estimates, accessed May 23, 2025.

38 "History of the Project," *Slave Voyages*, https://www.slavevoyages.org/about/about#history/1/en/, accessed May 23, 2025.

39 To view Kahn and Bouie's visualization and accompanying text, see Andrew Kahn and Jamelle Bouie, "The Atlantic Slave Trade in Two Minutes," *Slate*, September 16, 2021, https://slate.com/news-and-politics/2021/09/atlantic-slave-trade-history-animated-interactive.html, accessed May 23, 2025. Note that the visualization was created and released in 2016, and rereleased in 2021 with no changes that we could detect. In "New World: The Impact of Digitization on the Study of Slavery," literary scholar Britt Rusert helpfully summarizes the visualization's history: "The *Slate* map has at least two direct antecedents: a set of maps of the slave trade on the Trans-Atlantic Slave Trade Database website, and David Eltis and David Richardson's 2010 print volume, *Atlas of the Transatlantic Slave Trade*, whose maps were "constructed from a set of estimates derived from the Trans-Atlantic Slave Trade Database as it existed in January 2008—almost a year before the launch of the Voyages Web site." Rusert, "New World: The Impact of Digitization on the Study of Slavery," *American Literary History* 29.2 (2017): 274.

40 Rusert, "New World," 275.

41 By paraphrasing Rusert's argument as "another version" of the god trick, I'm not giving full justice to the richness of her argument. In her observation about "invisible hands," Rusert is at once referencing Adam Smith's formulation of the "invisible hand of the market," his famed articulation of the forces of free-market capitalism, and the literal invisible hands that seem to pluck the ships from the African coast and carry them across the Atlantic without ever being attached to the people responsible for setting that particular ship or the slave trade overall in motion. In the case of both free-market capitalism and the slave trade (which are themselves, in many ways, two halves of a whole), there emphatically are *people* responsible for the decisions that are presented by Smith and the visualization, respectively, as coming from "no body," to return to Haraway's phrase.

42 Johnson, "Markup Bodies," 41.

43 We are not the only design team to have considered similar questions. Celia E. Naylor, Kirsten Akey, Madiha Zahrah Choksi, Alex Gil, Moacir P. de Sa Pereira, and Monique S. J. Williams, in "Un(silencing) Slavery: Remembering the Enslaved at Rose Hall Plantation, Jamaica," similarly "aims to re-center the lives of the enslaved persons at Rose Hall Plantation by emphasizing the history of slavery and enslaved persons as that historical site." See the interactive visualization and "About" page at https://unsilencing-slavery.org/about, accessed May 26, 2025.

44 The canonical essay in this regard is Katherine McKittrick's "Mathematics Black Life," *Black Scholar* 44.2 (Summer 2014): 18, in which she comments directly on the "descriptive analytics of violence." More recently, Caitlin Rosenthal, in *Accounting for Slavery: Masters and Management* (Harvard UP, 2019) has returned to these numbers to support an argument about the management theory driving the plantation system, enlisting this violent data as evidence of the racism at the core of modern methods of industrial management. At the time that we began our design process, we did not consider the possibility that we might use this data as a form of "studying up" (cf. Laura Nader), but we note its continued possibilities in this regard.

45 See, for example, Stephanie Smallwood, *Saltwater Slavery: A Middle Passage from Africa to American Diaspora* (Harvard UP, 2007); Rusert, "New World"; Christina Sharpe, *In the Wake: On Blackness and Being* (Duke UP, 2016); Romi Morrison, "Gaps between the Digits: On the Fleshy Unknowns of the Human," *Information Design Journal* 25.1 (2019): 56–70; Tiffany Lethabo King, *The Black Shoals: Offshore Formations of Black and Native Studies* (Duke UP, 2019).

46 Morrison, "Gaps between the Digits," 56.

47 Orlando Patterson, *Slavery and Social Death: A Comparative Study* (Harvard UP, 1982).

48 We also appreciated the further connection to Spillers's theorization of the process of dehumanization as the conversion of bodies into flesh that the Middle Passage brought about. See Spillers, "Mama's Baby."

49 Equiano, *The Interesting Narrative*, 39, 45.

50 Kevin Quashie, *The Sovereignty of Quiet: Beyond Resistance in Black Culture* (Rutgers UP, 2012).

51 On the wide circulation and impact of the "Description," see Finley, *Committed to Memory*, 71–74. On Clarkson's *History*, see Finley, 88–92.

52 As Clarkson explains in the second chapter of the 1200-page tome, "I was in possession of more facts on this subject than any other person" (*History*, 1:31).

53 Daniel Rosenberg and Anthony Grafton, *Cartographies of Time: A History of the Timeline* (Princeton Architectural Press, 2012), 148.

54 Clarkson, *History*, 1:272.

55 Wood, 4.

56 Wood, 4.

57 As Rosenberg and Grafton explain, Strass "believed that a graphic representation of history held manifold advantages over a textual one: it revealed order, scale, and synchronism simply and without the trouble of memorization and calculation" (*Cartographies of Time*, 143).

58 Bell, quoted in Rosenberg and Grafton, *Cartographies of Time*, 147.

59 Rosenberg and Grafton, *Cartographies of Time*, 136.

60 Quoted in Rosenberg and Grafton, *Cartographies of Time*, 136.

61 William Playfair, introduction to *The Commercial and Political Atlas*, ed. Howard Wainer and Ian Spence (New York: Cambridge UP, 2023 [1801]), x. As we will learn in chapter 2, this is not the only instance of similar framing. Just as Playfair, in the subtitle to his *Atlas*, described his goal of providing a picture of British imports and exports in a "single view," so too did Clarkson describe how he sought to "bring [the history of abolition] before the reader, that he may comprehend the whole of it at a single view" (*History*, 1:259).

62 "Letter to the Committee for the Abolition of the Slave Trade," in Equiano, *The Interesting Narrative*, 205.

2 **Image:** William Playfair and the Politics of the Simple View

1 While there are no known examples of pie charts that predate Playfair's, there are several earlier entries in the bar chart category. In *A History of Data Visualization and Graphic Communication* (Harvard UP, 2021), Michael Friendly and Howard Wainer discuss a bar chart of water levels of the Seine, created by the French geographer Philippe Buache and his collaborator Guillaume de L'Isle, which dates to 1770. But because Buache and de L'Isle were using bars to represent water levels, Friendly and Wainer assert, their chart did not reflect the degree of abstraction of

Playfair's, which "use[d] the length of bars to show something intangible: imports and exports to Scotland that could be understood in a new way" (115). In her investigation of how data tables were "read" in the years before visualization, historian of science Lorraine Daston surfaces an even earlier bar chart attributed to Athanasius Kircher, from 1721, which shows temperatures in Berlin as bars. On the basis of the lack of response, even amongst the members of the Royal Society, Daston concludes that they "were more of a nuisance than a novelty," since early modern eyes had not yet been trained to interpret them (213). See Daston, "Super-Vision: Weather Watching and Table Reading in the Early Modern Royal Society and Académie Royale des Sciences," *Huntington Library Quarterly* 78.2 (Summer 2015): 187–215.

2 David R. Bellhouse, *The Flawed Genius of William Playfair: The Story of the Father of Statistical Graphics* (U of Toronto Press, 2023), xiii. The final section of this chapter discusses Playfair's reception history in more detail.

3 Howard Wainer and Ian Spence, "Introduction," in William Playfair, *The Commercial and Political Atlas and Statistical Breviary*, ed. Howard Wainer and Ian Spence (Cambridge UP, 2005), 9. Unless otherwise indicated, all direct quotations from the *Atlas* are transcribed from this edition.

4 Responsible for numerous failed money-making schemes, Playfair went through bankruptcy proceedings in the final years of the 1790s and never fully recovered financial solvency. Playfair's financial difficulties are discussed in more depth later in this chapter, and in Bellhouse, *The Flawed Genius*, 159–195.

5 My account of Neele and Playfair's distribution of tasks derives from Wainer and Spence, "Introduction."

6 My account of copperplate engraving derives from Wainer and Spence, "Introduction," as well as from Roger Gaskell, "Printing House and Engraving Shop: A Mysterious Collaboration," *BC* 53 (2004): 213–251. Interested readers may also consult "From Paper to Copper: The Engraver's Process," a demonstration by Andrew Stein Raftery, Associate Professor of Printmaking at the Rhode Island School of Design, viewable online at https://www.youtube.com/watch?v=fQvghHs15hA.

7 Playfair, *The Commercial and Political Atlas*, viii. Here, it is worth reiterating that the term "data visualization" did not exist in Playfair's time. Playfair himself never settled on a single term for his "invention," describing his approach variously in the *Atlas* as "applying lines to subjects of Finance" and applying "the principles of geometry to matters of Finance" (vii, viii). For the 1798 version of the *Atlas*, Playfair coined the phrase "lineal arithmetic," although this did not catch on (see *Lineal Arithmetic*, London, 1798).

8 Helen Kennedy, Rosemary Lucy Hill, Giorgia Aiello, and William Allen, "The Work that Visualization Conventions Do," *Information, Communication and Society* 19.6 (2016), 723. Also see Alberto Cairo, *How Charts Lie: Getting Smarter about Visual Information* (Norton, 2019).

9 Here I am encapsulating the major contribution of science and technology studies (STS) of the past several decades. Steven Shapin and Simon Schaffer's *Leviathan and the Air-Pump: Hobbes, Boyle, and the Experimental Life* (Princeton UP, 1985) lays strong conceptual groundwork for this claim. More recently, the essays included in *Social Studies of Scientific Imaging and Visualization*, ed. Joseph Dumit and Regula Valérie Burri (MIT Press, 2008), have applied this STS framework to visualization. 10 Quoted in Wainer and Spence, "Introduction," 14.

11 William Playfair, *The Commercial and Political Atlas*, 1st ed. (London, 1786), 7.

12 In both the preface and introduction to the third edition, Playfair includes quotations from the press coverage of earlier editions, as well as his own recollections of the book's early reception in England and France. See *The Commercial and Political Atlas*, iii–xv.

13 Playfair, *The Commercial and Political Atlas*, ix–x.

14 On Enlightenment empiricism and the senses, see chapter 1, note 7.

15 Playfair, *The Commercial and Political Atlas*, x.

16 Playfair, *The Commercial and Political Atlas*, xiv.

17 In the field of visualization research, scholars have explored questions around the "value" of visualization, albeit focused primarily on methods of evaluation that align with its specific contributions, such as efficiency, confidence, and insight. In

"The Value of Visualization," Jarke van Wijk lays out the rationale for this research program: "If we know to which actions the visualization should lead to, the next steps are assessment whether the knowledge derived from the visualization does indeed support the decision, and also, to assess the economic value of this decision. This is not easy, but one can try for instance to estimate how much time is saved, or try to quantify the consequences of a wrong decision" (*IEEE Visualization 2005*, Minneapolis, 79–86). Subsequent work by John Stasko and Emily Wall models how specific values—in particular, "time, insight, essence, and confidence"—might be assessed. See Stasko, "Value-Driven Evaluation of Visualizations," *BELIV '14: Proceedings of the Fifth Workshop on Beyond Time and Errors: Novel Evaluation Methods for Visualization*, 46–53); and Emily Wall, Meeshu Agnihotri, Laura Matzen, Kristin Divis, Michael Haass, Alex Endert, and John Stasko, "A Heuristic Approach to Value-Driven Evaluation of Visualizations," *IEEE Transactions on Visualization and Computer Graphics* 25.1 (2019): 491–500.

18 Wainer and Spence push this analysis even further, finding "minor, but careless, arithmetical errors in the tables" which "throws into doubt the accuracy of the remaining numbers in these and the other tables," as well as discrepancies in the visual "rendering of the raw data," especially in the third edition ("Introduction," 19). Their analysis focuses on the charts involving the West Indies and Jersey, Guernsey, and Alderney, but they arrive at a conclusion about the charts overall, which is that "some of the curves that connect the datapoints seem to have derived their shapes from Playfair's opinion of how the intervening data should look" (18).

19 Even this was an innovation. Prior to Playfair, time was for the most part not delimited in a linear way. Rather, significant events were what defined the scale. See Daniel Rosenberg and Anthony Grafton, *Cartographies of Time: A History of the Timeline* (Princeton Architectural Press, 2010), 136.

20 This phase, a variation on the "simple and complete idea" that Playfair describes in his introduction, derives from the working title of the *Atlas*, "The commercial and political atlas representing, by means of stained copper-plate charts, the exports, imports, and general trade of England, at a single view." This was the title Playfair employed in a draft version that he circulated to friends, including Watt, in 1785.

21 On the saga of Playfair's bankruptcy, see Bellhouse, *The Flawed Genius*, 159–195.

22 This topic will be discussed in more depth in the section of this chapter on Playfair's charts in the present, but as an opening salvo, consider the similarities between Playfair's views about the value of visualization and Ben Shneiderman's "visual information-seeking mantra": "overview first, zoom and filter, then details on demand." Shneiderman, "The Eyes Have It: A Task by Data Type Taxonomy for Information Visualizations," in *The Craft of Information Visualization* (Morgan Kaufmann, 2003), 364–371.

23 Playfair, *The Commercial and Political Atlas*, ix.

24 While the critical consensus is that Playfair possessed poor business acumen, this particular turn toward financial precarity was not entirely his own doing. Playfair had opened a metal factory outside of Paris, which involved significant personal investment, just as the French Revolution took hold. It was by all accounts on track to become profitable, but Playfair could not stop himself from pursuing additional business ventures. He was in the midst of planning for a second mill, as well as a separate project involving land investments in the newly constituted United States, when the economic instability brought about by the Revolution ratcheted up another notch. Lending in France collapsed and investors' priorities shifted. In addition, the land investment scheme (in which Playfair was one of several partners) became increasingly dubious; it involved only the right to purchase land, rather than the land itself, a fact that was not clearly communicated to its investors and was discovered only when some of them, fleeing revolutionary France, traveled to the US to remake homes there. Playfair also founded a bank during this time; it was also marked by some irregularities, but these would not have been fatal had the Revolution not then reached a violent peak that forced Playfair to flee Paris, withdrawing his stake in the bank in the process. On this episode, see Bellhouse, *The Flawed Genius*, 70–115.

25 Bellhouse speculates that this change of fortune was what prompted Playfair's turn to increasingly dubious schemes (*The Flawed Genius*, 123).

26 Playfair, *The Commercial and Political Atlas*, iii.

27 Bellhouse, *The Flawed Genius*, 179.

28 Playfair, *The Commercial and Political Atlas*, iv.

29 Playfair, *The Commercial and Political Atlas*, iv.

30 Playfair, *The Commercial and Political Atlas*, iv.

31 E.g., Manuel Lima, *Visual Complexity: Mapping Patterns of Information* (Princeton Architectural Press, 2011).

32 Playfair, *The Commercial and Political Atlas*, xiv.

33 Playfair, *The Commercial and Political Atlas*, xv. As Jenny Odell has observed, it is in fact capitalism that is responsible for the constant push toward more efficiency, which Playfair's charts support in form, content, and audience. See Odell's most recent works, *How to Do Nothing: Resisting the Attention Economy* (Melville House, 2019) and *Saving Time: Discovering Life beyond the Clock* (Penguin-Random House, 2023). Thanks to Alexander Sherman for these citations.

34 The introduction to this chapter references work on visualization conventions led by Helen Kennedy, who brings together semiotic analyses of contemporary visualizations with interviews with practicing designers about their aims. Meanwhile, visualization researchers Jessica Hullman and Nicholas Diakopoulos have explored the idea of visualization rhetoric, analyzing the various forms of visual rhetoric that accompany visualizations that appear in a narrative context, and that contribute to the meaning that is conveyed. See "Visualization Rhetoric: Framing Effects in Narrative Visualization," *IEEE Transactions on Visualization and Computer Graphics* 17.12 (2011). Finally (although far from exhaustively), Ha-Kyung Kong, Zicheng Liu, and Karrie Karahalious have explored the impact of visualization titles, finding additional evidence of how viewers presume neutrality objectivity. See "Frames and Slants in Titles of Visualizations on Controversial Topics," in *CHI '18: Proceedings of the 2018 CHI Conference on Human Factors in Computing Systems*.

35 Catherine D'Ignazio and Lauren F. Klein, "Feminist Data Visualization," *Workshop on Visualization for Digital Humanities (VIS4DH) 2016*, Baltimore, MD; and Catherine D'Ignazio and Lauren Klein, "On Rational, Scientific, Objective Viewpoints from Mythical, Imaginary, Impossible Standpoints," in *Data Feminism* (MIT Press, 2020), 73–96.

36 Playfair was himself not only a proponent but also a victim of this business-first thinking. He remained drawn to the power wielded by the elite, even as he was never able to experience its benefits. In the end, the simplifying design of his charts further consolidated this power, rather than open it up to be dispersed.

37 See Cairo, *How Charts Lie*, xii–xiv.

38 Bellhouse consults two contemporaneous published versions of the custom house data, Charles Whitworth's *State of the Trade of Great Britain in Its Imports and Exports* (1776) and David Macpherson's *Annals of Commerce* (1805), to confirm this claim (59). Consulting these books, in turn, takes us back to the original custom house records, which have now been digitized and made available online via the British National Archives, viewable at https://discovery.nationalarchives.gov.uk/browse/r/h/C67.

39 See Bellhouse, *The Flawed Genius*, 89, 143.

40 Playfair, *The Commercial and Political Atlas*, 20. He concludes by noting that "much attention has been lately paid to the subject, and most well informed men are of opinion that a change might easily be effected that would be productive of general advantage."

41 This explanation is eleven pages long, as compared to the West Indies' one and a half.

42 Playfair, *The Commercial and Political Atlas*, 28.

43 Jorge Camoes, "Excel Charts meet William Playfair," December 6, 2011, https://excelcharts.com/excel-charts-meet-william-playfair/, accessed July 10, 2025.

44 Jo Wood, "Playfair's Wheat and Wages," March 10, 2024, https://observablehq.com/@jwolondon/wheat-and-wages, accessed July 10, 2025.

45 Michael Bostock, Vadim Ogievetsky, and Jeffrey Heer, "D3: Data-Driven Documents," *IEEE Transactions on Visualization and Computer Graphics* 17.12 (2011): 2301–2309.

46 Jacques Bertin, *Sémiologie graphique. Les diagrammes, les réseaux, les cartes* (Gauthier-Villars, 1967). The first English edition appeared in 1983, translated by William J. Berg with a foreword by Howard Wainer, who declares Bertin's book to be "the most important work on graphics since the publication of Playfair's *Atlas*" (iv).

47 Leland Wilkinson, *The Grammar of Graphics* (Springer, 1999). Notably, Wilkinson's text features Minard's diagram on its cover.

48 Hadley Wickham, *ggplot2: Elegant Graphics for Data Analysis* (Springer-Verlag New York, 2016), https://ggplot2.tidyverse.org/.

49 A similar line of thinking is taken up in Arvind Satyanarayan, Bongshin Lee, Donghao Ren, Jeffrey Heer, John Stasko, John Thompson, Matthew Brehmer, and Zhicheng Liu, "Critical Reflections on Visualization Authoring Systems," *IEEE Transactions on Visualization and Computer Graphics* 26.1 (2020): 461–471. This builds on work by Alex Bigelow, Steven Drucker, Danyel Fisher, and Miriah Meyer, including "Reflections on How Designers Design with Visualization," in *Proceedings of the 2014 International Working Conference on Advanced Visual Interfaces (AVI '14)* (Association for Computing Machinery, 2014), 17–24, and "Iterating between Tools to Create and Edit Visualizations," *IEEE Transactions on Visualization and Computer Graphics* 23.1 (2017): 481–490.

50 Cairo, *How Charts Lie*. In addition, work led by Crystal Lee has documented how malicious actors actively exploit the "aura of objectivity" of standard visualization forms in order to convince their viewers of the truth of what is, in fact, falsified data. See Crystal Lee, Tanya Yang, Gabrielle D. Inchoco, Graham M. Jones, and Arvind Satyanarayan, "Viral Visualizations: How Coronavirus Skeptics Use Orthodox Data Practices to Promote Unorthodox Science Online," in *Proceedings of the 2021 CHI Conference on Human Factors in Computing Systems (CHI '21)* (Association for Computing Machinery, 2021), article 607, 1–18.

51 William Playfair, *Joseph and Benjamin, a Conversation. Translated from a French Manuscript* (London, 1787), 232. It is worth noting that Franklin would have likely appreciated Playfair's contributions. Franklin had previously commended Joseph Priestley for his ingenuity in creating his 1765 "Chart of Biography," going so far as to nominate him for membership in the Royal Society, and Priestley reciprocated by dedicating his 1769 "New Chart of History" to Franklin. Franklin himself also had a long history of prioritizing efficiency in written and graphic communication, e.g., "Extract from Thomas Jefferson's Anecdotes of Benjamin Franklin," 4 December 1818, https://tjrs.monticello.org/letter/27, accessed July 10, 2025.

52 Keynes writes: "[Jevons] was not the first to plot economic statistics in diagrams; some of his diagrams bear, indeed, a close resemblance to Playfair's, with whose work he seems to have been acquainted." See J. M. Keynes, "William Stanley Jevons 1835–1882: A Centenary Allocution on His Life and Work as Economist and Statistician," *Journal of the Royal Statistical Society* 99.3 (1936): 523.

53 Quoted in Erica Royston, "Studies in the History of Probability and Statistics: III. A Note on the History of the Graphical Presentation of Data," *Biometrika* 43.3/4 (December 1956): 241 n1.

54 In their preface to the modern edition of Playfair's *Atlas*, Wainer and Spence describe it as the "Bible" of contemporary visualization design (vi). I have previously noted how Wainer, in his introduction to Jacques Bertin's *Sémiologie graphique*, traces a direct line from Playfair to Bertin, but not how Bertin himself selects an 1805 chart of Playfair's as one of only two historical images he reproduces in his foundational text. Most recently, Michael Friendly, writing with Wainer in *A History of Data Visualization and Graphic Communication*, says of Playfair: "He can rightly be called the father of modern graphical methods, and it is only a slight stretch to consider his contributions to be the Big Bang of data graphics" (95).

55 The chronology that Tufte recounts draws largely from that established by Funkhouser in his doctoral thesis, "Historical Development of the Graphical Representation of Statistical Data," published in *Osiris* 3 (1936): 269–404. Funkhouser and his advisor, Helen M. Walker, had previously coauthored an essay, "Playfair and His Charts," *Economic Journal* 45.1 (1935): 103–109, which made the case for Playfair as "the apparent inventor of statistical graphs."

56 "Chartjunk" is just what it sounds like—unnecessary visual embellishment that clutters the visual field. See Edward R. Tufte, *The Visual Display of Quantitative Information*, 2nd ed. (Graphics Press, 2001), 107–121. For an extended (and justified!) critique of chartjunk, see Derya Akbaba, Jack Wilburn, Main T. Nance, and Miriah Meyer, "Manifesto for Putting 'Chartjunk' in the Trash 2021!," alt.VIS 2021.

57 Here I am glossing Donna Haraway's claim that "it is precisely in the politics and epistemology of partial perspectives that the possibility of sustained, rational, objective inquiry rests." See Haraway, "Situated Knowledges: The Science Question in Feminism and the Privilege of Partial Perspective," *Feminist Studies* 14.3 (1988): 584.

3 **People:** Shanawdithit and Our Place in the Colonial Frame

1 Deep gratitude is owed to Lori Temple and the management team at The Rooms for their feedback and fact-checking on earlier drafts of this chapter.

2 "Beothuk Lake" is now the official name of this body of water, the result of a process led by the Government of Newfoundland and Labrador in keeping with Canadian and First Nations naming protocols. At the time of the events narrated in this chapter, and until 2021, its official name was Red Indian Lake. This name was bestowed by the British in reference to the Beothuk practice of applying a red ochre dye to their bodies and possessions. The Beothuk were also called the Red Indians at this time. In an attempt to honor the outcome of the renaming process, as well as to reflect the revised names of the historical maps discussed in this chapter, I use the name "Beothuk Lake," employing its previous name only when quoting directly from historical sources.

3 Ingeborg Marshall, *A History and Ethnography of the Beothuk* (McGill-Queen's UP, 1996), 261–262. Here I should note that, while I have attempted to prioritize Indigenous sources and scholars throughout this chapter, Marshall's history remains the most comprehensive source of information about this earlier era. For an account of the later part of this history that centers Beothuk perspectives, see two young adult novels by Mi'Sel Joe (Miawpukek Mi'kmaq) and Sheila O'Neill (Qalipu Mi'kmaq), *My Indian* (Breakaway, 2021) and its sequel, *Suliewey* (Breakaway, 2023), which bring together Mi'kmaq oral history with many of the sources referenced in this chapter to construct a set of character-driven narratives of these events. For a critique of Marshall, see Fiona Polack, "Introduction: De-Islanding the Beothuk," in *Tracing Ochre: Changing Perspectives on the Beothuk*, ed. Fiona Polack (U of Toronto Press, 2018), 12–14. And a final note on my marking of Indigenous nation/affiliation: per convention, I include such information when discussing Indigenous people (both historical figures and scholars) in the text, but do not insert affiliations into bibliographic citations. For reasons of legibility, I have chosen to leave settler scholars unmarked, but direct readers to Max Liboiron (Métis/Michif) in *Pollution Is Colonialism* (Duke UP, 2021), 3 n10 for an important discussion about how this choice does imply a normative identity from which others diverge. Notably, Liboiron currently lives and works in St. John's, where this chapter is set.

4 Marshall, *A History and Ethnography of the Beothuk*, 14–16.

5 Polack, "Introduction," 5.

6 Polack, "Introduction," 5.

7 Polack, "Introduction," 5.

8 John Peyton, as related to James P. Howley and recorded by Howley as "Capture of Mary March (Demasduit) on Red Indian Lake, in the month of March 1819," in *The Beothucks or Red Indians: The Aboriginal Inhabitants of Newfoundland* (Cambridge UP, 1915), 96.

9 Howley, *The Beothucks*, 93.

10 Frank G. Speck, "The Story of Buchan's Expedition (Related by John Paul, of Badger's Brook, Newfoundland, 68 years of age in 1914, who heard it from his grandfather)" in *Beothuk and Micmac* (Museum of the American Indian–Heye Foundation, 1922), 49.

11 Marshall believes that Shanawdithit was around 20 years old at the time of her capture. Because Demasduit's capture took place three years earlier, she would then have been about 17 (*A History and Ethnography of the Beothuk*, 208).

12 As an example of the latter, see Penelope Myrtle Kelsey, *Reading the Wampum: Essays on the Hodinöhsö:ni' Visual Code and Epistemological Recovery* (Syracuse UP, 2019).

13 This claim reflects the foundational work of Linda Tuhiwai Smith (Māori iwi Ngāti Awa and Ngāti Porou) in *Decolonizing Methodologies: Research and Indigenous Peoples*, 2nd ed. (Zed, 2012) and, more recently, Stephanie Carroll et al., "The CARE Principles for Indigenous *Data Governance*," *Data Science Journal* 19 (2020): 1–12, as well as the overarching Indigenous protocols of respect, responsibility, and reciprocity.

14 See "Land Acknowledgment Statement," *Emory University*, https://www.emory.edu/home/explore/history/land-acknowledgment/index.html, accessed July 25, 2025. Here I employ the traditional spelling of "Muscogee," consistent with the preference of Mvskoke scholars including Laura Harjo, cited later in this chapter.

15 E.g., Eve Tuck and K. Wayne Yang, "Decolonization Is Not a Metaphor," *Decolonization: Indigeneity, Education and Society* 1.1 (2012): 1–40.

16 Here we are again guided by the work of Max Liboiron, whose thoughtful discussion of the range of expressions of anticolonial research has helped focus this chapter, as our project overall, on the goals of recognizing and resisting "the settler and colonial entitlement to Land and Indigenous cultures, concepts, knowledges, and life" that have long characterized academic research. See Liboiron, *Pollution Is Colonialism*, 26–27 and 133.

17 Liboiron, citing Eve Tuck (Unangax), writes strongly against reproducing "damage-centered research ... that operates, even benevolently, from a theory of change that establishes harm or injury in order to achieve reparation." Tuck, "Suspending Damage: A Letter to Communities," *Harvard Educational Review* 79.3 (2009): 413, cited in Liboiron, *Pollution Is Colonialism*, 34.

18 According to Peyton, Demasduit at one point attempted to crawl across the snow under the cover of night, trailing her deer-skin cloak behind her so as to erase her tracks. See "Capture of Mary March (Demasduit) on Red Indian Lake, in the month of March 1819" in Howley, *The Beothucks*, 94.

19 William Nugent Glascock, "Report of Capt. Glascock," July 20, 1819, reproduced in Howley, *The Beothucks*, 113–116.

20 William Epps Cormack, "Information obtained from Shanawdithit" (n.d.), reproduced in Howley, *The Beothucks*, 229.

21 William Epps Cormack, "History of the Red Indians of Newfoundland" (1829), reproduced in Howley, *The Beothucks*, 224.

22 Marshall, *A History and Ethnography of the Beothuk*, 217.

23 The numbered order appears to have been decided by Howley, who explains in his book that he "shall take them according to their dates," and labels them from one to five. Acknowledging that this may not have been the order in which Shanawdithit created them, or intended them to be viewed, The Rooms has recently (as of 2024) created updated titles for the maps that no longer include numbers, nor the designation of "sketch."

24 John Paul, "The Story of Buchan's Expedition," in Speck, *Beothuk and Micmac*, 49.

25 For more on this ill-fated expedition, see John Paul, "The Story of Buchan's Expedition," 49–51.

26 "Turtle Island" is a name used by a number of Indigenous groups to describe the North American continent. It derives from the creation stories told by the Lenape, Haudenosaunee, and other nations rooted along the Northeast Coast, in which the back of a turtle becomes the ground on top of which the landmass of earth rises from the sea. Per the topic of this chapter, it is also worth noting that the continent of North America resembles a turtle, with the United States and central Canada as the body, Alaska and the Yukon as one top flipper and Quebec and Newfoundland as the other, and Mexico and the rest of Central America as one of the bottom flippers and Florida as the other, and New Orleans as its tail.

27 Neither Peyton's account nor Cormack's annotations suggest any direct link between these dwellings and the specific events depicted, but the annotations of some of Shanawdithit's other maps suggest that this might be the case.

28 "Capture of Mary March (Demasduit) on Red Indian Lake, in the month of March 1819," in Howley, *The Beothucks*, 92.

29 William Epps Cormack, "Information obtained from Shanawdithit" (n.d.), reproduced in Howley, *The Beothucks*, 228.

30 In his introduction to "Capture of Mary March (Demasduit)," Howley asserts that Peyton was "the actual captor of the Beothuck woman" (*The Beothucks*, 91).

31 William Epps Cormack, "Cormack to Bishop of Nova Scotia," January 10, 1829, reprinted in Howley, *The Beothucks*, 210.

32 While it exceeds the scope of this chapter, it is interesting to observe that another one of Shanawdithit's maps contains annotations so "copious," as Howley himself described them, that they threaten to overrun the map itself (*The Beothucks*, 243). These annotations are primarily numerical, and as such can be placed in the parallel history of Indigenous data and the violence it often brought about. See the historical overview offered in Maggie Walter and Chris Anderson, *Indigenous Statistics: A Quantitative Research Methodology* (Routledge, 2013).

33 See, for example, Nihad M. Farooq, *Undisciplined: Science, Ethnography, and Personhood in the Americas* (NYU Press, 2016).

34 Ben Shneiderman, "The Eyes Have It: A Task by Data Type Taxonomy for Information Visualizations," *Proceedings of the 1996 IEEE Symposium on Visual Languages*, Boulder, CO, 1996, 336–343.

35 See note 13 above. As an example of these principles in practice, see Margaret W. Pearce and Stephen J. Hornsby, "Making the *Coming Home* Map," *Cartographica* 55.3 (2020): 170–176, as well as Pearce's other work, including the project discussed later in this chapter.

36 On Indigenous and decolonial data visualization, see January O'Connor, Mark Parman, Nicole R. Bowman, and Stephanie Evergreen, "Decolonizing Data Visualization: A History and Future of Indigenous Data Visualization," *Journal of MultiDisciplinary Evaluation* 19.44 (2023): 62–79; Jeff Doctor, Jeff Ward, Trevor Jang, and Jen Polack, "Decolonizing Digital: Data's Role in Indigenous Data Sovereignty," https://animikii.com/insights/decolonizing-digital-data-s-role-in-indigenous-data-sovereignty, accessed July 24, 2025, and their book *#DataBack: Asserting and Supporting Indigenous Data Sovereignty* (Animikii, 2023); BIPOC Design History, "(re) Creating Turtle Island: Native American Design Remembered through History," https://bipocdesignhistory.com/products/re-creating-turtle-island-native-american-design-through-remembered-history, accessed July 24, 2025; and work envisioned by students at Leeward Community College and the University of Hawai'i Mānoa, under the guidance of Kamuela Enos, viz. "Students Showcase Projects Melding Indigenous, Contemporary Sciences," *University of Hawai'i News*, September 30, 2024, https://www.hawaii.edu/news/2024/09/30/student-projects-indigenous-contemporary-sciences/, accessed July 24, 2025.

37 Mishuana Goeman, *Mark My Words: Native Women Mapping Our Nations* (U of Minnesota Press, 2013), 4.

38 Goeman, *Mark My Words*, 4.

39 Goeman, *Mark My Words*, 4.

40 Thanks to Derya Akbaba for pushing me to extend this insight to the epistemology of visualization.

41 Interestingly, similar questions are beginning to be asked at the end phases of visualization projects, e.g., Derya Akbaba, Devin Lange, Michael Correll, Alexander Lex, and Miriah Meyer, "Troubling Collaboration: Matters of Care for Visualization Design Study," *Proceedings of the 2023 Conference on Human Factors in Computing (CHI '23)*, Hamburg, Germany, April 2023.

42 Margaret Wickens Pearce and Renee Pualani Louis, "Mapping Indigenous Depth of Place," *American Indian Culture and Research Journal* 32:3 (2008): 110.

43 Interestingly, this claim that mapping is best defined in terms of process is not limited to Indigenous contexts. See Matthew Edney, *Cartography: The Ideal and Its History* (U of Chicago Press, 2019), for a recent example of this view. Here we might also consider Laura Harjo's critique of how viewing the land and its features as "assets" reinforces a static and ownership-driven conceptualization of space. See Harjo, *Spiral to the Stars: Mvskoke Tools of Futurity* (U of Arizona Press, 2019).

44 "Mary March" was the name given to Demasduit by the British upon her capture. They chose the last name to mark the month of her capture. I have chosen not to reproduce this name in this text, with the exception of direct quotations such as this.

45 Pearce and Louis, "Mapping Indigenous Depth of Place," 110–111.

46 Harjo, *Spiral to the Stars*, 28; Edward R. Tufte, *The Visual Display of Quantitative Information*, 2nd ed. (Graphics Press, 2001), 40.

47 Matthew Sparke, "Between Demythologizing and Deconstructing the Map: Shawnawdithit's New-found-land and the Alienation of Canada," *Cartographica* 32.1 (Spring 1995): 12.

48 This observation paraphrases a point made by Sparke in another essay, "Mapped Bodies and Disembodied Maps: (Dis)placing Cartographic Struggle in Colonial Canada," in *Places through the Body*, ed. Heidi J. Nast and Steve Pile (Routledge, 1998), 310.

49 Howley, "Shanawdithit's drawings," in *The Beothucks*, 238.

50 Gerald Vizenor, *Manifest Manners: Narratives on Postindian Survivance* (U of Nebraska Press, 1994).

51 Cormack, quoted in Nicholas Chare, "In Her Hands: Affect, Encounter and Gesture of Wit(h)nessing in Shanawdithit's Drawings," *Parallax* 26.3 (2020): 294.

52 Chare, "In Her Hands," 294.

53 At this juncture we would do well to recall the arguments made throughout *The Imperial Map: Cartography and the Mastery of Empire*, ed. James Akerman (U of Chicago Press, 2009), each of which acknowledge the power that maps accord while also challenging binary framings of colonial vs. Indigenous (and East vs. West) maps. Thanks to Bill Rankin for alerting me to these valuable essays.

54 Lisa Brooks, *The Common Pot: The Recovery of Native Space in the Northeast* (U of Minnesota Press, 2008), 234.

55 For a detailed account of the map's genesis, see Mark Warhus, *Another America: Native American Maps and the History of Our Land* (St. Martin's Press, 1997), 1–3.

56 Cathy O'Neil, *Weapons of Math Destruction: How Big Data Increases Inequality and Threatens Democracy* (Crown, 2016).

57 Susan Schulten, *Mapping the Nation: History and Cartography in Nineteenth-Century America* (U of Chicago Press, 2012), 24.

58 Emma Willard, *History of the United States, or The Republic of America*, 5th ed. (New York, 1834), vii, cv.

59 As Paul Kahn has observed, this map and its framing was inspired by a map that appeared in Willard's earlier global historical atlas, *Willard's Ancient Geography, Ancient Atlas to Accompany Universal Geography* (1827), which was in turn inspired by Matthew Carey's *Complete Genealogical, Historical, Chronological, and Geographical Atlas; Being a General Guide to History, Both Ancient and Modern ... According to the Plan of Le Sage, Greatly Improved. The Whole Forming a Complete System of History and Geography* (1820). See "Inclusion and Exclusion in Emma Willard's Maps of History," *Nightingale: Journal of the Data Visualization Society,* March 30, 2023, https://nightingaledvs.com/inclusion-and-exclusion-in-emma-willard-maps-of-history/, accessed July 25, 2025.

60 Schulten, *Mapping the Nation*, 25.

61 Kahn calls out several places in subsequent maps where Indigenous nations do reappear: "Indigenous people reappear as insets labeled *Places Mentioned in the History of the Pequod War* in the *Third Map of 1643* and *Places Mentioned in the History of King Philip's War* in the *Fourth Map of 1692*. ... The *Fifth Map of 1755* shows the names of the indigenous people who controlled the interior of the southern colonies—Tuscarora, Cawataba, Congaree, Yamasee, Cherokee, and Apalachee—but the area west of the coastal rivers remains empty. It is only in the *Sixth Map of 1765* that [Willard] actually includes an indigenous nation with territory, the *Country of the Six Nations ... [and in]* the *Ninth Map of 1826, referred to in the History as the Map of the Present Day*, many of the previously excluded nations represented in the De L'Isle map of New France reappear. ... But the most interesting detail of the Map of 1826 is the depiction of the Far West in the upper left. The North West Territory is labeled, *Those lands possessed in common by the Sioux, Chippewas, Winnebago, and Sauks,* an assertion that would have been news to the people who lived there. The western boundary of this territory is the upper Mississippi. Beyond that is a space with no boundaries labeled *Sioux Indians or Naudowessies*."

62 On the "thematic map," its origins, and its limitations, see Bill Rankin, "Layers in the Service of the State," in *Radical Cartography* (Viking, 2025), 63–70.

63 Rankin, "Layers in the Service of the State," 61.

64 Warhus, *Another America*, 9–11.

65 I have chosen not to reproduce the text of this note. Interested viewers may consult the original image, viewable via the British Library at: https://www.imagesonline.bl.uk/asset/3043/.

66 The emphasis here is on the preservation of the map itself. Written accounts of such maps are plentiful and date back centuries. See Warhus, *Another America*, 9–11.

67 Sarah Sense, "Power Lines," https://sarahsense.com/Artists/11571/website_statements_2021-22.pdf, accessed July 25, 2025.

68 John Peyton, as related to James P. Howley and recorded by Howley in *The Beothucks*, 96.

69 This analysis could be pushed further still. In Peyton's account, his group's surveillance of the Beothuk in advance of their attack on the winter camp structures his entire narrative, even as it contains details that stretch back decades. To wit, the narrative is titled "Capture of Mary March (Demasduit) on Red Indian Lake, in the month of March 1819."

70 Though born in Brooklyn, Speck was sent to Connecticut as a child and raised by a family friend, a Mohegan woman named Fidelia Fielding, from whom he learned the Mohegan language; evidently, he spoke it fluently. This likely granted him increased access and trust to the Native peoples he studied. See National Museum of the American Indian, "Frank Speck," https://americanindian.si.edu/collections-search/edan-record/ead_component%3Asova-nmai-ac-001-ref15769, accessed July 25, 2025.

71 Speck, *Beothuk and Micmac*, 23.

72 Speck, *Beothuk and Micmac*, 23.

73 As Max Liboiron writes, and I confirmed during my short visit, "I can't talk about Land in Newfoundland and Labrador if I don't talk about the weather. Weather [and oil prices] are what shape life here." Liboiron, *Pollution Is Colonialism*, 30.

74 In her work with the Penobscot Nation to create a new map of their traditional place names, Margaret Pearce describes her process, as an outside collaborator, of coming to realize how her own experience and perspective—her own "vantage point," so to speak—was required in order to assemble the place names and the stories that she'd learned through the design process into a meaningful conceptual whole. Very crucially, she realized, this conceptual whole would and could not be the same for all viewers, as they each had different relationships to the land itself. See Pearce, "The Last Piece Is You," *Cartographic Journal* 51 (2014): 107–122.

75 Only after completing this visualization did we learn of the work of Pedro Cruz, who uses the visual metaphor of tree rings to anchor a series of visualizations on US immigration data. "Simulated Dendrochronology of US Immigration, 1790–2016," https://pmcruz.com/dendrochronology/, accessed July 25, 2025.

4 **Knowledge:** Elizabeth Palmer Peabody and the Value of Epistemology

1 An earlier version of this chapter appears as "What Data Visualization Reveals: Elizabeth Palmer Peabody and the Work of Knowledge Production," *Harvard Data Science Review* 4.2 (2022).

2 For more on Peabody's biography, see Bruce A. Ronda, *Elizabeth Palmer Peabody: A Reformer on Her Own Terms* (Harvard UP, 1999), Megan Marshall, *The Peabody Sisters: Three Women Who Ignited American Romanticism* (New York: Mariner, 2006), and Monika Elbert, Julie E. Hall, and Katharine Rodier, eds., *Reinventing the Peabody Sisters* (U of Iowa Press, 2006).

3 Elizabeth Palmer Peabody, *The Polish-American System of Chronology, Reproduced, with Some Modifications, from General Bem's Franco-Polish Method* (Boston, 1850), 3.

4 The term "feminism" began to coalesce in the second half of the nineteenth century—first in France, and then in the United States, as historian Karen Offen has explored. While some credit the French utopian philosopher (and major influence on the transcendentalist movement) Charles Fourier with coining the term, this has not been borne out by the research. See Nancy Cott, *The Grounding of Modern Feminism* (Yale UP, 1987), and Karen Offen, "Defining Feminism: A Comparative Historical Approach," *Signs* 14.1 (1988): 119–157.

5 On Antoni Jażwiński and the Polish system, see Hunter Dukes and Adam Green, "Visualizing History: The Polish System," *Public Domain Review*, May 5, 2021, and Daniel Rosenberg and Anthony Grafton, *Cartographies of Time: A History of the Timeline* (Princeton Architectural Press, 2010), 203–209. While it exceeds the scope of this chapter, it is worth noting the particular appeal for Peabody of the Polish origins of this method. Peabody had a particular interest in the ideas coming out of central and eastern Europe as a result of her ongoing engagement in theories of language. She had sustained correspondence with the Hungarian philologist Charles Kraitsir throughout the 1840s, and he traveled to Boston to meet with her in 1847. Peabody went on to support the (failed) Hungarian revolution of 1848, which she viewed as allied with the aims of American independence. Among her central concerns with standard accounts of US history was that they were written from the perspective of England. She would have likely seen an anticolonial stance as inherent in

Bem's system. For more on her relationship with Kraitsir, see Ronda, *Elizabeth Palmer Peabody*, 225–226. For more on her support for the Hungarian revolution and her views on history, see Nina Baym, "The Ann Sisters: Elizabeth Peabody's Millennial Historicism," *American Literary History* 3.1 (1991): 25–45.

6 Rosenberg and Grafton, *Cartographies of Time*, 205.

7 Ronda, *Elizabeth Palmer Peabody*, 227.

8 Elizabeth Palmer Peabody, *Chronological History of the United States: Arranged with Plates on Bem's Principle* (New York, 1856), 9.

9 Peabody, *Chronological History*, 9.

10 Peabody, *The Polish-American System*, 10.

11 Peabody, *Chronological History*, 10.

12 Scholars now date the death of Pocahontas to five years earlier, in 1617. She had been abroad in England with her husband (the British settler John Rolphe) and their child, where she served as the main draw in a campaign to attract new investment in the Jamestown colony. She had just begun the long ship voyage home, not yet out of the Thames, when she became sick and died shortly after being taken off the boat. This is another story that exceeds the scope of this chapter, but it is important to acknowledge the power relations that framed Pocahontas's marriage to Rolphe, as well as how her story has been shaped over centuries to fit various colonial and national origin stories. See Rayna Green, "The Pocahontas Perplex: The Image of Indian Women in American Culture," *Massachusetts Review* 16.4 (Autumn 1975): 698–714, and more recently Chris Finley, "Decolonizing the Queer Native Body (and Recovering the Native Bull-Dyke)," in *Queer Indigenous Studies: Critical Interventions in Theory, Politics, and Literature*, ed. Qwo-Li Driskill, Chris Finley, Brian Joseph Gilley, and Scott Lauria Morgensen (U of Arizona Press, 2011), 31–42.

13 Peabody, *Chronological History*, 5.

14 For more on the role of mnemonics in nineteenth-century US history education, see "Mnemonics, Methods, and Memory: Organizing and Encoding the American Story," in Barry Joyce, *The First US History Textbooks: Constructing and Disseminating the American Tale in the Nineteenth Century* (Lexington Books, 2015), 61–100.

15 Peabody, *Chronological History*, 7.

16 Nikole Hannah-Jones et al., "The 1619 Project," *New York Times Magazine,* August 14, 2019, https://www.nytimes.com/interactive/2019/08/14/magazine/1619-america-slavery.html.

17 Jake Silverstein, "Why We Published the 1619 Project," *New York Times Magazine*, December 20, 2019, https://www.nytimes.com/interactive/2019/12/20/magazine/1619-intro.html.

18 Katherine McKittrick, *Dear Science and Other Stories* (Duke UP, 2021), 3.

19 Among McKittrick's points in *Dear Science* is that her theory of relational knowledge stands apart from Haraway's: "A black sense of place is not a standpoint or a situated knowledge; it is a location of difficult encounter and relationality," she explains, choosing not to include Haraway in her own genealogy of thought (106). She further observes how Haraway developed her theory of situated knowledges through her analysis of works by the Nigerian novelist Buchi Emecheta, whom Haraway should have more fully recognized as her source of inspiration (130).

20 With her students, McKittrick has explored what this liberatory project might look like in online form. Inspired by the work of the philosopher Sylvia Wynter, McKittrick and her students seek to create "small-scale aesthetic projects" in visual, audio, and multimedia form that exist "in tandem with, yet imagine futures outside, colonial logics." See *The Revolutionary Demand for Happiness* at https://revolutionarydemandforhappiness.com/, accessed July 30, 2025.

21 Elizabeth Palmer Peabody, *Universal History; Arranged to Illustrate Bem's Charts of Chronology* (New York, 1859), v.

22 In Peabody's own words, the Polish-American system "does not pretend to be what an outline can never be, namely: a perfect frame work for history." See Elizabeth Palmer Peabody "Bem's Method of History: An Appeal to Teachers and Boards of Education," *Indiana School Journal* 1.2 (1856): 325.

23 As evident in the pedagogy surrounding her charts, Peabody believed that self-directed play mattered as much as formal instruction, and that knowledge was derived not from any external authority but instead from a focused and sustained analysis of "the self-activity of the mind." See Elizabeth Palmer Peabody, "The Origin and Growth of the Kindergarten," *Education* 2 (1882): 507–527. Here it is also worth noting that, while Peabody was the sister-in-law of famed educational innovator and advocate Horace Mann, she did not maintain a close relationship with him after he married her sister Mary. In fact, there was significant tension between Elizabeth and Mary over the intensity of Elizabeth's intellectual relationship with Mann. Only after Mann's death in 1859 did the sisters' shared interest in theories (and practices) of education become rekindled. For more on this new phase of Peabody's educational theories, developed in conversation with her sister, see Ronda, *Elizabeth Palmer Peabody*, 269–317.

24 Ronda, *Elizabeth Palmer Peabody*, 227.

25 Peabody, *Chronological History*, 7.

26 Caldas, a Colombian naturalist who roomed with Humboldt in Quito for eight months while the latter created his famed visualizations of Chimborazo, was in fact the person who first created the forms for which Humboldt would soon be renowned. See Jorge Cañizares-Esguerra, "Screwing Humboldt and His Hagiographers," *Medium,* September 27, 2019, https://jorgecanizaresesguerra.medium.com/screwing-in-two-positions-82c2cc5b09db, and in a more scholarly register, Mauricio Nieto Olarte, *La obra cartográfica de Francisco José de Caldas* (Universidad de los Andes, 2006).

27 For an extended analysis of Willard's charts, see Susan Schulten, "Emma Willard's Maps of Time," *Public Domain Review* (2020), https://publicdomainreview.org/essay/emma-willard-maps-of-time/.

28 Peabody, "Bem's Method of History," 326.

29 Susan Schulten, *Mapping the Nation: History and Cartography in Nineteenth-Century America* (U of Chicago Press, 2012), 31.

30 On positivism as the dominant epistemology of visualization, and feminism as an alternative, see Derya Akbaba, Lauren Klein, and Miriah Meyer, "Entanglements for Visualization: Changing Research Outcomes through Feminist Theory," *IEEE Transactions on Visualization and Computer Graphics* (VIS) 31.1 (2024): 1279–1289.

31 As an example of how domain experts might be enabled to contribute their own knowledge to a visualization system, see Haihan Lin, Derya Akbaba, Miriah Meyer, and Alexander Lex, "Data Hunches: Incorporating Personal Knowledge into Visualizations," IEEE Transactions on Visualization and Computer Graphics (VIS) 29.1 (2022): 504–514.

32 Peabody, *Chronological History*, 9.

33 Elizabeth Palmer Peabody, "My Experience as a Teacher," *American Journal of Education* 32 (1882): 785.

34 Ronda, *Elizabeth Palmer Peabody*, 236.

35 To view a portfolio of Kelly Dobson's data visceralization work, see https://web.media.mit.edu/~monster/. For an extended discussion of data visceralization in relationship to data visualization, see Catherine D'Ignazio and Lauren F. Klein, *Data Feminism* (MIT Press, 2020), 73–96.

36 Jessica Hullman, "The Purpose of Visualization Is Insight, Not Pictures: An Interview with Ben Shneiderman," *ACM Interactions*, August 19, 2019, https://interactions.acm.org/blog/view/the-purpose-of-visualization-is-insight-not-pictures-an-interview-with-ben.

37 Thanks to RJ Andrews for sending the email that alerted me to this fact. The mural chart, "Christian Era," associated with her *Universal History* (1850), can now be viewed online via the David Rumsey Historical Map Collection, image number 15895000.jp2.

38 In his biography of Peabody, Ronda speculated that they "must have been much larger than even folio size" (*Elizabeth Palmer Peabody*, 234). The chart in the David Rumsey collection measures 176 by 148 centimeters.

39 Ronda, *Elizabeth Palmer Peabody*, 227.

40 Thanks to Granville Ganter for sharing these photos and allowing me to reproduce them here. For Ganter's work on Clarke, who created her timelines with the help of her

sister, also named Elizabeth, see Granville Ganter, "Mistress of Her Art: Anne Laura Clarke, Traveling Lecturer of the 1820s," *New England Quarterly* 87.4 (2014): 709–746.

41 Papers discussing this process include: Caroline Foster et al., "The Shape of History: Reimagining Nineteenth-Century Data Visualization," *Digital Humanities 2017* (Association of Digital Humanities Organizations, 2017); and Alexandra Beall et al., "Reimagining Elizabeth Palmer Peabody's Lost 'Mural Charts,'" *Digital Humanities 2018* (Association of Digital Humanities Organizations, 2018).

42 The touch matrix was prototyped by Angela Vujic and Maninder Japra. The design was refined and further implemented by Courtney Allen, Alexandra Beall, and Gwendolyn McGuire.

43 The fabric topper was designed and fabricated by Sarah Schoemann. The LED system was prototyped by Svyatoslav Kucheryavykh. The design was further refined by Courtney Allen and Alexandra Beall. The final system was designed and implemented by Courtney Allen, Gwendolyn McGuire, and Sarah Schoemann. Noah Sutter wrote the underlying code.

44 Neither focus nor fatigue are limited to the physical realm, of course. Chapter 6 addresses these considerations with respect to the making of the visualizations and other digital components of this project.

45 Elizabeth Peabody to Samuel Gray Ward, September 1850, quoted in Ronda, *Elizabeth Palmer Peabody*, 235.

46 The process of acquiring, restoring, and preserving historical artifacts is of course far from magic to those who participate in it. Alex Clausen, of Ruderman Antique Maps, reported in an email how the chart arrived in poor condition, and required significant restoration before it was resold to David Rumsey.

47 The initial listing that struck the eye of Barry Ruderman and Alex Clausen of Ruderman Antique Maps can be found here: https://www.invaluable.com/auction-lot/antique-wall-maps-charts-lot-of-four-2432-c-cc645e1b12?objectID=164057415. Thanks to Barry Ruderman and Alex Clausen for sharing this information with me.

48 For an overview of this gendered history of work, see "Show Your Work," in D'Ignazio and Klein, *Data Feminism*, 173–202.

49 While not explored in this chapter, the history of computing provides another key example, one with which—in the object-oriented way in which Peabody structures her event data—Peabody might also be engaged. On the history of women in computing, see Jennifer Light, "When Computers Were Women," *Technology and Culture* 40.3 (1999): 455–483; Nathan Ensmenger, *The Computer Boys Take Over: Computers, Programmers, and the Politics of Technical Expertise* (MIT Press, 2010); and Mar Hicks, *Programmed Inequality: How Britain Discarded Women Technologists and Lost Its Edge in Computing* (MIT Press, 2017).

50 E.g., William Ferris, ed., *Afro-American Folk Art and Crafts* (G. K. Hall, 1983); and S. J. Markowitz, "The Distinction between Art and Craft," *Journal of Aesthetic Education* 28.1 (Spring 1994): 55–70.

51 E.g., R. Kalina, "Gee's Bend Modern," *Art in America* 10 (2003): 104–149; and S. A. Duncan, "From Cloth to Canvas: Reinventing Gee's Bend Quilts in the Name of Art," *Museum Anthropology* 28.1 (March 2005): 27–28.

52 Ryan Leahy, "The Alabama Women Who Made Their Quilts a Part of Modern Art," *Artsy* (2018), https://www.artsy.net/article/artsy-editorial-alabama-women-made-quilts-modern-art

53 Elsa Barkley Brown, "African-American Women's Quilting: A Framework for Conceptualizing and Teaching African-American Women's History," *Signs* 14.4 (1989): 926.

54 Barkley Brown, "African-American Women's Quilting," 927.

55 Barkley Brown, "African-American Women's Quilting," 928.

56 Edward Tufte, *Seeing with Fresh Eyes: Meaning, Space, Data, Truth* (Graphics Press, 2020), 29.

57 Barkley Brown, "African-American Women's Quilting," 926.

58 Dan Jutan, the project team member responsible for converting Peabody's event data into tabular form, observed that Native American groups are at times represented by other colors in later charts, as the conflicts between them could not be captured by the English vs. Americans distinction.

5 **Change:** Lula Iola Mack, W. E. B. Du Bois, and Visualization for Liberation

1 "Table of Temperature and Rainfall, Week Ended December 25, 1899," *Public Health Reports (1896–1970)*, January 12, 1900, 73–74.

2 David Levering Lewis, *W. E. B. Du Bois: A Biography* (Henry Holt, 2009), 154. The first day of the 1899–1900 winter term is listed as December 25, 1899 in the *Catalogue of the Officers and Students of Atlanta University, 1898–1899* (Atlanta UP, 1899), 2.

3 Per the 1898–1899 *Catalogue*, Mack was awarded "high honor" after her sophomore year in recognition of "maintain[ing] a certain average rank" (24). Mack would be awarded "honor" along with a classmate, Henry Napoleon Lee, at her own graduation. See *Catalogue of the Officers and Students of Atlanta University, 1900–1901* (Atlanta UP, 1901), 25.

4 As listed in the 1899–1900 *Catalogue,* the textbook employed for the course was Richmond Mayo-Smith's *Statistics and Sociology* (1895), which when published became the first textbook on quantitative methods of social science research. See *Catalogue of the Officers and Students of Atlanta University, 1899–1900* (Atlanta UP, 1900), 7.

5 *Catalogue of the Officers and Students of Atlanta University, 1899–1900* , 14.

6 On Calloway's role in championing the exhibit, see *A Small Nation of People: W. E. B. Du Bois and African American Portraits of Progress*, ed. David Levering Lewis and Deborah Willis (Harper Collins, 2005), 16–17.

7 Eugene F. Provenzo Jr., *W. E. B. Du Bois' Exhibit of American Negroes: African Americans at the Beginnings of the Twentieth Century* (Rowman & Littlefield, 2013), xii.

8 Whitney Battle-Baptiste and Britt Rusert, *W. E. B. Du Bois's Data Portraits: Visualizing Black America* (Princeton Architectural Press, 2018), 9.

9 W. E. B. Du Bois, *A Soliloquy on Viewing My Life from the Last Decade of Its First Century* (Oxford, 2014 [1968]), 220. As mentioned in the introduction, visualization designer Jason Forrest believes that there is evidence of an additional chart, bringing the total to 64. See Forrest, "Discovering an Unknown Chart from W. E. B. Du Bois's 'The Exhibition of American Negroes," *Nightingale: The Journal of the Data Visualization Society*, January 1, 2019, https://medium.com/nightingale/discovering-a-new-chart-from-w-e-b-du-boiss-the-exhibition-of-american-negros-part-6-66e7b340b90c, accessed August 18, 2025.

10 The numbers derive from "Paris 1900: City of Entertainment," *Portland Art Museum* (2019), https://portlandartmuseum.org/event/paris-1900-city-of-entertainment/, accessed August 18, 2025. It is also important to note that the exhibit was not installed in the main US exhibition space, its "national building." Rather, in the Palace of Social Economy and Congresses, it "shared a small space with maps detailing US resources, models of New York City tenements, and information on libraries, railroad pensions, and labor unions." Lewis and Willis, *A Small Nation of People*, 17.

11 Battle-Baptiste and Rusert, *W. E. B. Du Bois's Data Portraits*. The images published in the book were based on a 2014 effort by the Library of Congress to make the charts available online. See Jeff Bridgers, "Du Bois's American Negro Exhibit for the 1900 Paris Exposition," *Library of Congress Blog*, February 28, 2014, https://blogs.loc.gov/picturethis/2014/02/du-boiss-american-negro-exhibit-for-the-1900-paris-exposition/?loclr=blogpic. The full collection is available under the heading "Charts and Graphs Showing the Condition of African Americans at the Turn of the Century Exhibited at the Paris Exposition Universelle in 1900," *Library of Congress* (2014), https://www.loc.gov/pictures/item/2005679642/, accessed August 18, 2025.

12 W. E. B. Du Bois, "The American Negro at Paris," *American Monthly Review of Reviews* 12.5 (November 1900): 576.

13 E.g., W. E. B. Du Bois, *The Souls of Black Folk* (Modern Library, 2003 [1903]).

14 Rogers is credited by name in the *Atlanta University Bulletin* 106 (January 1900): 1; and "Atlanta University Exhibit at Paris," *Atlanta Journal*, February 22, 1900. The *Catalogue of the Officers and Students of Atlanta University, 1899–1900* confirms this role, listing Rogers's occupation as "Work on Paris Exhibit" (31). With that said, as graphic designer (and *Data by Design* team member) Silas Munro explains, "based on the volume of the designs, each piece's complexity and detail,

and the compressed project timeline, it seems implausible that Rogers and Du Bois worked alone to complete the project." Silas Munro, "Introduction to the Plates," in Battle-Baptiste and Rusert, *W. E. B. Du Bois's Data Portraits*, 50.

15 Tré Seals, "Du Bois," *Vocal Type*, https://www.vocaltype.co/history-of/du-bois.

16 Another likely source was federal geographer Henry Gannett's 1895 study *Statistics of the Negroes in the United States* (Baltimore, 1894). Gannett also served as chief geographer of the US Censuses of 1880, 1890, and 1900, as discussed in more detail below.

17 In his third and final autobiography, written late in life, Du Bois recalls only, "I got a couple of my best students and put a series of facts into charts," resulting in the "most interesting set of drawings" that were displayed in Paris. Du Bois, *A Soliloquy*, 18.

18 For a rich theorization of archival silence, see Michel-Rolph Trouillot, *Silencing the Past: Power and the Production of History* (Beacon, 1997).

19 Those interested in how I theorize my own approach to fragmentary archives may wish to read the introduction to my first book, "No Eating in the Archive," in *An Archive of Taste: Eating and Aesthetics in the Early United States* (U of Minnesota Press, 2020), 1–33.

20 *Catalogue of the Officers and Students of Atlanta University, 1899–1900*, 27.

21 *Catalogue of the Officers and Students of Atlanta University, 1899–1900*, 7.

22 In all other years, this work took the form of the Atlanta University Studies, discussed in depth later in this chapter. This public-facing work was to enable "graduate study of the social problems in the South by the most approved scientific methods of Atlanta, Fisk, and other institutions," as stated in the course description in the *Catalogue of the Officers and Students of Atlanta University, 1899–1900*, 14.

23 Reportage in the January 1900 *Atlanta University Bulletin* documents how Du Bois "[gave] up the routine of school work, except the class in Sociology, for three months" to prepare for the exhibit, further suggesting that the sociology course (and the students in it) were central to the work. *Atlanta University Bulletin* 106 (January 1900), 2.

24 Mabel O. Wilson, "The Cartography of W. E. B. Du Bois's Color Line," in Battle-Baptiste and Rusert, *Du Bois's Data Portraits*, 39.

25 For an overview of the history of lynching in the United States, see the Equal Justice Initiative's comprehensive report, *Lynching in America: Confronting the Legacy of Racial Terror*, 3rd ed., https://lynchinginamerica.eji.org/report/, accessed August 18, 2025.

26 On counterdata, see Seyi Olojo, "Counterdata," in *Keywords for a Datafied State* (Data & Society, 2024), 170–181.

27 Overviews of this era in the history of statistics, the culmination of its initial rise, include Theodore M. Porter, *The Rise of Statistical Thinking, 1820–1900* (Princeton UP, 1986), Ian Hacking, *The Taming of Chance* (Cambridge UP, 2014), and more recently Chris Wiggins and Matthew L. Jones, *How Data Happened: A History from the Age of Reason to the Age of Algorithms* (Norton, 2023). For books focusing on the history of eugenics and its ties to statistics, see Banu Subramaniam, *Ghost Stories for Darwin: The Science of Variation and the Politics of Diversity* (U of Illinois Press, 2014), and more recently Anita Chan, *Predatory Data: Eugenics in Big Tech and Our Fight for an Independent Future* (U of California Press, 2024). Note that Du Bois was himself not immune to ties to so-called "positive eugenics." (This was the goal of bettering the human population by encouraging the intentional reproduction of those deemed "fit," as opposed to "negative eugenics," which discouraged the reproduction of those deemed "unfit.") Du Bois's colleague Jane Addams, one of the co-founders of Chicago's Hull House (whose own social surveys had served as a model for his own), was a leading proponent of this view, as was Margaret Sanger—the founder of Planned Parenthood—and many other reformers of the early twentieth century.

28 Autumn Womack, *The Matter of Black Living: The Aesthetic Experiment of Racial Data, 1880–1930* (U of Chicago Press, 2022), 21. It should be noted that Wells and Du Bois did not enjoy an easy relationship. Du Bois notoriously excluded Wells from his list of founders of the NAACP, and while he claimed that Wells had chosen not to be named, she contended that she had been deliberately excluded. If this were the case, it would have been in keeping with a trend of Du Bois turning away from Black women activists as he sought the collaboration of their white counterparts. On Wells and the NAACP, see Wells, *Crusade for Justice: The Autobiography of Ida B. Wells*, ed. Alfreda Duster (U of Chicago Press, 1970), 322. On Du Bois and Black women, see Farah Jasmine Griffin, "Black Feminists and Du Bois: Respectability, Protection, and Beyond," *Annals of the American Academy of Political and Social Science* 568.1 (2000): 28–40, and Britt Rusert, "W. E. B. Du Bois and the Origins of Visual Sociology: The 1900 Paris Exposition and Beyond," in *Oxford Handbook of W. E. B. Du Bois*, ed. Aldon D. Morris et al., 371–390. Thanks to Britt Rusert for encouraging me to elaborate this line of inquiry.

29 Womack, *The Matter of Black Living*, 21.

30 W. E. B. Du Bois, "The American Negro in Paris," *The American Monthly Review of Reviews* 22.5 (November 1900): 577.

31 Thanks to Rahul Bhargava for this observation, which eluded this Georgia resident's eye!

32 Lewis, *W. E. B. Du Bois: A Biography*, 139–140.

33 Gannett, *Statistics of the Negroes in the United States*.

34 Gannett, *Statistics of the Negroes in the United States*, plate VII.

35 For more on this particular comparison, see Ben Schmidt, "Canonization," in *Creating Data: The Invention of Information in the American State, 1850–1950*, https://creatingdata.benschmidt.org/Dubois_and_Gannett, accessed August 18, 2025.

36 David Walker was an early nineteenth-century writer and activist who famously advocated that Black Americans purchase a copy of Thomas Jefferson's *Notes on the State of Virginia,* which contains some of Jefferson's most racist views, so as to better be able to refute it. "Unless we try to refute Mr. Jefferson's arguments respecting us, we will only establish them," Walker wrote in 1829. See David Walker, *Appeal to the Colored Citizens of the World* (Boston, 1829). For commentary on Walker, see Gene Andrew Jarrett, "'To Refute Mr. Jefferson's Arguments Respecting Us': Thomas Jefferson, David Walker, and the Politics of Early African American Literature," *Early American Literature* 46.2 (2011): 291–318 and, in a graphic design context, Marcy J. Dinius, "Look!! Look!!! At This!!!!': The Radical Typography of David Walker's *Appeal*," *PMLA* 126.1 (January 2011): 55–72.

37 Du Bois, "The American Negro at Paris," 576.

38 The frame is noted in the *Atlanta University Bulletin* article reporting on the exhibit: "One of the most remarkable bits of work there, is a carved wooden frame for one of the charts made entirely by an unskilled, illiterate old Freedman, living in Atlanta." See "The Georgia Negro Exhibit for the Paris Exposition," *Atlanta University Bulletin* 110 (May 1900): 2.

39 On the "Black Codes," see Battle-Baptiste and Rusert, *W. E. B. Du Bois's Data Portraits*, 21–22, and Provenzo, *W. E. B. Du Bois' Exhibit*, 7.

40 The photo is credited to Thomas Askew, a prominent Black photographer who lived in Atlanta and who took many of the Atlanta-based photographs included in the exhibit. See Deborah Willis, "The Sociologist's Eye: W. E. B. Du Bois and the Paris Exposition," in Lewis and Willis, *A Small Nation of People,* 51–78.

41 The women's dorm is described in the *Catalogue of the Officers and Students of Atlanta University, 1899–1900*, 22. Its 45 residents, including Mack, then age 20, are documented in the 1900 US Federal Census, viewable at https://www.ancestry.com/search/collections/7602/records/8895499, accessed August 22, 2025.

42 Du Bois, "The American Negro in Paris," 577. It's also worth noting that photographs were increasingly featured in atlases of the time, as Ben Schmidt observed during this book's community review. Schmidt points to the *Atlas of Asiatic Russia* (1914) that is "filled with photographs—many ethnographic—in the first half," as a key example of this practice.

43 Du Bois, *The Souls of Black Folk*, 8; quoted in Shawn Michelle Smith, *Photography on the Color Line: W. E. B. Du Bois, Race, and Visual Culture* (Durham: Duke UP, 2004), 25. Lewis and Willis, *A Small Nation of People*, referenced above, also explores the exhibit's photographs in detail.

44 Historian of photography Deborah Willis identifies one of these photographers as the prominent Black portraitist Thomas Askew, who shot the photo of the Atlanta University women. The other photographers remain unknown. See "The Sociologist's Eye:

W. E. B. Du Bois and the Paris Exposition," in Lewis and Willis, *A Small Nation of People,* 51–78.

45 In their discussion of the charts, Battle-Baptiste and Rusert elaborate this line of thinking, emphasizing how Du Bois understood double consciousness as a "kind of 'second sight' that might be transformed from a curse into a 'gift' that offered a unique and superior perspective on turn-of-the-century race relations, sociability, and even existence itself." Battle-Baptiste and Rusert, *W. E. B. Du Bois's Data Portraits*, 15.

46 W. E. B. Du Bois, *Dusk of Dawn: An Essay toward an Autobiography of a Race Concept*, ed. Irene Diggs (Transaction, 1984 [1940]), 51.

47 Du Bois, *Dusk of Dawn*, 67. This powerful line and its variants, what is safe to call a motif of Du Bois's oeuvre, have been analyzed by generations of scholars. For a summary of this work, see Benjamin Murphy, "Multiplied without Number": Lynching, Statistics, and Visualization in Ida B. Wells, Mark Twain, and W. E. B. Du Bois," *American Literature* 93.2 (2021): 205–207.

48 Because of the biased nature of the legal record, it is difficult to determine what exactly took place. My account, which is intentionally scant for this reason, derives from Darren E. Grem, "Sam Jones, Sam Hose, and the Theology of Racial Violence," *Georgia Historical Quarterly* 90.1 (2006): 35–61.

49 Du Bois, *Dusk of Dawn*, 67.

50 My interpretation of this incident has been strongly influenced by Bryan Wagner's analysis in *Disturbing the Peace: Black Culture and the Police Power after Slavery* (Harvard UP, 2009), 180–183. Britt Rusert discusses this incident in relation to the charts in "W. E. B. Du Bois and the Origins of Visual Sociology."

51 Du Bois, *Dusk of Dawn*, 67–68.

52 This both echoes Du Bois's earlier assertion, as documented by Autumn Womack, that "there is only one sure basis of social reform and that is Truth—a careful, detailed knowledge of the essential facts of each problem," and signals his departure from it. Quoted in Womack, *The Matter of Black Living*, 41.

53 This was the legislation that, as Whitney Battle-Baptiste and Britt Rusert explain, "stretch[ed] from the slave codes of the colonial and antebellum period to the segregationist policies and laws of the present," and which attempted to control and constrain all aspects of Black life (*W. E. B. Du Bois's Data Portraits*, 19). Du Bois's transcription of the Black Codes filled three large manuscript volumes. For a sustained meditation on the legacy of the Black Codes and their relation to contemporary digital life, see the special issue of *The Black Scholar* on "Black Code," edited by Jessica Marie Johnson and Mark Anthony Neal, and in particular the introduction, "Wild Seed in the Machine," *Black Scholar* 47.3 (2017): 1–2.

54 The *Atlanta University Catalogues* between the years of 1901 and 1914, when they stopped publishing individual alumni information, list locations in Georgia, Alabama, and Tennessee.

55 Marggette Wilkins is listed as Mack's daughter in the 1920 US Census. (She herself was recorded as Lula Brown, having remarried at that point.) Marggette's name was listed as Marguerite in the 1930 US Census. See "Lula Brown in the 1920 United States Federal Census" and "Lula I Brown in the 1930 United States Federal Census," *Ancestry.com*, accessed August 19, 2025.

56 *Catalogue of the Officers and Students of Atlanta University, 1909–1910* (Atlanta UP, 1909), 36.

57 Westmoreland's occupation remains consistent until at least 1914, the last year in which individual alumni information was reported in the annual catalogue.

58 Simon may have been using skills he learned at Atlanta University, as the Atlanta University Studies were published in-house. On the history of Black printing in relation to Du Bois, see Elizabeth McHenry, *To Make Negro Literature: Writing, Literary Practice, and African American Authorship* (Duke UP, 2021).

59 I was unable to find a "Henry Napoleon Lee" in any of federal censuses for which he was alive. There were several listings for "Henry Lee," including some born in Atlanta in the late 1800s, aligning with a 1900 college graduation date. But I could not find enough conclusive evidence that any single listing was Henry Napoleon Lee.

60 The exact date, August 27, 1963, was famously the same that hundreds of thousands of Black Americans and their allies would begin assembling on the Washington Mall to hear Martin Luther King Jr. deliver his now-canonical "I Have a Dream" speech.

61 In the summer of 1910, Du Bois resigned from his position at Atlanta University to take up the role as founding editor of *The Crisis*, the monthly journal of the NAACP, which required that he move to New York. He nonetheless continued to edit the Atlanta University Studies until 1914. He would return to Atlanta University in 1934.

62 Lewis, Morris, Rusert, and Womack all make versions of this claim.

63 W. E. B. Du Bois, "Sociology Hesitant," *boundary 2* 27.3 (Fall 2000): 42. In this posthumous essay, believed to be authored around 1905, Du Bois also advances his support for interdisciplinary scholarship, writing that "mutual understanding must come under a working hypothesis which will give scope to Historian as well as Biologist," and that to explain statistical findings, "we must assume Law and Chance working in conjunction—Chance being the scientific side of inexplicable Will. Sociology, then, is the Science that seeks the limits of Chance in human conduct" (44).

64 This is perhaps best exemplified by the Du Bois Challenge, an annual event sponsored by the Data Visualization Society aimed at "celebrating [Du Bois's] data visualization legacy" by recreating his charts with contemporary visualization libraries and tools. For more on this event, see Anthony Starks, "The #DuBois Challenge," *Nightingale: Journal of the Data Visualization Society*, February 1, 2022, https://nightingaledvs.com/the-dubois-challenge/, accessed August 19, 2025.

65 Mona Chalabi, "WEB Du Bois: Retracing His Attempt to Challenge Racism with Data," *Guardian*, February 14, 2017, https://www.theguardian.com/world/2017/feb/14/web-du-bois-racism-data-paris-african-americans-jobs, accessed August 19, 2025.

66 Frederick F. Wherry et al., "Comparing and Calculating Your Student Loan Debt," https://www.dignityanddebt.org/projects/student-debt-calculator/, accessed August 19, 2025.

67 The project is also notable for how it acknowledges the importance of change through policy: "Visualizations can render data to make it meaningful for people around the world. Furthermore, they are able to reveal what might be either changed or strengthened through policy and individual action." Wherry et al., "Comparing and Calculating Your Student Loan Debt."

68 In *Data Feminism*, Catherine D'Ignazio and I share an essay by Candice Lanius, "Fact Check: Your Demand for Statistical Proof Is Racist," in which she summarizes the ample research on how those in positions of power accept anecdotal evidence from those like themselves, but demand endless statistics from minoritized groups. In those cases, she argues convincingly, more data will never be enough" (58).

69 Mimi Ọnọụha, "In Absentia" (2019), https://mimionuoha.com/in-absentia, accessed August 19, 2025.

70 In relaying this anecdote, Battle-Baptiste and Rusert also observe its "close proximity to the hold of the ship" (*W. E. B. Du Bois's Data Portraits*, 17).

71 *Catalogue of the Officers and Students of Atlanta University, 1898–1899*, 43.

72 This choice seems particularly resonant in light of Du Bois's flawed gender politics, which resulted in his largely viewing Black women as best suited for domestic rather than intellectual roles. See Griffin, "Black Feminists and Du Bois," and Rusert, "W. E. B. Du Bois and the Origins of Visual Sociology."

73 Saidiya Hartman, *Wayward Lives, Beautiful Experiments: Intimate Histories of Social Upheaval* (Norton, 2019), xiii. Later in the book, Hartman in fact speculates about another woman photographed for the Paris Exposition, as she does—very briefly—about Du Bois's charts. See *Wayward Lives*, 17–19 for the photograph and 109–113 for the charts.

74 A web compilation of the Atlanta University Studies, including information and links to scanned versions, can be found on the "Atlanta University Publications" page of WEBDuBois.org, edited by Robert W. Williams. See http://www.webdubois.org/wdb-AtlUniv.html. On the significance of the studies for Du Bois's pioneering contributions to the field of sociology, see Aldon Morris, *The Scholar Denied: W. E. B. Du Bois and the Birth of Modern Sociology* (U of California Press, 2015). On the history of Black printing and Du Bois's role in it, see McHenry, *To Make Negro Literature*.

75 For example, "Religion of American Negroes" and "Statistics of Negro Church Organizations" employ numbers very similar to those provided in table 1 of the third study, *Some Efforts of American Negroes for Their Own Social Betterment* (Atlanta University, 1898); and "Negro Business Men in the United States" takes its data directly from table 2 of the fourth study, *The Negro in Business* (Atlanta University, 1899), 7.

76 Morris, *The Scholar Denied*, 71.

77 In the absence of the Du Bois-Atlanta School, the University of Chicago would claim this honor with work beginning around 1915. See Morris, *The Scholar Denied*, 109–118.

78 The official title was *The College-Bred Negro* (Atlanta University, 1900).

79 As Du Bois explains in the report, "A list of living colored college graduates was obtained and a blank with the following questions was sent them. About 800 answers to these blanks were received." While he does not specify the total number of living graduates, the number of graduates, living and dead, was 3,856. *The College-Bred Negro American* (Atlanta University, 1910), 11–12.

80 Per her alumni listing in the *Catalogue of the Officers and Students of Atlanta University, 1909–1910*, 36; I could not find her listed in the 1910 US Census there.

81 *The College-Bred Negro American*, 47.

82 This number was determined by subtracting the 163 Atlanta University graduates from the total number of 3,856 initially provided by Du Bois. See note 79.

83 *The College-Bred Negro American*, 46.

84 Tommaso Elli, Adam Bradley, et al., "Tied in Knots: A Case Study on Anthropographic Data Visualization about Sexual Harassment in the Academy," *2020 IEEE VIS Arts Program (VISAP)*, Salt Lake City, UT, USA, 2020, 29–44. Also see the website viewable at https://tiedinknots.io/#/.

85 It is likely that Mack completed at least some of the surveys sent by Du Bois, since she consistently responded to Atlanta University alumni questionnaires for several decades. From this information, we know that, in 1910, Mack had moved back to Alabama and from there to Virginia, where she stayed for at least two years. It is unknown what happened to Wilkins; the 1918 alumni directory reports Mack having moved again, to Urbana, Ohio, where she was once again working as a teacher, remarried to a man named John Anderson Brown. According to Google Street View, their foursquare-style house on Urbana's South Main Street, where Mack lived until the final years of her life, is still standing today. The exact address, 927 S Main St, Urbana, OH 43078, is reported in the *General Catalogue of Atlanta University, 1867–1918*, ed. Myron Adams (Atlanta UP, 1918), 59. In the obituary for John Brown, her husband, it mentions that he died in the house. Mack is also reported as living there in the 1920 and 1930 US Censuses. In her obituary, dated September 20, 1940, it reports that she lived in Bellefontaine, a small city about 20 miles away. "Miss Lulu Brown," *The Lima News*, September 20, 1940, 6.

86 While Clarkson's advocacy efforts were oriented toward the abolition of the slave *trade*, slavery itself would not be abolished in England until 1834, and in the United States until 1865. (Both countries banned the slave trade—the actual trafficking of people on ships—in 1807.)

6 **Labor:** Data by Design What We Learn from Visualization Work

1 Andy Hertzfeld, "-2000 Lines of Code," *Folklore.org* (February 1982), https://folklore.org/Negative_2000_Lines_Of_Code.html?sort=date, accessed August 30, 2025.

2 In her study of Indigenous women's impact on electronic manufacturing, Lisa Nakamura likewise highlights how digital work can literally be understood as "the work of the hands and its digits" (932), again accentuating the human involvement in the creation of digital technologies that often go under- (if not fully un-) recognized. Nakamura, "Indigenous Circuits: Navajo Women and the Racialization of Early Electronic Manufacture," American Quarterly 66.4 (2014): 919–941, https://dx.doi.org/10.1353/aq.2014.0070.

3 E.g., Jessica Bruder, "These Workers Have a New Demand: Stop Watching Us," *Nation*, May 27, 2015; Ifeoma Ajunwa, *The Quantified Worker: Law and Technology in the Modern Workplace* (Cambridge UP, 2023); and more recently Milagros Miceli et al., "The Data-Workers Inquiry," https://data-workers.org/, accessed August 30, 2025.

4 For a more in-depth examination of this phase of the project in particular, see our collaborative essay "What Data Do and Do Not Represent: Visualizing the Archive of Slavery" in *IEEE Computer Graphics and Applications* 45.3 (May-June 2025): 45–57, doi: 10.1109/MCG.2025.3553412.

5 Tina Campt, *Listening to Images* (Duke UP, 2017), 9.

6 Here I use "grammar" in its basic sense. For the use of the term in a visualization context, see note 12.

7 I often think of Bruce Fink's understanding of language as a force that we use and that also "uses" us. See Fink, *The Lacanian Subject: Between Language and Jouissance* (Princeton UP, 1997), 14.

8 Campt, *Listening to Images*, 116, emphasis in original.

9 Campt, *Listening to Images*, 116.

10 Toni Morrison, "Cinderella's Stepsisters," in *The Source of Self-Regard* (Knopf, 2019), 111.

11 Campt, *Listening to Images*, 8–9.

12 Leland Wilkinson's *Grammar of Graphics* (2001), also referenced in the introduction, has strongly influenced the design of many widely used contemporary visualization libraries, such as ggplot (the "g"s are for "grammar" and "graphics") and Vega-Lite, as well as the commercial visualization platform Tableau.

WORKS CITED

Acuña, René, ed. *Relaciones geográficas del siglo XVI*. UNAM, 1984.

Adams, Myron, ed. *General Catalogue of Atlanta University, 1867–1918*. Atlanta UP, 1918.

Ajunwa, Ifeoma. *The Quantified Worker: Law and Technology in the Modern Workplace*. Cambridge UP, 2023.

Akbaba, Derya, Lauren Klein, and Miriah Meyer. "Entanglements for Visualization: Changing Research Outcomes through Feminist Theory." *IEEE Transactions on Visualization and Computer Graphics* 31.1 (2025): 1279–1289.

Akbaba, Derya, Devin Lange, Michael Correll, Alexander Lex, and Miriah Meyer. "Troubling Collaboration: Matters of Care for Visualization Design Study." *Proceedings of the 2023 Conference on Human Factors in Computing (CHI '23)*, Hamburg, Germany, April 2023.

Akbaba, Derya, Jack Wilburn, Main T. Nance, and Miriah Meyer. "Manifesto for Putting 'Chartjunk' in the Trash 2021!" alt.VIS 2021.

Akerman, James, ed. *The Imperial Map: Cartography and the Mastery of Empire*. U of Chicago Press, 2009.

Andrews, RJ. "Don't Say 'Data Visualizations.'" Chartography, January 28, 2025. https://www.chartography.net/p/dont-say-data-visualizations, accessed May 16, 2025.

"Atlanta University Exhibit at Paris." *Atlanta Journal*, February 22, 1900.

Atlanta University. *Catalogue of the Officers and Students of Atlanta University, 1898–1899*. Atlanta UP, 1899.

Atlanta University. *Catalogue of the Officers and Students of Atlanta University, 1899–1900*. Atlanta UP, 1900.

Atlanta University. *Catalogue of the Officers and Students of Atlanta University, 1900–1901*. Atlanta UP, 1901.

Atlanta University. *Catalogue of the Officers and Students of Atlanta University, 1909–1910*. Atlanta UP, 1910.

"Atlanta University dormitory in the 1900 United States Federal Census." Ancestry.com. https://www.ancestry.com/search/collections/7602/records/8895399. Accessed August 22, 2025.

Barkley Brown, Elsa. "African-American Women's Quilting: A Framework for Conceptualizing and Teaching African-American Women's History." *Signs* 14.4 (1989): 921–929.

Bartram, Lyn, Michael Correll, and Melanie Tory. "Untidy Data: The Unreasonable Effectiveness of Tables." *IEEE Transactions on Visualization and Computer Graphics* 28.1 (2002): 686–696.

Battle-Baptiste, Whitney, and Britt Rusert. *W. E. B. Du Bois's Data Portraits: Visualizing Black America*. Princeton Architectural Press, 2018.

Baucom, Ian. *Specters of the Atlantic: Finance Capital, Slavery, and the Philosophy of History*. Duke UP, 2005.

Baym, Nina. "The Ann Sisters: Elizabeth Peabody's Millennial Historicism." *American Literary History* 3.1 (1991): 25–45.

Beall, Alexandra, Courtney Allen, Angela Vujic, and Lauren F. Klein. "Reimagining Elizabeth Palmer Peabody's Lost 'Mural Charts.'" *Digital Humanities 2018*. Association of Digital Humanities Organizations, 2018.

Bellhouse, David R. *The Flawed Genius of William Playfair: The Story of the Father of Statistical Graphics*. U of Toronto Press, 2023.

Benjamin, Ruha. *Imagination: A Manifesto*. Norton, 2024.

Bernier, Celeste-Marie. "The Slave Ship Imprint: Representing the Body, Memory, and History in Contemporary African American and Black British Painting." *Callaloo* 37.4 (2014): 990–1022.

Bertin, Jacques. *Sémiologie graphique. Les diagrammes, les réseaux, les cartes*. Gauthier-Villars, 1967. Translated by William J. Berg as *Semiology of Graphics*, with a forward by Howard Wainer (U of Wisconsin Press, 1983).

Bhargava, Rahul. *Community Data: Creative Approaches to Empowering People with Information*. Oxford UP, 2024.

Bigelow, Alex, Steven Drucker, Danyel Fisher, and Miriah Meyer. "Iterating between Tools to Create and Edit Visualizations." *IEEE Transactions on Visualization and Computer Graphics* 23.1 (2017): 481–490.

Bigelow, Alex, Steven Drucker, Danyel Fisher, and Miriah Meyer. "Reflections on How Designers Design with Visualization." In *Proceedings of the 2014 International Working Conference on Advanced Visual Interfaces (AVI '14)*, 17–24. Association for Computing Machinery, 2014.

BIPOC Design History. "(re) Creating Turtle Island: Native American Design Remembered through History." https://bipocdesignhistory.com/products/re-creating-turtle-island-native-american-design-through-remembered-history, accessed March 17, 2026.

Bissell, Evan, Erik Loyer, Tanya Orellana, Lisa Nowlain, and Josh Begley. *The Knotted Line*. 2014. https://knottedline.com/

Borowski, Susan. "The Multi-Layered Wisdom of Edmond Halley." American Association for the Advancement of Science (blog), October 12, 2012. https://www.aaas.org/taxonomy/term/10/multi-layered-wisdom-edmond-halley, accessed March 11 2026.

Bostock, Michael, Vadim Ogievetsky, and Jeffrey Heer. "D3: Data-Driven Documents." *IEEE Transactions on Visualization and Computer Graphics* 17.12 (2011): 2301–2309.

Bradshaw, Noel-Ann. "Florence Nightingale (1820–1910): An Unexpected Master of Data." *Patterns* 8.1/2 (2020).

Bridgers, Jeff. "Du Bois's American Negro Exhibit for the 1900 Paris Exposition." Library of Congress Blog, February 28, 2014. https://blogs.loc.gov/picturethis/2014/02/du-boiss-american-negro-exhibit-for-the-1900-paris-exposition/?loclr=blogpic.

Brody, Howard, Michael Russell Rip, Peter Vinten-Johansen, Nigel Paneth, and Stephen Rachman. "Map-Making and Myth-Making in Broad Street: The London Cholera Epidemic, 1854." *The Lancet* 356.9223 (2000): 64–68.

Brooks, Lisa. *The Common Pot: The Recovery of Native Space in the Northeast*. U of Minnesota Press, 2008.

Browne, Simone. *Dark Matters: On the Surveillance of Blackness*. Duke UP, 2015.

Bruder, Jessica. "These Workers Have a New Demand: Stop Watching Us." The *Nation*, May 27, 2015

Burke, Christopher, and Gunther Sandner. *History and Legacy of Isotype*. Bloomsbury, 2024.

Buurma, Rachel Sagner, and Laura Heffernan. "Search and Replace: Josephine Miles and the Origins of Distant Reading." *Modernism/Modernity* 3.1 (2018).

Cairo, Alberto. *The Functional Art: An Introduction to Information Graphics and Visualization*. New Riders, 2012.

Cairo, Alberto. *How Charts Lie: Getting Smarter about Visual Information*. Norton, 2019.

Camoes, Jorge. "Excel Charts Meet William Playfair." December 6, 2011. https://excelcharts.com/excel-charts-meet-william-playfair/, accessed July 10, 2025.

Campt, Tina. *Listening to Images*. Duke UP, 2017.

Cañizares-Esguerra, Jorge. "Screwing Humboldt and His Hagiographers." *Medium*, September 27, 2019. https://jorgecanizaresesguerra.medium.com/screwing-in-two-positions-82c2cc5b09db.

Card, Stuart K., Jock D. Mackinlay, and Ben Shneiderman. *Readings in Information Visualization: Using Vision to Think*. Morgan Kaufman, 1999.

Cardoso Llach, Daniel. *Builders of the Vision Software and the Imagination of Design*. Routledge, 2015.

Carey, Matthew. *Complete Genealogical, Historical, Chronological, and Geographical Atlas; Being a General Guide to History, Both Ancient and Modern … According to the Plan of Le Sage, Greatly Improved. The Whole Forming a Complete System of History and Geography*. Philadelphia, 1820.

Carretta, Vincent. *Equiano, the African: Biography of a Self-Made Man*. U of Georgia Press, 2005.

Carroll, Stephanie, Ibrahim Garba, Oscar L. Figueroa-Rodríguez, Jarita Holbrook, Raymond Lovett, Simeon Materechera, Mark Parsons, Kay Raseroka, Desi Rodriguez-Lonebear, Robyn Rowe, Rodrigo Sara, Jennifer D. Walker, Jane Anderson, and Maui Hudson. "The CARE Principles for Indigenous Data Governance." *Data Science Journal* 19 (2020): 1–12.

Caso, Alfonso. *Los calendarios prehispánicos*. Universidad Nacional Autónoma de México, 1967.

Caso, Alfonso. "El mapa de Teozacoalco." *Cuadernos Americanos* 47.5 (1949): 145–181.

Chalabi, Mona. "WEB Du Bois: Retracing His Attempt to Challenge Racism with Data." *Guardian*, February 14, 2017. https://www.theguardian.com/world/2017/feb/14/web-du-bois-racism-data-paris-african-americans-jobs.

Chan, Anita. *Predatory Data: Eugenics in Big Tech and Our Fight for an Independent Future*. U of California Press, 2024.

Chare, Nicholas. "In Her Hands: Affect, Encounter and Gesture of Wit(h)nessing in Shanawdithit's Drawings." *Parallax* 26.3 (2020): 294.

"Chart, n." *The Oxford English Dictionary*, September 2025. https://doi.org/10.1093/OED/4890659491.

"Charts and Graphs Showing the Condition of African Americans at the Turn of the Century Exhibited at the Paris Exposition Universelle in 1900." Library of Congress, 2014. https://www.loc.gov/pictures/item/2005679642/.

Chassot, Joanne. *Ghosts of the African Diaspora: Re-visioning History, Memory, and Identity*. Dartmouth College Press, 2018.

Cifor, Marika, Patricia Garcia, TL Cowan, Jasmine Rault, Tonya Sutherland, Anita Say Chan, Jennifer Rode, Anna Lauren Hoffmann, Niloufar Salehi, and Lisa Nakamura. *Feminist Data Manifest-No*. 2019. https://www.manifestno.com/, accessed May 20, 2025.

Clarkson, Thomas. *The History of the Rise, Progress, and Accomplishment of the Abolition of the African Slave-Trade by the British Parliament*. Vol. 2. London, 1808.

Cline, Howard F. "The Relaciones Geográficas of the Spanish Indies, 1577–1586." *Hispanic American Historical Review* 44.3 (1964): 341–374.

Cormack, William Epps. "Cormack to Bishop of Nova Scotia." January 10, 1829. Reproduced in James P. Howley, *The Beothucks or Red Indians: The Aboriginal Inhabitants of Newfoundland*, 210. Cambridge UP, 1915.

Cormack, William Epps. "History of the Red Indians of Newfoundland." 1829. Reproduced in James P. Howley, *The Beothucks or Red Indians: The Aboriginal Inhabitants of Newfoundland*, 222–224. Cambridge UP, 1915.

Cormack, William Epps. "Information Obtained from Shanawdithit." Reproduced in James P. Howley, *The Beothucks or Red Indians: The Aboriginal Inhabitants of Newfoundland*, 225–229. Cambridge UP, 1915.

Cott, Nancy. *The Grounding of Modern Feminism*. Yale UP, 1987.

Cruz, Petro. "Simulated Dendrochronology of US Immigration, 1790–2016." https://pmcruz.com/dendrochronology/, accessed July 25, 2025.

Cugoano, Quobna Ottobah. *Thoughts and Sentiments on the Evil of Slavery*. Penguin, [1791] 1999.

Daston, Lorraine. "Super-Vision: Weather Watching and Table Reading in the Early Modern Royal Society and Académie Royale des Sciences." *Huntington Library Quarterly* 78.2 (Summer 2015): 187–215.

"Declaration of Independence: A Transcription." National Archives. https://www.archives.gov/founding-docs/declaration-transcript.

D'Ignazio, Catherine, and Lauren F. Klein. *Data Feminism*. MIT Press, 2020.

D'Ignazio, Catherine, and Lauren Klein. "Feminist Data Visualization." *Workshop on Visualization for Digital Humanities (VIS4DH)* 2016, Baltimore, MD.

Dinius, Marcy J. "'Look!! Look!!! At This!!!!': The Radical Typography of David Walker's Appeal." *PMLA* 126.1 (January 2011): 55–72.

Dobson, Kelly. Data Visceralization Portfolio. https://web.media.mit.edu/~monster/, accessed March 17, 2026.

Doctor, Jeff, Jeff Ward, Trevor Jang, and Jen Polack. "Decolonizing Digital: Data's Role in Indigenous Data Sovereignty." *Animikii*. https://animikii.com/insights/decolonizing-digital-data-s-role-in-indigenous-data-sovereignty, accessed July 24, 2025.

Doctor, Jeff, Jeff Ward, Trevor Jang, and Jen Polack. *#DataBack: Asserting & Supporting Indigenous Data Sovereignty*. Animikii, 2023.

Dragicevic, Pierre, Yvonne Jansen, and Andrew Vande Moere. "Data Physicalization." In *Springer Handbook of Human Computer Interaction*, edited by Jean Vanderdonckt, Philippe Palanque, and Marco Winckler. Springer, 2021.

Driskill, Qwo-Li, Chris Finley, Brian Joseph Gilley, and Scott Lauria Morgensen, eds. *Queer Indigenous Studies: Critical Interventions in Theory, Politics, and Literature*. U of Arizona Press, 2011.

Du Bois, W. E. B. "The American Negro at Paris." *American Monthly Review of Reviews* 22.5 (November 1900): 575–577.

Du Bois, W. E. B. *The College-Bred Negro*. Atlanta University, 1900.

Du Bois, W. E. B. *The College-Bred Negro American*. Atlanta University, 1910.

Du Bois, W. E. B. *Dusk of Dawn: An Essay toward an Autobiography of a Race Concept*. Edited by Irene Diggs. Transaction, 1984.

Du Bois, W. E. B. "The Georgia Negro Exhibit for the Paris Exposition." *Atlanta University Bulletin* 110 (May 1900): 2.

Du Bois, W. E. B. *The Negro in Business*. Atlanta University, 1899.

Du Bois, W. E. B. "Sociology Hesitant." *boundary 2* 27.3 (Fall 2000): 37–44.

Du Bois, W. E. B. *A Soliloquy on Viewing My Life from the Last Decade of Its First Century*. Oxford, [1968] 2014.

Du Bois, W. E. B. *Some Efforts of American Negroes for Their Own Social Betterment*. Atlanta University, 1898.

Du Bois, W. E. B. *The Souls of Black Folk*. Modern Library, [1903] 2003.

Dukes, Hunter, and Adam Green. "Visualizing History: The Polish System." *Public Domain Review*, May 5, 2021.

Dumit, Joseph and Regula Valérie Burri, eds. *Social Studies of Scientific Imaging and Visualization*. MIT Press, 2008.

Duncan, S.A. "From Cloth to Canvas: Reinventing Gee's Bend Quilts in the Name of Art." *Museum Anthropology* 28.1 (March 2005): 27–28.

Edney, Matthew. *Cartography: The Ideal and Its History*. U of Chicago Press, 2019.

Ehrenreich, Barbara, and Deirdre English. *Witches, Midwives, and Nurses: A History of Women Healers*. Feminist Press, [1973] 2010.

Elbert, Monika, Julie E. Hall, and Katharine Rodier, eds. *Reinventing the Peabody Sisters*. U of Iowa Press, 2006.

Elli, Tommaso, Adam Bradley, Christopher Collins, Uta Hinrichs, Zachary Hills, and Karen Kelsky. "Tied in Knots: A Case Study on Anthropographic Data Visualization about Sexual Harassment in the Academy." 2020 IEEE VIS Arts Program (VISAP), Salt Lake City, UT, USA, 2020, 29–44. https://tiedinknots.io/#/.

Emory University. "Land Acknowledgment Statement." https://www.emory.edu/home/explore/history/land-acknowledgment, accessed July 25, 2025.

Ensmenger, Nathan. *The Computer Boys Take Over: Computers, Programmers, and the Politics of Technical Expertise*. MIT Press, 2010.

Equal Justice Initiative. *Lynching in America: Confronting the Legacy of Racial Terror*. 3rd edition. https://lynchinginamerica.eji.org/report/.

Equiano, Olaudah. *The Interesting Narrative of the Life of Olaudah Equiano, or Gustavus Vassa, the African, Written by Himself*. Edited by Werner Sollors. Norton, [1789] 2001.

Farooq, Nihad M. *Undisciplined: Science, Ethnography, and Personhood in the Americas*. NYU Press, 2016.

Ferris, William, ed. *Afro-American Folk Art and Crafts*. G. K. Hall, 1983.

Festa, Lynn. *Sentimental Figures of Empire in Eighteenth-Century Britain and France*. Johns Hopkins UP, 2006.

Fink, Bruce. *The Lacanian Subject: Between Language and Jouissance*. Princeton UP, 1997.

Finley, Cheryl. *Committed to Memory: The Art of the Slave Ship Icon*. Princeton UP, 2018.

Finley, Chris. "Decolonizing the Queer Native Body (and Recovering the Native Bull-Dyke)." In *Queer Indigenous Studies: Critical Interventions in Theory, Politics, and Literature*, edited by Qwo-Li Driskill, Chris Finley, Brian Joseph Gilley, and Scott Lauria Morgensen, 31–42. U of Arizona Press, 2011.

Forrest, Jason. "Discovering an Unknown Chart from W. E. B. Du Bois's 'The Exhibition of American Negroes.'" *Nightingale: The Journal of the Data Visualization Society*, January 1, 2019. https://medium.com/nightingale/discovering-a-new-chart-from-w-e-b-du-boiss-the-exhibition-of-american-negros-part-6-66e7b340b90c.

Foster, Caroline, Adam Hayward, Svyatoslav Kucheryavykh, Angela Vujic, Maninder Japra, Shivani Negi, and Lauren Klein. "The Shape of History: Reimagining Nineteenth-Century Data Visualization." *Digital Humanities 2017*. Association of Digital Humanities Organizations, 2017.

Foucault, Michel. "Nietzsche, Genealogy, History." In *Language, Counter-Memory, Practice: Selected Essays and Interviews*, edited by D. F. Bouchard, 139–164. Cornell UP, 1977.

Friendly, Michael. "The Golden Age of Statistical Graphics." *Statistical Science* 23.4 (2008): 502–535.

Friendly, Michael. "Visions and Re-Visions of Charles Joseph Minard." *Journal of Educational and Behavioral Statistics* 27.1 (2002): 31–51.

Friendly, Michael, and Daniel J. Denis. "Varieties of Data Visualization." *Milestones in the History of Thematic Cartography, Statistical Graphics, and Data Visualization*. https://www.datavis.ca/milestones/index.php?page=varieties+of+data+visualization, accessed May 16, 2025.

Friendly, Michael, Matthew Sigal, and Derek Harnanansingh. "The Milestones Project: A Database for the History of Data Visualization." https://datavis.ca/papers/MilestonesProject.pdf, accessed January 25, 2026.

Friendly, Michael, and Howard Wainer. *A History of Data Visualization and Graphic Communication*. Harvard UP, 2021.

Funkhouser, Howard Grey. "Historical Development of the Graphical Representation of Statistical Data." *Osiris* 3 (1936): 269–404.

Funkhouser, Howard Grey, and Helen M. Walker. "Playfair and His Charts." *Economic Journal* 45.1 (1935): 103–109.

Gannett, Henry. *Statistics of the Negroes in the United States*. Baltimore, 1894.

Ganter, Granville. "Mistress of Her Art: Anne Laura Clarke, Traveling Lecturer of the 1820s." *New England Quarterly* 87.4 (2014): 709–746.

Garth, Christoph, Min Chen, Alex Endert, Petra Isenberg, Alexander Lex, Shixia Liu, and Anders Ynnerman. "Things are Changing in 2021: The New VIS Conference." IEEE VIS (blog), October 7, 2020. https://ieeevis.org/year/2024/blog/things-are-changing-2021, accessed January 15, 2026.

Garvey, Ellen Gruber. "'facts and FACTS': Abolitionists' Database Innovations." In *"Raw Data" Is an Oxymoron*, edited by Lisa Gitelman, 89–102. MIT Press, 2013.

Gaskell, Roger. "Printing House and Engraving Shop: A Mysterious Collaboration." *BC* 53 (2004): 213–251.2

Glascock, William Nugent. "Report of Capt. Glascock." July 20, 1819. Reproduced in James P. Howley, *The Beothucks or Red Indians: The Aboriginal Inhabitants of Newfoundland,* 113–116. Cambridge UP, 1915.

Goeman, Mishuana. *Mark My Words: Native Women Mapping Our Nations*. U of Minnesota Press, 2013.

Green, Rayna. "The Pocahontas Perplex: The Image of Indian Women in American Culture." *Massachusetts Review* 16.4 (Autumn 1975): 698–714.

Grem, Darren E. "Sam Jones, Sam Hose, and the Theology of Racial Violence." *Georgia Historical Quarterly* 90.1 (2006): 35–61.

Griffin, Farah Jasmine. "Black Feminists and Du Bois: Respectability, Protection, and Beyond." *Annals of the American Academy of Political and Social Science* 568.1 (2000): 28–40.

Hacking, Ian. *The Taming of Chance*. Cambridge UP, 1990.

Hall, Peter A., and Patricio Dávila. *Critical Visualization: Rethinking the Representation of Data*. Bloomsbury, 2022.

Halpern, Orit. *Beautiful Data: A History of Vision and Reason since 1945*. Duke UP, 2015.

Hannah-Jones, Nikole, et al. "The 1619 Project." *New York Times Magazine*, August 14, 2019. https://www.nytimes.com/interactive/2019/08/14/magazine/1619-america-slavery.html.

Haraway, Donna. "Situated Knowledges: The Science Question in Feminism and the Privilege of Partial Perspective." *Feminist Studies* 14.3 (1988): 575–599.

Harjo, Laura. *Spiral to the Stars: Mvskoke Tools of Futurity*. U of Arizona Press, 2019.

Harley, J. B., and David Woodward, eds. *History of Cartography*. 6 vols. Chicago: U of Chicago Press.

Hartman, Saidiya. *Scenes of Subjection: Terror, Slavery, and Self-Making in Nineteenth-Century America*. Oxford UP, 1997.

Hartman, Saidiya. *Wayward Lives, Beautiful Experiments: Intimate Histories of Social Upheaval*. Norton, 2019.

"Hear Me Now: The Black Potters of Old Edgefield, South Carolina." Exhibition. High Museum of Art, Atlanta, February 16–May 12, 2024. https://high.org/exhibition/hear-me-now, accessed May 16, 2025.

Hertzfeld, Andy. "-2000 Lines of Code." *Folklore.org*, February 1982. https://folklore.org/Negative_2000_Lines_Of_Code.html, accessed August 30, 2025.

Hicks, Mar. *Programmed Inequality: How Britain Discarded Women Technologists and Lost Its Edge in Computing*. MIT Press, 2017.

Hidalgo, Alex. *Trail of Footprints: A History of Indigenous Maps from Viceregal Mexico*. U of Texas Press, 2019.

Hochschild, Adam. *Bury the Chains: Prophets and Rebels in the Fight to Free an Empire's Slaves*. Houghton Mifflin, 2005.

Howley, James P. *The Beothucks or Red Indians: The Aboriginal Inhabitants of Newfoundland*. Cambridge UP, 1915.

Hullman, Jessica. "The Purpose of Visualization Is Insight, Not Pictures: An Interview with Ben Shneiderman." *ACM Interactions*, August 19, 2019. https://interactions.acm.org/blog/view/the-purpose-of-visualization-is-insight-not-pictures-an-interview-with-ben, accessed May 16, 2025.

Hullman, Jessica, and Nicholas Diakopoulos. "Visualization Rhetoric: Framing Effects in Narrative Visualization." *IEEE Transactions on Visualization and Computer Graphics* 17.12 (2011): 2231–2240.

Hume, David. "On National Characters." In *Political Essays*, edited by Knud Haakonssen, 78–92. Cambridge UP, 2012.

Jansen, Maarten E. R. G. N., and Gabina Aurora Pérez Jiménez. "Chronological Correlations in Aztec and Mixtec History." In *Mesoamerican Manuscripts: New Scientific Approaches and Interpretations*, edited by Maarten E. R. G. N. Jansen, Virginia Lladò-Buisán, and Ludo Snijders, 341–379. Brill, 2019.

Jarrett, Gene Andrew. "'To Refute Mr. Jefferson's Arguments Respecting Us': Thomas Jefferson, David Walker, and the Politics of Early African American Literature." *Early American Literature* 46.2 (2011): 291–318.

Jefferson, Thomas. "Extract from Thomas Jefferson's Anecdotes of Benjamin Franklin." December 4, 1818. https://tjrs.monticello.org/letter/27, accessed July 10, 2025.

Jefferson, Thomas. *The Writings of Thomas Jefferson*. Vol. 1, *Memoir, Correspondence, and Miscellanies from the Papers of Thomas Jefferson*. Edited by Thomas Jefferson Randolph. Gray and Bowen, 1830.

Joe, Mi'Sel, and Sheila O'Neill. *My Indian*. Breakaway, 2021.

Joe, Mi'Sel, and Sheila O'Neill. *Suliewey*. Breakaway, 2023.

Johnson, Jessica Marie. "Markup Bodies: Black [Life] Studies and Slavery [Death] Studies at the Digital Crossroads." *Social Text* 36.4 (2017): 57–79.

Johnson, Jessica Marie, and Mark Anthony Neal. "Wild Seed in the Machine." *Black Scholar* 47.3 (2017): 1–2.

Joyce, Barry. *The First US History Textbooks: Constructing and Disseminating the American Tale in the Nineteenth Century*. Lexington Books, 2015.

Kahn, Andrew, and Jamelle Bouie. "The Atlantic Slave Trade in Two Minutes." September 16, 2021. https://slate.com/news-and-politics/2021/09/atlantic-slave-trade-history-animated-interactive.html, accessed May 23, 2025.

Kahn, Paul. "Inclusion and Exclusion in Emma Willard's Maps of History." *Nightingale: Journal of the Data Visualization Society*, March 30, 2023. https://nightingaledvs.com/inclusion-and-exclusion-in-emma-willard-maps-of-history/, accessed July 25, 2025.

Kalina, Richard. "Gee's Bend Modern." *Art in America* 10 (2003): 104–149.

Kamen, Henry. *Philip of Spain*. Yale UP, 1997.

Kelsey, Penelope Myrtle. *Reading the Wampum: Essays on the Hodinöhsö:ni' Visual Code and Epistemological Recovery*. Syracuse UP, 2019.

Kennedy, Helen, Rosemary Lucy Hill, Giorgia Aiello, and William Allen. "The Work that Visualisation Conventions Do." *Information, Communication and Society* 19.6 (2016): 715–735.

Keynes, J. M. "William Stanley Jevons 1835–1882: A Centenary Allocution on His Life and Work as Economist and Statistician." *Journal of the Royal Statistical Society* 99.3 (1936): 516–548.

King, Tiffany Lethabo. *The Black Shoals: Offshore Formations of Black and Native Studies*. Duke UP, 2019.

Kirschenbaum, Matthew. "The .txtual Condition: Digital Humanities, Born-Digital Archives, and the Future Literary." *Digital Humanities Quarterly* 7.1 (2013).

Klein, Lauren. *An Archive of Taste: Race and Eating in the Early United States*. U of Minnesota Press, 2020.

Klein, Lauren. "The Image of Absence: Archival Silence, Data Visualization, and James Hemings." *American Literature* 85.4 (2013): 661–688.

Klein, Lauren. "What Data Visualization Reveals: Elizabeth Palmer Peabody and the Work of Knowledge Production." *Harvard Data Science Review* 4.2 (2022).

Kong, Ha-Kyung, Zicheng Liu, and Karrie Karahalious. "Frames and Slants in Titles of Visualizations on Controversial Topics." In *Proceedings of the 2018 CHI Conference on Human Factors in Computing Systems (CHI '18)*, article 437, 1–12. Association for Computing Machinery, 2018.

Korsmeyer, Carolyn. "The Hierarchy of the Senses." In *Making Sense of Taste: Food and Philosophy*, 20–53. Cornell UP, 2015.

Kraak, Menno-Jan, and Ferjan Ormeling. *Cartography: Visualization of Spatial Data*. 4th ed. Routledge, 2023.

Latour, Bruno. *We Have Never Been Modern*. Harvard UP, 1993.

Leahy, Ryan. "The Alabama Women Who Made Their Quilts a Part of Modern Art." *Artsy*, 2018. https://www.artsy.net/article/artsy-editorial-alabama-women-made-quilts-modern-art.

Lee, Crystal, Tanya Yang, Gabrielle D. Inchoco, Graham M. Jones, and Arvind Satyanarayan. "Viral Visualizations: How Coronavirus Skeptics Use Orthodox Data Practices to Promote Unorthodox Science Online." In *Proceedings of the 2021 CHI Conference on Human Factors in Computing Systems (CHI '21)*, article 607, 1–18. Association for Computing Machinery, 2021.

Lewis, David Levering. *W. E. B. Du Bois: A Biography*. Henry Holt, 2009.

Lewis, David Levering, and Deborah Willis, eds. *A Small Nation of People: W. E. B. Du Bois and African American Portraits of Progress*. Harper Collins, 2005.

Li, Shiyao, Margy Adams, Tanvi Sharma, Jay Varner, and Lauren Klein. "What Data Do and Do Not Represent: Visualizing the Archive of Slavery." *IEEE Computer Graphics and Applications* 45.3 (May–June 2025): 45–57. https://doi.org/10.1109/MCG.2025.3553412.

Liboiron, Max. *Pollution Is Colonialism*. Duke UP, 2021.

Light, Jennifer. "When Computers Were Women." *Technology and Culture* 40.3 (1999): 455–483.

Lima, Manuel. *Visual Complexity: Mapping Patterns of Information*. Princeton Architectural Press, 2011.

Lin, Haihan, Derya Akbaba, Miriah Meyer, and Alexander Lex. "Data Hunches: Incorporating Personal Knowledge into Visualizations." *IEEE Transactions on Visualization and Computer Graphics* 29.1 (2022): 504–514.

Loukissas, Yanni. *All Data Are Local: Thinking Critically in a Data-Driven Society*. MIT Press, 2019.

Loukissas, Yanni. "The Life and Death of Data." https://lifeanddeathofdata.org/, accessed July 10, 2025.

"Lula Brown in the 1920 United States Federal Census." Ancestry.com. Accessed August 19, 2025.

"Lula I Brown in the 1930 United States Federal Census." Ancestry.com. Accessed August 19, 2025.

MacEachren, Alan M., and D. R. Fraser Taylor, eds. *Visualization in Modern Cartography*. Pergamon, 1994.

Marey, Étienne-Jules. *The Graphic Method.* Translated by Georges Hattab and RJ Andrews. Visionary Press, 2022.

Markowitz, Sally. J. "The Distinction between Art and Craft." *Journal of Aesthetic Education* 28.1 (Spring 1994): 55–70.

Marshall, Ingeborg. *A History and Ethnography of the Beothuk*. McGill-Queen's UP, 1996.

Marshall, Megan. *The Peabody Sisters: Three Women Who Ignited American Romanticism*. Mariner, 2006.

Mayo-Smith, Richmond. *Statistics and Sociology*. 1895.

McHenry, Elizabeth. *To Make Negro Literature: Writing, Literary Practice, and African American Authorship*. Duke UP, 2021.

McKittrick, Katherine. *Dear Science and Other Stories*. Duke UP, 2020.

McKittrick, Katherine. "Mathematics Black Life." *Black Scholar* 44.2 (Summer 2014): 16–28.

Mei, Xiyao, Yu Zhang, Chaofan Yang, Rui Shi, and Xiaoru Yuan. "ZuantuSet: A Collection of Historical Chinese Visualizations and Illustrations." arXiv preprint, February 26, 2025. https://arxiv.org/pdf/2502.19093

Miceli, Milagros, et al., *The Data-Workers Inquiry*. https://data-workers.org/, accessed August 30, 2025.

"Miss Lulu Brown." *The Lima News*, September 20, 1940, 6.

Morgan, Jennifer. *Reckoning with Slavery: Gender, Kinship, and Capitalism in the Early Black Atlantic*. Duke UP, 2021.

Morgenstern, Michelle, Amy Fox, Graham M. Jones, and Arvind Satyanarayan. "Visualization Vibes: The Socio-Indexical Function of Visualization Design." *IEEE Transactions on Visualization and Computer Graphics* 32.1 (2026): 1284–1294.

Morris, Aldon. *The Scholar Denied: W. E. B. Du Bois and the Birth of Modern Sociology*. U of California Press, 2015.

Morrison, Romi. "Gaps between the Digits: On the Fleshy Unknowns of the Human." *Information Design Journal* 25.1 (2019): 56–70.

Morrison, Toni. "Cinderella's Stepsisters." In *The Source of Self-Regard*. Knopf, 2019.

Mundy, Barbara. *The Mapping of New Spain: Indigenous Cartography and the Maps of the Relaciones Geográficas*. U of Chicago, 2000.

Munzner, Tamara. *Visualization Analysis and Design*. CRC Press, 2014.

Murphy, Benjamin. "'Multiplied without Number': Lynching, Statistics, and Visualization in Ida B. Wells, Mark Twain, and W. E. B. Du Bois." *American Literature* 93.2 (2021): 205–207.

Nakamura, Lisa. "Indigenous Circuits: Navajo Women and the Racialization of Early Electronic Manufacture." *American Quarterly* 66.4 (December 2014): 919–941.

Nast, Heidi J., and Steve Pile, eds. *Places through the Body*. Routledge, 1998.

National Museum of the American Indian. "Frank Speck." https://americanindian.si.edu/collections-search/edan-record/ead_component%3Asova-nmai-ac-001-ref15769, accessed July 25, 2025.

Naylor, Celia E., Kirsten Akey, Madiha Zahrah Choksi, Alex Gil, Moacir P. de Sa Pereira, and Monique S. J. Williams. "Un(silencing) Slavery: Remembering the Enslaved at Rose Hall Plantation, Jamaica." https://unsilencing-slavery.org/about, accessed May 26, 2025.

Neurath, Otto. *From Hieroglyphics to Isotype: A Visual Autobiography*. Hyphen Press, 2010.

Nieto Olarte, Mauricio. *La obra cartográfica de Francisco José de Caldas*. Universidad de los Andes, 2006.

O'Connor, January, Mark Parman, Nicole R. Bowman, and Stephanie Evergreen. "Decolonizing Data Visualization: A History and Future of Indigenous Data Visualization." *Journal of Multidisciplinary Evaluation* 19.44 (2023): 62–79.

Odell, Jenny. *How to Do Nothing: Resisting the Attention Economy*. Melville House, 2019.

Odell, Jenny. *Saving Time: Discovering Life beyond the Clock*. Penguin-Random House, 2023.

Offen, Karen. "Defining Feminism: A Comparative Historical Approach." *Signs* 14.1 (1988): 119–157.

Offenhuber, Dietmar. *Autographic Design: The Matter of Data in a Self-Inscribing World*. MIT Press, 2023.

Oldfield, John. *Popular Politics and British Anti-Slavery: The Mobilization of Public Opinion against the Slave Trade, 1787–1807*. Frank Cass, 1998.

Olojo, Seyi. "Counterdata." In *Keywords of the Datafied State*, edited by Jenna Burrell, Ranjit Singh, and Patrick Davison, 170–181. Data & Society, 2024.

O'Neil, Cathy. *Weapons of Math Destruction: How Big Data Increases Inequality and Threatens Democracy*. Crown, 2016.

Ọnụọha, Mimi. "In Absentia." 2019. https://mimionuoha.com/in-absentia.

"Our Graduates." *Atlanta University Bulletin* 106 (January 1900): 1.

"Paris 1900: City of Entertainment." Portland Art Museum, 2019. https://portlandartmuseum.org/event/paris-1900-city-of-entertainment/.

Patterson, Orlando. *Slavery and Social Death: A Comparative Study*. Harvard UP, 1982.

Paul, Katherine Anne, Richard A. Pegg, and Katie E. Osborne. "A Jain Cosmographic Map of the Rose-Apple Continent (Jambudvipa)." Norman B. Leventhal Map & Education Center, Boston Public Library, https://www.leventhalmap.org/projects/jain-map-chat/, accessed March 17, 2026.

Peabody, Elizabeth Palmer. "Bem's Method of History: An Appeal to Teachers and Boards of Education." *Indiana School Journal* 1.2 (1856).

Peabody, Elizabeth Palmer. *Chronological History of the United States: Arranged with Plates on Bem's Principle*. New York, 1856.

Peabody, Elizabeth Palmer. "My Experience as a Teacher." *American Journal of Education* 32 (1882).

Peabody, Elizabeth Palmer. "The Origin and Growth of the Kindergarten." *Education* 2 (1882): 507–527.

Peabody, Elizabeth Palmer. *The Polish-American System of Chronology, Reproduced, with Some Modifications, from General Bem's Franco-Polish Method*. Boston, 1850.

Peabody, Elizabeth Palmer. *Universal History; Arranged to Illustrate Bem's Charts of Chronology*. New York, 1859.

Pearce, Margaret Wickens. "The Last Piece Is You." *Cartographic Journal* 51 (2014): 107–122.

Pearce, Margaret Wickens, and Stephen J. Hornsby. "Making the Coming Home Map." *Cartographica* 55.3 (2020): 170–176.

Pearce, Margaret Wickens, and Renee Pualani Louis. "Mapping Indigenous Depth of Place." *American Indian Culture and Research Journal* 32:3 (2008): 110.

Peyton, John. "Capture of Mary March (Demasduit) on Red Indian Lake, in the month of March 1819." In James P. Howley, *The Beothucks or Red Indians: The Aboriginal Inhabitants of Newfoundland*. Cambridge UP, 1915.

Phillipsen, Ivan. "The Natural Wonders of Oaxaca, Mexico." *Wild Latitudes*. https://wildlatitudes.com/natural-wonders-of-oaxaca-mexico/, accessed May 16, 2025.

Phipps, Elena. *Cochineal Red: The Art History of a Color*. Metropolitan Museum of Art, 2010.

Playfair, William. *The Commercial and Political Atlas*. 1st edition. London, 1786.

Playfair, William. *The Commercial and Political Atlas*. 3rd edition. Edited by Howard Wainer and Ian Spence. New York: Cambridge UP, [1801] 2023.

Playfair, William. *Joseph and Benjamin, a Conversation. Translated from a French Manuscript*. London, 1787.

Playfair, William. *Lineal Arithmetic*. London, 1798.

Polack, Fiona. "Introduction: De-Islanding the Beothuk." In *Tracing Ochre: Changing Perspectives on the Beothuk*, edited by Fiona Polack. U of Toronto Press, 2018.

Porter, Theodore M. *The Rise of Statistical Thinking, 1820–1900*. Princeton UP, 1986.

Poskett, James. *Horizons: A Global History of Science*. Penguin, 2022.

Provenzo, Eugene F., Jr. *W. E. B. Du Bois' Exhibit of American Negroes: African Americans at the Beginnings of the Twentieth Century*. Rowman & Littlefield, 2013.

Quashie, Kevin. *The Sovereignty of Quiet: Beyond Resistance in Black Culture*. Rutgers UP, 2012.

Raftery, Andrew Stein. "From Paper to Copper: The Engraver's Process." YouTube video. https://www.youtube.com/watch?v=fQvghHs15hA.

Rankin, Bill. *Radical Cartography*. Viking, 2025.

Ravina, Mark. "Algorithmic Maps and the Political Geography of Early-Modern Japan." *Cultural Analytics* 8.3, https://doi.org/10.22148/001c.84860.

Rediker, Marcus. *The Slave Ship: A Human History*. Penguin, 2007.

Rendgen, Sandra. *History of Information Graphics*. Taschen, 2019.

Rendgen, Sandra. *The Minard System: The Complete Statistical Graphics of Charles-Joseph Minard*. Princeton Architectural Press, 2018.

Riskin, Jessica. *Science in the Age of Sensibility: The Sentimental Empiricists of the French Enlightenment*. U of Chicago Press, 2002.

Robinson, Cedric. *Black Marxism: The Making of the Black Radical Tradition*. U of North Carolina Press, 1993.

Ronda, Bruce A. *Elizabeth Palmer Peabody: A Reformer on Her Own Terms*. Harvard UP, 1999.

Rosenberg, Daniel, and Anthony Grafton. *Cartographies of Time: A History of the Timeline*. Princeton Architectural Press, 2012.

Rosenthal, Caitlin. *Accounting for Slavery: Masters and Management*. Harvard UP, 2019.

Royston, Erica. "Studies in the History of Probability and Statistics: III. A Note on the History of the Graphical Presentation of Data." *Biometrika* 43.3/4 (December 1956): 241–247.

Ruan, Felipe E. "Prudent Deferment: Cosmographer-Chronicler Juan López de Velasco and the Historiography of the Indies." *Americas* 74.1 (2017): 27–55.

Rusert, Britt. "New World: The Impact of Digitization on the Study of Slavery." *American Literary History* 29.2 (2017): 267–286.

Rusert, Britt. "W. E. B. Du Bois and the Origins of Visual Sociology: The 1900 Paris Exposition and Beyond." In *Oxford Handbook of W. E. B. Du Bois*, edited by Aldon D. Morris et al., 371–390.

Satyanarayan, Arvind, Bongshin Lee, Donghao Ren, Jeffrey Heer, John Stasko, John Thompson, Matthew Brehmer, and Zhicheng Liu. "Critical Reflections on Visualization Authoring Systems." *IEEE Transactions on Visualization and Computer Graphics* 26.1 (2020): 461–471.

Schiff, Robyn. *Information Desk: An Epic*. Penguin, 2023.

Schmidt, Ben. "Canonization." In *Creating Data: The Invention of Information in the American State, 1850–1950*. https://creatingdata.benschmidt.org/Dubois_and_Gannett.

Schuller, Kyla. *The Biopolitics of Feeling: Race, Sex, and Science in the Nineteenth Century*. Duke UP, 2018.

Schulten, Susan. "Emma Willard's Maps of Time." *Public Domain Review*, 2020. https://publicdomainreview.org/essay/emma-willard-maps-of-time/.

Schulten, Susan. *Mapping the Nation: History and Cartography in Nineteenth-Century America*. U of Chicago Press, 2012.

Seals, Tré. "Du Bois." *Vocal Type*. https://www.vocaltype.co/history-of/du-bois

Sense, Sarah. "Power Lines." https://sarahsense.com/Artists/11571/website_statements_2021-22.pdf, accessed July 25, 2025.

Shapin, Steven, and Simon Schaffer. *Leviathan and the Air-Pump: Hobbes, Boyle, and the Experimental Life*. Princeton UP, 1985.

Shapley, Deborah. "The Da Vinci of Data." *New York Times*, March 30, 1998.

Sharpe, Christina. *In the Wake: On Blackness and Being*. Duke UP, 2016.

Sherman, Alex. Comment on open review of *Data by Design*. January 13, 2025. https://hyp.is/_GCQpNH6Ee-wcGtkgLC9FQ/dataxdesign.io/chapters/intro.

Shimada, Izumi, ed. *The Inka Empire: A Multidisciplinary Approach*. U of Texas Press, 2015.

Shneiderman, Ben. "The Eyes Have It: A Task by Data Type Taxonomy for Information Visualizations." *Proceedings of the 1996 IEEE Symposium on Visual Languages*, Boulder, CO, 1996, 336–343.

Shoemaker, David E. "El Mapa de Teozacoalco." Foundation for the Advancement of Mesoamerican Studies, 1999. http://www.famsi.org/reports/98032/index.html.

Silverstein, Jake. "Why We Published the 1619 Project." *New York Times Magazine*, December 20, 2019. https://www.nytimes.com/interactive/2019/12/20/magazine/1619-intro.html.

Slave Voyages. "History of the Project." https://www.slavevoyages.org/about/about#history, accessed May 23, 2025.

Slave Voyages. "Trans-Atlantic Slave Trade—Estimates." https://www.slavevoyages.org/assessment/estimates, Accessed May 23, 2025.

Smallwood, Stephanie. *Saltwater Slavery: A Middle Passage from Africa to American Diaspora*. Harvard UP, 2007.

Smith, Linda Tuhiwai. *Decolonizing Methodologies: Research and Indigenous Peoples*. 2nd edition. Zed, 2012.

Smith, Shawn Michelle. *Photography on the Color Line: W. E. B. Du Bois, Race, and Visual Culture*. Duke UP, 2004.

Solano, Francisco de, ed. *Cuestionarios para la formación de las relaciones geográficas de Indias: Siglos XVI/XIX*. Consejo Superior de Investigaciones Científicas, 1988.

Sparke, Matthew. "Between Demythologizing and Deconstructing the Map: Shawnawdithit's New-found-land and the Alienation of Canada." *Cartographica* 32.1 (Spring 1995): 1–21.

Sparke, Matthew. "Mapped Bodies and Disembodied Maps: (Dis)placing Cartographic Struggle in Colonial Canada." In *Places through the Body*, edited by Heidi J. Nast and Steve Pile. Routledge, 1998.

Speck, Frank G. *Beothuk and Micmac*. Museum of the American Indian–Heye Foundation, 1922.

Spillers, Hortense. "Mama's Baby, Papa's Maybe: An American Grammar Book." *Diacritics* 17.2 (Summer 1987): 65–81.

Starks, Anthony. "The #DuBois Challenge." *Nightingale: Journal of the Data Visualization Society*, February 1, 2022. https://nightingaledvs.com/the-dubois-challenge/.

Stasko, John. "Value-Driven Evaluation of Visualizations." *BELIV '14: Proceedings of the Fifth Workshop on Beyond Time and Errors: Novel Evaluation Methods for Visualization*, 46–53.

Strogatz, Steven. "From a Swinging Chandelier to Global Positioning Systems." *American Scientist* 109.2 (2021): 106–109.

Stuart, Alanna, Alyssa Vernon, Alice Grandoit-Šutka, DJ Lynnée Denise, fields harrington, Geoffrey Webster, Jeden Tolentino, Katherine McKittrick, Krista Franklin, Mark Campbell, Paul Akpomuje, Safia Siad, and Tina Munroe. *The Revolutionary Demand for Happiness*. https://revolutionarydemandforhappiness.com, accessed July 30, 2025.

"Students Showcase Projects Melding Indigenous, Contemporary Sciences." *University of Hawai'i News*, September 30, 2024. https://www.hawaii.edu/news/2024/09/30/student-projects-indigenous-contemporary-sciences, accessed July 25, 2025.

Subramaniam, Banu. *Ghost Stories for Darwin: The Science of Variation and the Politics of Diversity*. U of Illinois Press, 2014.

"Table of Temperature and Rainfall, Week Ended December 25, 1899." *Public Health Reports (1896–1970)*, January 12, 1900, 73–74.

Terraciano, Kevin. *The Mixtecs of Colonial Oaxaca: Ñudzahui History, Sixteenth through Eighteenth Centuries*. Stanford UP, 2002.

Thornton, Russell. "Native American Demographic and Tribal Survival into the Twenty-first Century." *American Studies* 46.3/4 (2005): 23–38.

Trouillot, Michel-Rolph. *Silencing the Past: Power and the Production of History*. Beacon Press, 1997.

Tuck, Eve. "Suspending Damage: A Letter to Communities." *Harvard Educational Review* 79.3 (2009): 409–427.

Tuck, Eve, and K. Wayne Yang. "Decolonization Is Not a Metaphor." *Decolonization: Indigeneity, Education and Society* 1.1 (2012): 1–40.

Tufte, Edward R. *Seeing with Fresh Eyes: Meaning, Space, Data, Truth*. Graphics Press, 2020.

Tufte, Edward R. *The Visual Display of Quantitative Information*. 2nd edition. Graphics Press, [1983] 2001.

Tukey, John. *Exploratory Data Analysis*. Addison-Wesley, 1977.

van Wijk, Jarke. "The Value of Visualization." *IEEE Visualization 2005*, Minneapolis, 79–86.

Vizenor, Gerald. *Manifest Manners: Narratives on Postindian Survivance*. U of Nebraska Press, 1994.

Wagner, Bryan. *Disturbing the Peace: Black Culture and the Police Power after Slavery*. Harvard UP, 2009.

Wainer, Howard, and Ian Spence. "Introduction." In William Playfair, *The Commercial and Political Atlas* and *Statistical Breviary*, edited by Howard Wainer and Ian Spence. New York: Cambridge UP, 2005.

Walker, David. *Appeal to the Colored Citizens of the World*. Boston, 1829.

Wall, Emily, Meeshu Agnihotri, Laura Matzen, Kristin Divis, Michael Haass, Alex Endert, and John Stasko. "A Heuristic Approach to Value-Driven Evaluation of Visualizations." *IEEE Transactions on Visualization and Computer Graphics* 25.1 (2019): 491–500.

Walter, Maggie, and Chris Anderson. *Indigenous Statistics: A Quantitative Research Methodology*. Routledge, 2013.

Warhus, Mark. *Another America: Native American Maps and the History of Our Land*. St. Martin's Press, 1997.

Wells, Ida B. *Crusade for Justice: The Autobiography of Ida B. Wells*. Edited by Alfreda Duster. U of Chicago Press, 1970.

Wherry, Frederick F., Jeffrey Himpele, et al. "Comparing and Calculating Your Student Loan Debt." https://www.dignityanddebt.org/projects/student-debt-calculator/.

Wickham, Hadley. *ggplot2: Elegant Graphics for Data Analysis*. Springer-Verlag, 2016. https://ggplot2.tidyverse.org/

Wiggins, Chris, and Matthew L. Jones. *How Data Happened: A History from the Age of Reason to the Age of Algorithms*. Norton, 2023.

Wilkinson, Leland. *The Grammar of Graphics*. Springer, 1999.

Willard, Emma. *History of the United States, or The Republic of America*. 5th edition. New York, 1834.

Willard, Emma. *Willard's Ancient Geography, Ancient Atlas to Accompany Universal Geography*. Hartford, 1827.

Williams, Robert W., ed. "Atlanta University Publications." WEBDuBois.org. http://www.webdubois.org/wdb-AtlUniv.html.

Womack, Autumn. *The Matter of Black Living: The Aesthetic Experiment of Racial Data, 1880–1930*. U of Chicago Press, 2022.

Wood, Jo. "Playfair's Wheat and Wages." March 10, 2024. https://observablehq.com/@jwolondon/wheat-and-wages, accessed July 10, 2025.

Wood, Marcus. *Blind Memory: Visual Representations of Slavery, 1780–1865*. Manchester UP, 2000.

Wood, Stephanie, ed. *The Mapas Project*. https://mapas.wired-humanities.org, accessed January 14, 2026.

Wynter, Sylvia. "1492: A New World View." In *Race, Discourse, and the Origin of the Americas: A New World View*, edited by Vera Lawrence Hyatt and Rex Nettleford, 5–57. Smithsonian Institution Press, 1994.

Yannakakis, Yanna. *Since Time Immemorial: Native Custom and Law in Colonial Mexico*. Duke UP, 2023.

INDEX

A

Abolition. *See also* US Civil War
- abolitionists, 19, 49–74, 142, 181 (*see also specific names*)
- resistance and, 43, 60
- visualization and, 43, 46, 49–74, 181, 203

Abstraction
- vs. detail, 44, 50, 75, 89, 186
- geometric, 139, 165
- insight and, 51–56
- Minard and, 171
- missing data and, 147
- as tool of oppression, 224
- in visualization design, 43, 222

Accessibility, 51, 104, 207, 223. *See also* Alt text
Ac ko mok ki, 122–123
Activism, 60, 169–205. *See also* Abolition
Adams, Margy, 12, 51, 55, 120, 130–131, 160, 206–207, 210, 213–218, 223, 226, 235
AI (artificial intelligence), 134, 210, 237
Algorithmic justice, 13, 122
Alt text, 51, 223
American Antiquarian Society, 147–150, 218, 236–237
Anticolonialism, 105, 115, 189
Atkinson, Bill, 207
Atlanta, 9, 185–186, 189
Atlanta University, 21, 172–189, 194–204
- Atlanta University Studies, 186, 195, 198–199

Attribution, 3, 46, 173, 199, 210
AWS (Amazon Web Services), 209

B

Bacon, Francis, 31
Bainbrigge, Captain William, 127–129
Bar charts, 9, 21, 77, 80, 93, 180, 189, 191, 193, 210–211
Barkley Brown, Elsa, 163–165, 214
Battle-Baptiste, Whitney, 171
Bem, Józef, 140–144
Beothuk, 19, 44, 103–134
Bertin, Jacques, 18, 24, 94
Big Tech, 229
Black codes, 183, 186
Black data, 60, 147, 163–165, 169–204
Black feminist theory, 55–56, 63, 146–147, 164, 195, 223–226
Blackness, 146–147
Black studies, 55–56, 63, 146–147, 186, 199. *See also specific scholars*
Bouie, Jamelle, 60
Boyle, Robert, 30, 38, 51
Broligarchy. *See* Big Tech
Browne, Simone, 60, 63

C

Cabot, John, 103, 123
Caldas, Francisco José de, 149, 154
Campt, Tina, 218–226
Capitalism. *See also* Big Tech; Colonialism: resource extraction and
- data visualization and, 11, 13, 47, 229
- design work and, 220
- history of, 11
- labor and, 25, 43–45, 63, 161, 163, 166, 180, 207–208
- land and, 19, 21, 32, 38, 43, 105, 122, 126, 146, 180, 193
- people as products in, 55, 63, 89
- racial capitalism, 43

Card, Stuart, 24–25
Cartography. *See* Maps
Chalabi, Mona, 25–28, 189–191
Charles II (king of England), 32
Charts. *See* Data visualization
Chatbots, 229
Chitimacha, 128–129
Choctaw, 128–129
Chronological charts, 24, 44–45, 70–74, 137–166
Clark, William, 122
Clarke, Anne Laura, 158–159
Clarkson, Thomas, 49–51, 57–60, 70–74, 83, 88, 139, 163, 203
Collaboration, 11–13, 45–46, 198, 219–222, 230
Colonialism, 10–13, 44, 47, 50, 103–134, 229
- anticolonialism, 105, 115, 189
- decolonialism, 104
- knowledge extraction and, 32, 105, 114–115, 129, 134–135
- resource extraction and, 19, 32, 38, 43, 103, 232
- violence in, 105, 119, 130, 229

Commercial and Political Atlas (Playfair), 9, 34, 43, 77–84, 89, 91, 98, 152
Computer science, 11, 18, 24–25, 45, 64, 163, 204
Consent, 50–51, 103
Cormack, William Epps, 107, 112–119, 126, 129, 133, 139
Counterdata, 57, 178
Counterhistories, 42, 107, 117, 137
Countervisualization, 172, 179, 183
Craft, 163–164
Credit. *See* Attribution
CSS, 209
Cugoano, Quobna Ottobah, 50, 56

D

D3 (data-driven documents), 44, 94–97, 209
Data access, 51, 115
Data centers, 229
Data collection, 33, 46, 57, 169, 178–183, 198, 230–231
Data colonialism, 105, 114, 135
Data physicalization, 25, 44, 159
Data stewardship, 104, 134–135
Data-veillance, 208
Data visceralization, 159. *See also* Data physicalization
Data visualization. *See also* Capitalism; Colonialism; Design process: knowledge production and; Labor; Occlusion; Quilting
as art, 18, 25, 28
bearing witness in, 44, 168, 193
best practices in, 13, 42, 44, 46, 81, 114, 137, 147, 172, 193, 226
blank space in, 56, 85, 100, 104
clarity and, 13, 24, 61, 86, 98, 114–115, 161, 230
collective knowledge and, 13, 38, 99, 134, 159, 214, 226, 230–232
definition of, 24–25, 28–29
demographics and, 33, 88
digital tools and, 18, 97, 134, 209, 214, 230
distance and, 104, 122, 131
efficiency of, 18, 44, 50, 56, 81, 105
in epidemiology, 32, 34
ethical clarity in, 13, 38, 43, 63
false insights and, 59, 95
"golden age" of, 32
as grammar, 94, 224–226
illusion of objectivity in, 21, 44, 80, 87–89, 101, 191, 223
intentional practice and, 12–13, 21, 29, 38, 44–48, 67–71, 104, 134, 203, 209, 235
metaphor in, 71–73, 134, 210, 218, 231
narrative and, 11, 47, 50, 100, 146–147, 152, 164–166
as power, 13–24, 29, 42–50, 57, 73, 81, 87–89, 98–100, 149, 162, 172, 185, 222, 226–237
scale and, 23, 33, 44, 53–60, 73, 85–86, 94, 104, 137, 171, 179, 214
social justice in, 46, 171
story of, 9–13, 30–43, 164–166, 171, 191, 194–223, 232
as superpower, 57
values in, 38, 172, 180, 220
David Rumsey Map Collection, 23, 41, 72, 80–81, 124–125, 140, 153, 157, 159–162
Demasduit, 104–114, 117, 133
Descartes, René, 31–32
Design bias, 222
Design process
abstraction and, 32
appeal to senses in, 147, 223
of authors, 55, 63, 67, 74, 93–96, 161, 192, 195, 208–226, 231
authorship and, 165
context and, 47, 185
data loss and, 56, 114–115
embodiment and, 45, 147
empowerment and, 167
engraving and, 80
expertise in, 161
innovation and, 28
knowledge production and, 12, 45–47, 50, 118, 165, 209, 224
liberation and, 74
maps in, 114–115
materials and, 80–82, 93
participation and, 148
people and, 214–218
recreation and, 94
results and, 12
simplicity in, 88
unidirectional, 129
vectorizing and, 95–96
Digital photography, 134
D'Ignazio, Catherine, 57, 235
Dignity + Debt Network, 191–192
DOGE, 207
DOM (document object model), 94
Du Bois, W. E. B. *See also* Exposition Universelle
Atlanta University Studies, 186, 195, 198–199
double consciousness in, 183–185
Du Bois-Atlanta School of Sociology, 198
epistemological evolution of, 44, 165–166, 185–186
labor and, 44, 166, 198, 219, 224, 230
liberation and, 172, 189, 193
pedagogy of, 44, 169, 198
"red ray," 186
responsibility of designers, 19
second sight in, 184
spiral chart form, 179–180, 191–193
students of, 44, 165, 169–173, 178–204

E

Elford, William, 51, 55–59
Emory University, 11, 230
Center for Digital Scholarship, 11, 209, 214, 221, 235
Digital Humanities Lab, 11–12, 235

Empiricism, 10, 31–33, 38, 44, 47, 50, 83
Engraving, 51, 54, 77, 83–88, 95
Enlightenment, 43, 83
Epistemology
alternative frameworks in, 100, 134, 147, 165
assumptions in, 93–97
complex data in, 59, 86
feminism and, 139, 149, 166
future and, 44
history of, 30
interaction and, 148
knowledge production and, 44, 83, 97
Mvskoke, 119–120
pluralism, 185–186
politics and, 38
positivism and, 152
Equiano, Oladuah, 49–56, 66–67, 71–74
Eugenics, 33, 98, 179–181
Exposition Universelle (Paris, 1900), 44, 165, 185, 224
Palace of Social Economy and Congresses, 171, 185

F

Fang, Zhou, 12, 175, 177, 214–217, 241
Fidler, Peter, 122
Figma, 95–97, 209–210, 213–214, 219
Finley, Cheryl, 51
Fisk, Harold, 62–67
Floor Chart project, 160–161
Folk art. *See* Craft
Forrest, Jason, 170
Franklin, Benjamin, 98
French Revolution, 86
Friendly, Michael, 32, 59, 236
Fu, Jianing, 12, 214–217, 235
Fugitive Slave Act, 149
Funkhouser, Howard Grey, 24, 30, 98

G

Galilei, Galileo, 30–31, 38
Gannett, Henry, 181–184
Gates, Theaster, 189
Gee's Bend quilters, 163–165, 214
Gender, 28, 38, 53–55, 139, 159–163
Georgia, 89, 172–175, 178–180, 183–189
Georgia Institute of Technology, 9, 11, 161, 214, 218, 221, 235
GitHub, 208–214, 224
God trick, 55–61, 88, 95, 147
Goeman, Mishuana, 115, 126
Google Docs, 210, 214
Grafton, Anthony, 73, 140

H

Halley, Edmond, 32, 36
Hannah-Jones, Nikole, 146
Haraway, Donna, 55–60, 147
Harjo, Laura, 118–119, 235
Hartman, Saidiya, 63, 195
Hawthorne, Julian, 148, 159
Hayward, Adam, 235
Hertzfeld, Andy, 207
Higher education, 232
Hose, Sam, 186, 189, 193
Howley, James P., 107, 114–120, 130–133
HTML, 209
Hudson's Bay Company, 122
Humanities, 9–13, 46–47, 103, 173, 223–226, 230–237
Humboldt, Alexander von, 149, 152, 154, 157
Hume, David, 10, 30
Huygens, Christiaan, 32

I

Indigenous peoples. *See individual peoples by name*
Innu, 104

J

Jacobinism, 87
Jambudvīpan, Jain diagram of, 41–42
Jamestown settlers, 142–146
Jażwiński, Antoni, 139–140
Jefferson, Thomas, 10–13, 181
Jevons, William Stanley, 77, 98–99
Jim Crow laws, 178
Johnson, Jessica Marie, 55, 63, 235
Joseph II (emperor), 98
Jutan, Dan, 12, 145, 215–217, 221, 235

K

Kahn, Andrew, 60
Kames, Henry Home, Lord, 10
Khipu, 38–39
Kikaigon, 127
Ku Klux Klan, 178

L

Labor
  collaboration in, 45, 63, 198, 208–210, 219–225, 230
  design work in, 43
  enslaved and, 89
  gender division in, 163
  history, 28
  measurement of, 207, 218
  students and, 11–13, 30, 44, 115, 146–152, 159–165, 170–192, 213–214, 224, 230–232
  test subjects and, 148
  value of, 166, 226
Lambert, Johann, 30–31
Lee, Henry Napoleon, 173, 187–189, 199–200
Lenape, 104, 122
Lewis, Meriwether, 122
Li, Shiyao, 12, 64, 66, 209, 214–217, 222, 235
Li, Yang, 12, 214–217, 235
Liberation, 46–47, 74, 168, 172, 189, 193, 225–226
Library Company of Philadelphia, 22, 77, 79, 84, 91, 94, 148, 157, 237
Library of Congress, 138, 169–170, 173–179, 181–182, 184, 196, 198
Lincoln, Abraham, 137, 157
Line graphs, 20, 87–88, 171, 180, 189, 191. *See also* Time series charts
Locke, John, 30, 51, 83
Louis XVI (king of France), 87
Louis, Renee Pualani, 118
Loukissas, Yanni, 18, 236, 238
Lupi, Giorgia, 26, 28
Lynching, 178–179, 186, 189, 193

M

Mack, Lula Iola, 45, 168–169, 183, 186–189, 194–195, 199
Mackinlay, Jock D., 24–25
Maps. *See also* Treemaps
  Cartesian, 118, 152
  colonialism and, 15, 19, 21, 32–43, 104–105, 114–122, 129–134, 147
  constellations in, 30–31, 118–119, 126, 218
  directional, 127, 129, 134
  flow maps, 86, 118–120, 171
  Indigenous cartography, 103–119, 122–123, 127–129
  "Mapa de Teozacoalco," 15, 17–21, 30, 33, 38, 46
  Mercator projection, 178
  Nuwuvi Salt Song Trail, 118
  performative, 118
  sand paintings, 118
  slave trade and, 60–61, 70, 73
  space-time in, 104, 118–119, 126
  stone carvings as, 118
  symbolic components of, 161
  tattoos as, 118
  topographical detail in, 19, 119
  trade winds in, 32, 36–37
  verbal, 118
Marey, Étienne-Jules, 24
Marie Séraphique (ship), 54–56, 61
McKittrick, Katherine, 146–147, 164
Meta, 230
Microsoft, 92–93, 230
Mi'kmaq, 104
Minard, Charles, 32, 35, 86, 118–120, 171
Missing data, 71, 75, 98, 134, 139, 147, 163, 167, 214
Modernity, 38, 42–43
Mola, Anna, 12, 97, 197–198, 215–217, 235
Mondrian, Piet, 163
Morris, Aldon, 198
Morrison, Romi, 63
Morrison, Toni, 225
Munro, Silas, 12, 55, 215–217, 219, 235
Murray, Daniel A. P., 169–171, 183
Mvskoke (Muscogee), 104, 118

N

NAACP, 189
Napoleon Bonaparte, 32, 35, 86
National Endowment for the Humanities (NEH), 230
  Office of Digital Humanities (ODH), 230, 236
Network diagrams, 21, 215–218
Neutrality, 13, 21, 47, 57, 59, 63, 80, 88, 223
Nightingale, Florence, 32–36
Nonosabasut, 104, 112

O

Objectivity, 21, 44, 55–60, 76, 80–81, 86–89, 98, 139, 191, 223, 226
Observable Plot, 92–94
Occlusion, 67
Ojibwe, 127
O'Neil, Cathy, 122
Ọnọụha, Mimi, 192–193
Oppression, 13, 45, 168, 178, 193, 229
Optical character recognition (OCR), 203
Orangi, Morgan, 235

Oresme, Nicole, 30–31
Organizing, 25, 28, 232

P

p5, 209, 219
Parrey (British naval captain), 51, 57
Participatory learning, 148, 152, 159
Peabody, Elizabeth Palmer, 44–45, 136–166, 185, 214, 218, 221
Pearce, Margaret Wickens, 118
Pearson, Karl, 98
Penn family, 122
Periscopic, 86–88
Peyton, John Jr., 103–133
Philip II (king of Spain), 15, 21, 38
Photoshop, 95
Pie charts, 9, 28–29, 77, 194–203
Playfair, William, 9–12, 32–34, 44, 73–100, 105, 122, 139, 149, 152, 161–163, 171, 189, 191, 210, 214. *See also Commercial and Political Atlas*
  education of, 9–11
  innovations of, 9, 32, 77
  untold stories in, 11
Plot, Robert, 31–32, 51
Polack, Fiona, 103
Polish-American system, 44, 137, 140, 146–150. *See also* Peabody, Elizabeth Palmer
Posavec, Stefanie, 26, 28
Powhatan, 143
Priestley, Joseph, 22–24, 152
Princeton University
  Special Collections, 72, 148
  VizE Lab for Ethnographic Data Visualization, 191–192
Progressive disclosure, 49–55, 103–115

Q

Quantitative information. *See also* Data collection; Data visualization
  abolition and, 51, 57
  colonialism and, 33
  Du Bois and, 189, 199, 203
  empiricism and, 50
  Halley and, 33
  Oresme and, 31
  power of, 52
  in relation to qualitative, 51, 57, 203
  Tufte and, 9, 24, 98
Quashie, Kevin, 67, 225
QuickDraw, 207
Quilting, 159–165, 212–214, 231

R

Racism, 45, 171–172, 178, 184, 189, 193, 230
Rankin, Bill, 126, 236
Reconstruction, 43, 46, 178, 183
Red ray, 186
Re-Indigenization, 129
Resistance, 43, 60–67, 98, 119, 135–137, 222, 230
Rogers, William Andrew, 169–175, 178, 180, 185–204
Ronda, Bruce, 140
Rooms, The (St. John's, Newfoundland), 109, 117, 131, 133, 236
Rosenberg, Daniel, 73, 140, 236
Royal Society, 31, 36, 51
Ruest, Annina, 28–29
Rumsey, David. *See* David Rumsey Map Collection
Rusert, Britt, 61, 171, 235

S

Schulten, Susan, 123, 126, 152
Scientific Revolution, 30
Sectarianism, 149
Self-governance, 42
Sense, Sarah, 128–129
Sensitive images, 51, 55
Shanawdithit, 43, 102–135, 137, 139, 147
Sharma, Tanvi, 12–13, 51, 65–66, 95, 106–107, 116–117, 132–133, 144–145, 160, 174–177, 188, 195, 197–198, 200–201, 203, 206–207, 209, 212–213, 215–217, 219–220, 235
Shneiderman, Ben, 24–25, 159, 214
Siksika, 122–123
Simon, Edward Lee, 173, 178, 188–190, 193–204
1619 Project, 146
Slack, 207
Slavery. *See also* Abolition
  capitalism and, 11, 43, 49
  Middle Passage, 52–53, 55, 60–63, 67, 210, 224
  SEAST in, 49, 50, 55, 57, 74, 77, 80, 89, 208
  slave ships, 43–44, 49–60, 71–74, 142, 204
  trade in, 43–74
  US Civil war and, 43, 46, 122, 137, 148, 178, 183
  Voyages Database and, 60–69
Small multiples, 193, 214
Smith, Adam, 10
Smith, Shawn Michelle, 183
Snow, John, 32–34

Sparke, Matthew, 199
Spence, Ian, 10, 236
Spillers, Hortense, 53–56, 224–226
Spiral chart. *See* Du Bois, W. E. B.: Du Boisian spiral
Spreadsheets, 10, 195
Stack, 208–210, 214, 218
Statistical atlases, 98–99, 181–182. *See also Commercial and Political Atlas*
Statistical transformation. *See* Wilkinson, Leland
Statistics, 32, 33, 98, 172, 175, 179, 180–184
Stereoscopes, 185
SVG, 209

T

Tech stack. *See* Stack
Teedyuscung, 122
Thoreau, Henry David, 137
Time series charts, 44, 80, 93, 214. *See also* Line graphs
Tongva, 104
Transcendentalism, 137
Treemaps, 214. *See also* Shneiderman, Ben
Trump, Donald J., 229–230, 237
Tufte, Edward, 9–12, 24, 32–33, 35, 56, 98, 118–119, 164
Turner Contemporary, 163
Twillingate (Newfoundland), 105–107

U

US Bureau of Labor Statistics, 180
US Census, 24, 172, 181, 189
US Civil War, 43, 46, 122, 137, 148, 178, 183

V

Varner, Jay, 12, 51, 66, 95–97, 145, 175, 197–198, 206, 209–218, 221, 235
Vector-based paths, 95
Vibe coding, 210
Viégas, Fernanda Bertini, 18, 57
Visualization. *See* Data visualization
Vocal Type, 172

W

Wainer, Howard, 10, 30, 59
Wattenberg, Martin, 18, 236
Weaving patterns, 128–129
Wells, Ida B., 178–179
Westmoreland, William George, 173, 178, 188–189, 194, 199–203
Whitney Museum of American Art, 163
Wikipedia, 28
Wilberforce, William, 59
Wilkins, F. H., 186
Wilkinson, Leland, 94, 226
Willard, Emma Hart, 122–126, 149–157, 165
Wilson, Mabel O., 178
Womack, Autumn, 179
Wood, Jo, 92–93
Wood, Marcus, 56, 59, 71
Wu, Shirley, 25–26

Y

Yale University, Beinecke Library, 58, 148
Yang, Nicholas, 12, 198, 215–217, 245
Yannakakis, Yanna, 20, 236
Yuhuiayu (community), 19–20

Z

Zotero, 210, 214

KEY

- Explainer Image
- Visualization
- Annotated Visualization
- Visualization by DxD Team
- Screenshot
- Photo
- Diagram

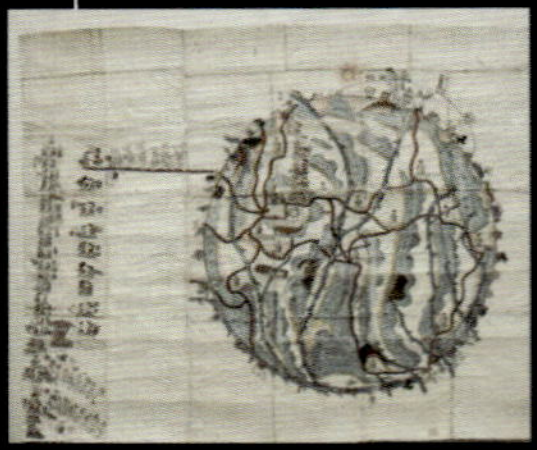

Mapa de Teozacoalco
FIGURE 0.1 PG 1

FIGURE 0.4A PG 20

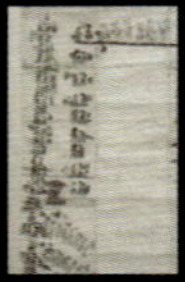

FIGURE 0.4B PG 20

Wind Map
FIGURE 0.2 PG 18

The Wind Map in the "Energy" exhibition at MoMA
FIGURE 0.3 PG 18

A Chart of Biography
FIGURE 0.5 PG 22

A New Chart of History
FIGURE 0.6 PG 23

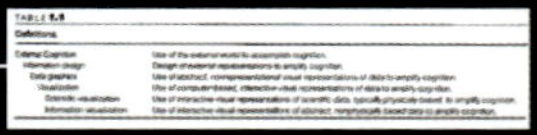

Definitions of Visualization
FIGURE 0.7 PG 24

US Workers on Strike
FIGURE 0.8 PG 25

Hong Kong Artists, Women
FIGURE 0.9 PG 26

A Week of Clocks
FIGURE 0.10A PG 25

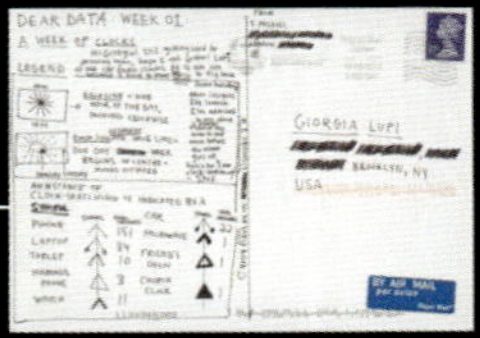

A Week of Clocks
FIGURE 0.10B PG 26

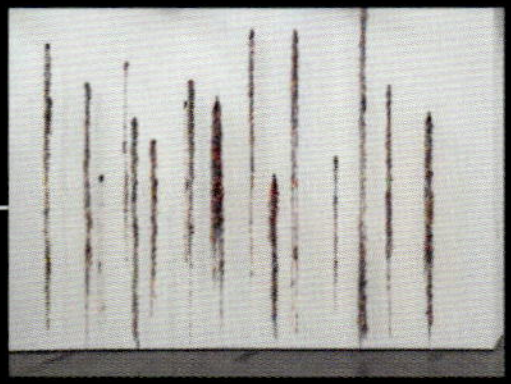

Border Quipu / Quipu Fronterizo
FIGURE 0.11 PG 26

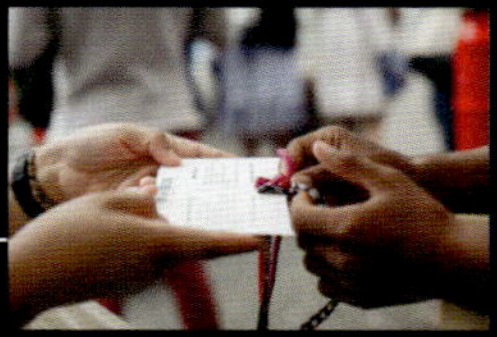

Border Quipu / Quipu Fronterizo
FIGURE 0.12 PG 27

A Piece of the Pie Chart
FIGURE 0.13 PG 27

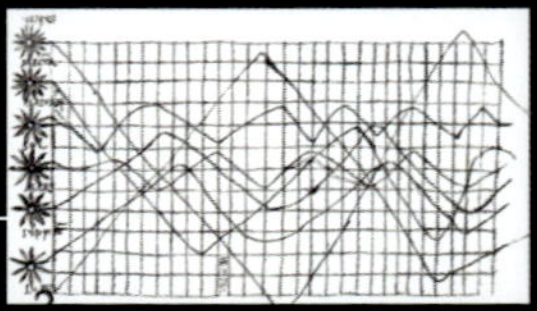

The Changing Locations of Constellations in the Night Sky
FIGURE 0.14 PG 30

Relationships between Quantities
FIGURE 0.15 PG 30

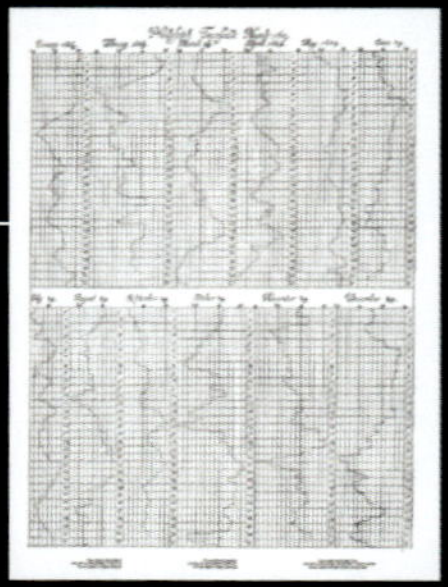

History of the Weather
FIGURE 0.16 PG 30

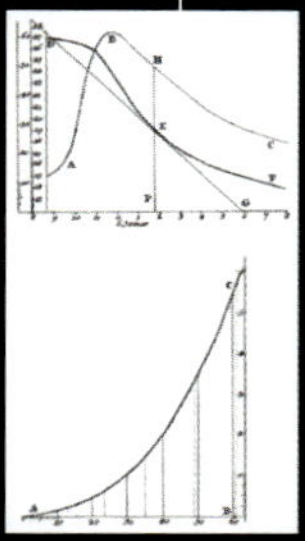

Essai d'hygrométrie ou sur la mesure de l'humidité

FIGURE 0.17 PG 30

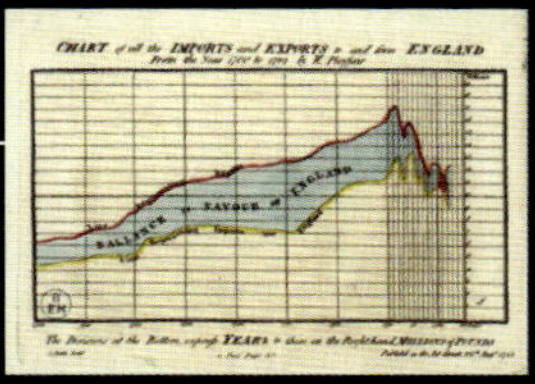

Chart of All the Imports and Exports to and from England from the Year 1700 to 1782

FIGURE 0.18 PG 34

Report on the Cholera Outbreak in the Parish of St. James, Westminster, during the Autumn of 1854

FIGURE 0.19 PG 34

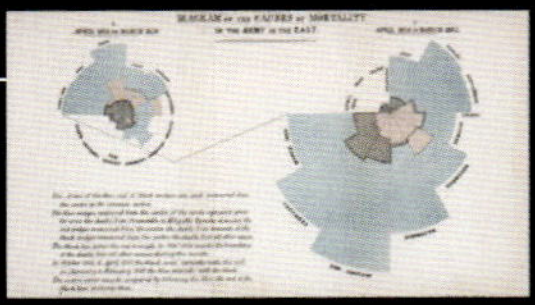

Diagram of the Causes of Mortality in the Army in the East

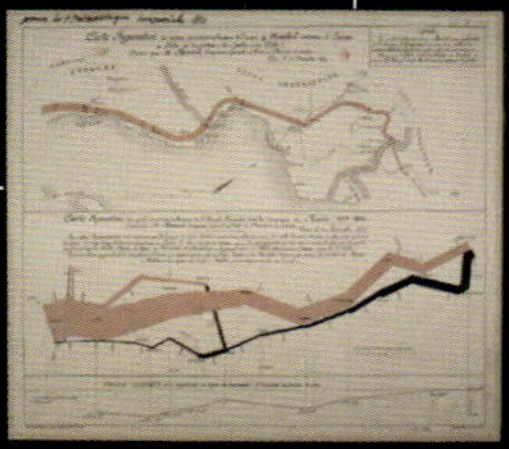

Carte Figurative des pertes successives en hommes de l'armée française dans la campagne de Russie 1812–1813 comparées à celle d'Hannibal durant la 2ème Guerre Punique

FIGURE 0.21 PG 35

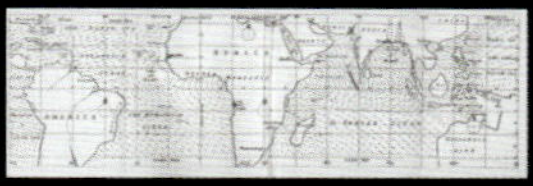

Map of the Trade Winds

FIGURE 0.22 PG 36–37

Inkan Khipu

FIGURE 0.23 PG 39

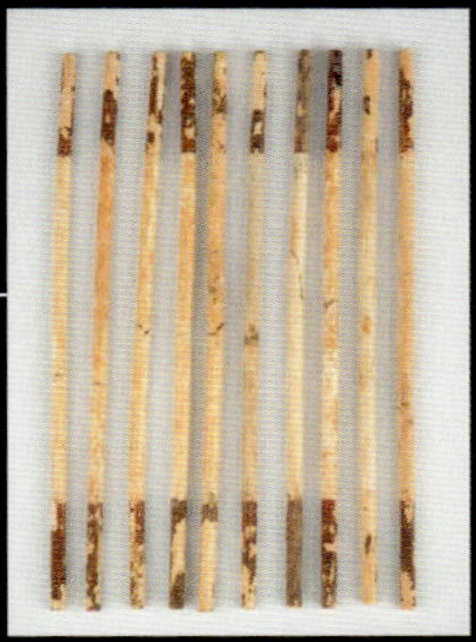

Chinese Counting Rods

Proto-Cuneiform Tablet

FIGURE 0.25 PG 39

Shogunal Map

FIGURE 0.26 PG 40

Jain Diagram of Jambudvīpa

FIGURE 0.27 PG 41

Portrait of Thomas Clarkson
FIGURE 1.1 PG 49

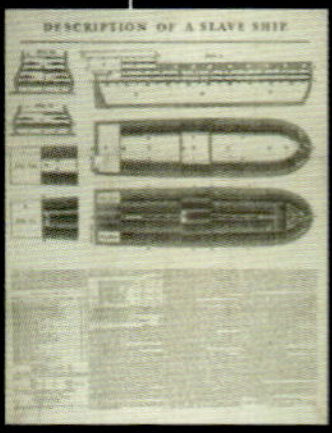

Description of a Slave Ship
FIGURE 1.5 PG 58

Portrait of Olaudah Equiano
FIGURE 1.2 PG 49

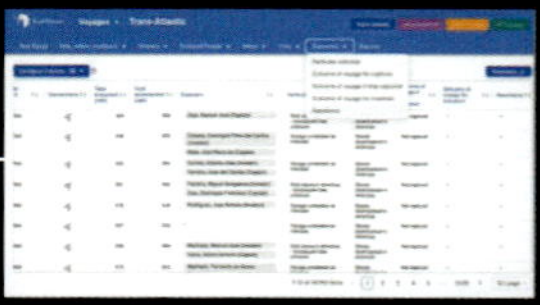

Web Interface of the Voyages Database
FIGURE 1.6 PG 61

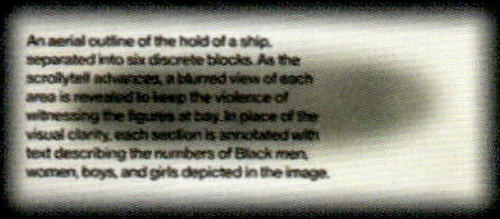

Example of sensitive image and alt text
FIGURE 1.3A PG 51

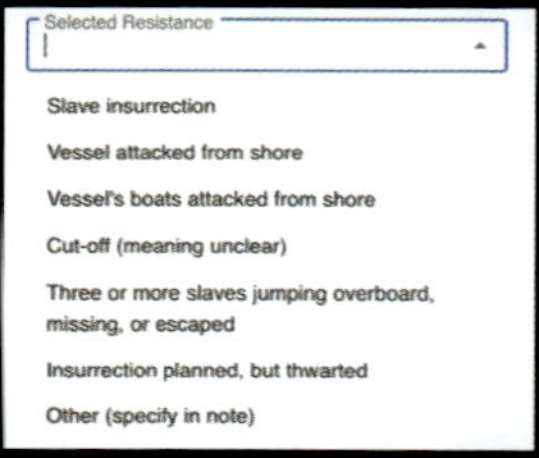

FIGURE 1.7 PG 61

Plan of an African Ship's Lower Deck with Negroes in the Proportion of Only One to a Ton
FIGURE 1.3B PG 52-53

Geological Investigation of the Alluvial Valley of the Lower Mississippi River
FIGURE 1.8 PG 62

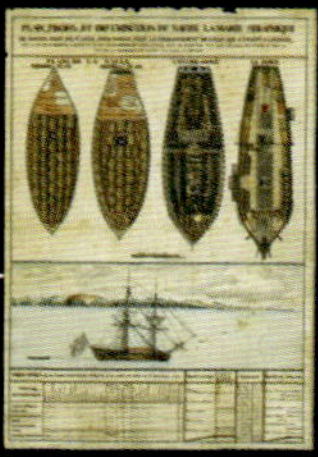

Plan, profil, et distribution du navire la Marie Séraphique de Nantes
FIGURE 1.4 PG 54

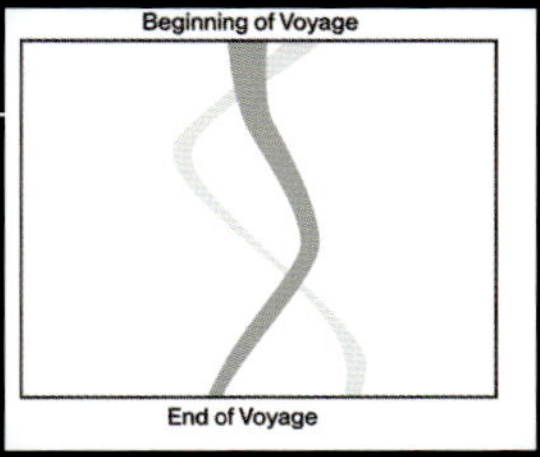

FIGURE 1.9A PG 64

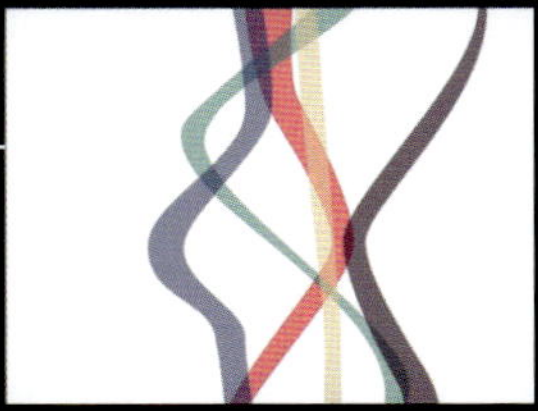

FIGURE 1.9B PG 64

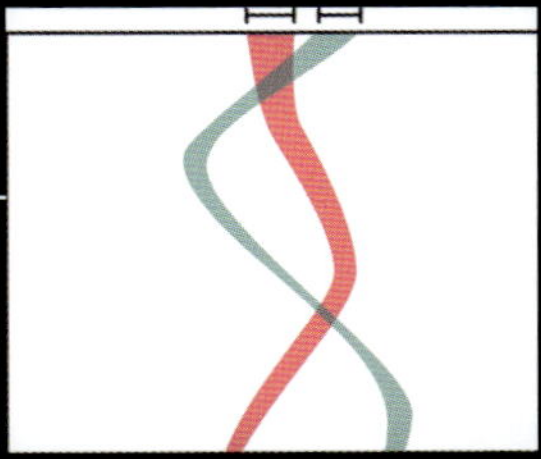

FIGURE 1.9C PG 64

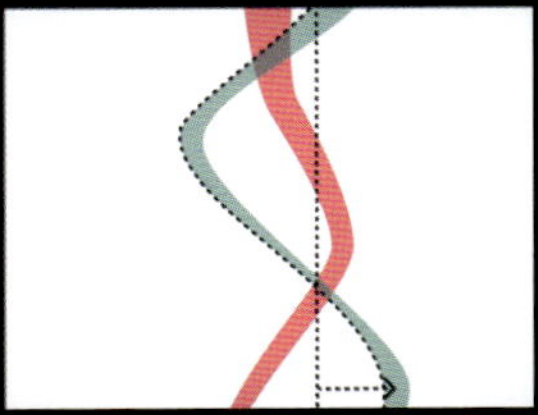

FIGURE 1.9D PG 65

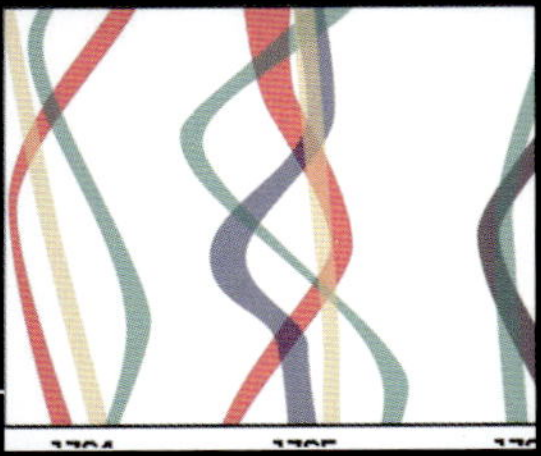

FIGURE 1.9E PG 65

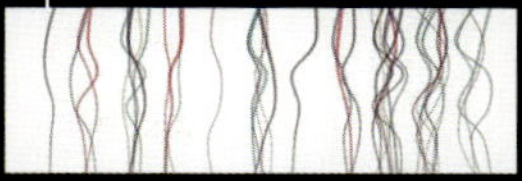

Visualizing Resistance, Phase 1

FIGURE 1.10 PG 66

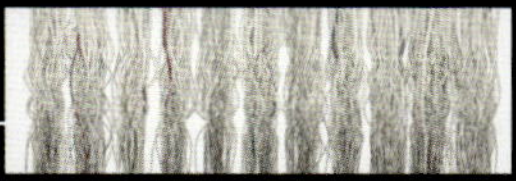

Visualizing Resistance, Phase 2

FIGURE 1.11 PG 66

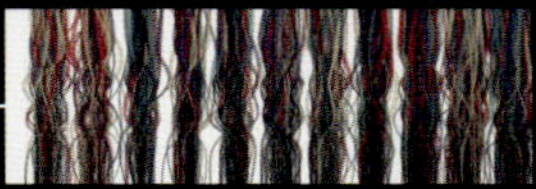

Visualizing Resistance, Phase 3

FIGURE 1.12 PG 66

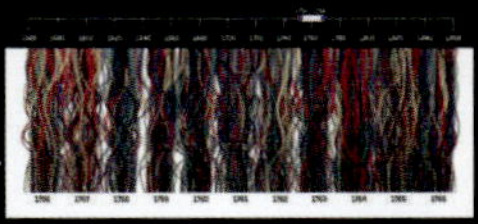

Visualizing Resistance

FIGURE 1.13 PG 68–69

From The History of the Rise, Progress, and Accomplishment of the Abolition of the African Slave-Trade by the British Parliament

FIGURE 1.14 PG 70

Strom der Zeiten

FIGURE 1.15 PG 72

Stream of Time

FIGURE 1.16 PG 72

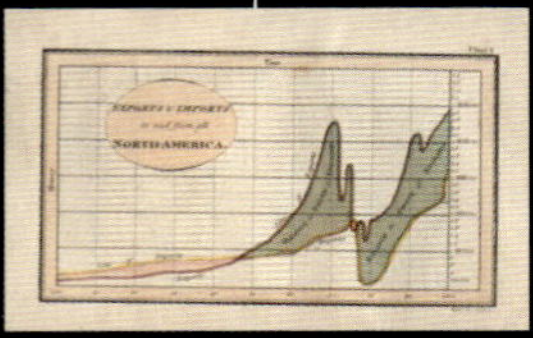

Exports & Imports to and from all North-America

FIGURE 2.1 PG 78–79

FIGURE 2.1B PG 77

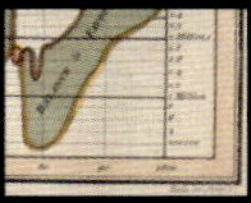

FIGURE 2.1B PG 77

FIGURE 0.4A PG 20

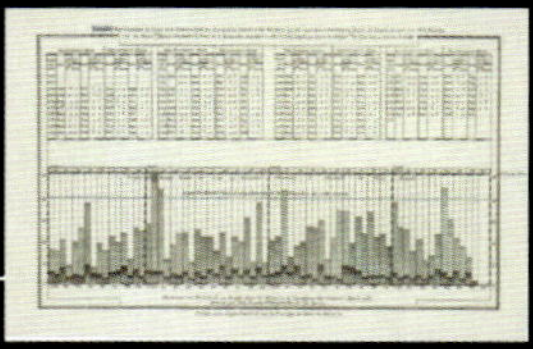

Profils représentants la crue et la diminution des eaux de la Seine et des rivières qu'elle reçoit dans le Paris-haut au dessus de Paris

FIGURE 2.2 PG 80

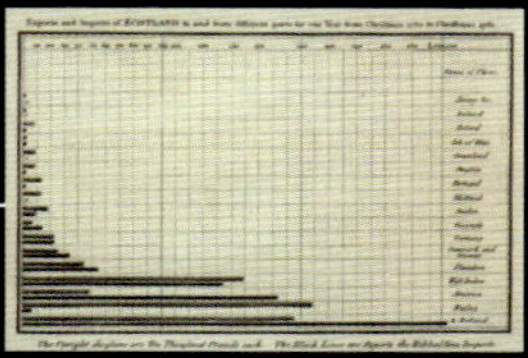

Exports and Imports of Scotland to and from different parts for one Year from Christmas 1780 to Christmas 1781

FIGURE 2.3 PG 80

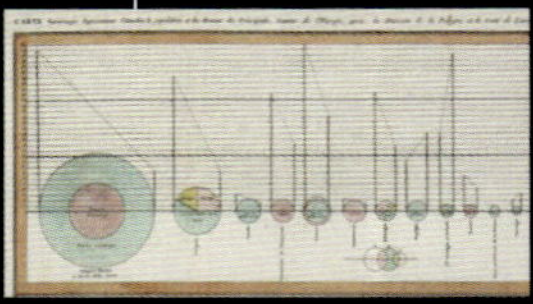

Chart Representing the Extent, Population & Revenues, of the Principal Nations in Europe, after the Division of Poland & Treaty of Luneville

FIGURE 2.4 PG 81

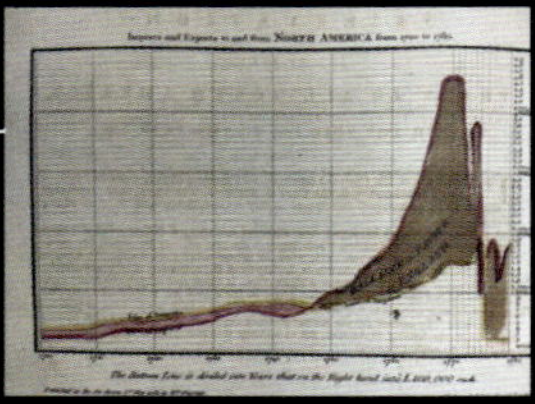

Imports and Exports to and from North America from 1700 to 1780

FIGURE 2.5 PG 82

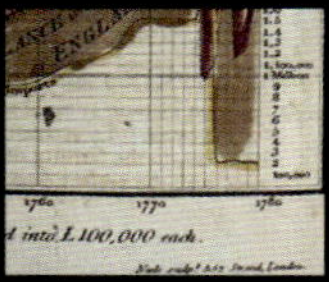

FIGURE 2.8A PG 85

PLATE V. AMERICA.

Years.	Imports.	Exports.	Balance.
1770,	1,480,000	4,550,000	3,070,000
1771,	1,430,000	4,630,000	3,200,000
1772,	1,445,000	3,600,000	2,155,000
1773,	1,465,000	2,465,000	1,000,000
1774.	1,435,000	3,840,000	2,300,000
1775,	2,065,000	985,000	1,080,000
[illegible]	245,000	1,190,000	945,000
1777,	230,000	1,820,000	1,650,000
1778,	265,000	1,150,000	885,000
1779,	295,000	1,370,000	1,075,000
1780,	300,000	1,805,000	1,505,000
1781,	385,000	1,545,000	1,160,000
1782,	295,000	905,000	610,000

America

FIGURE 2.6 PG 84

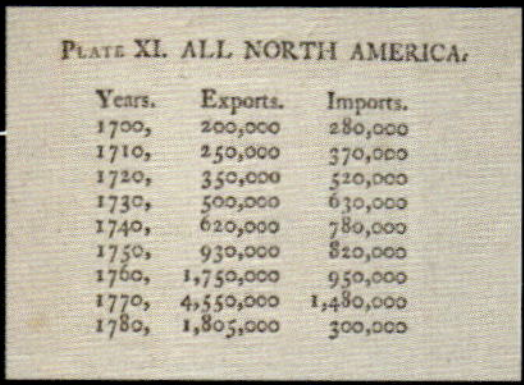

PLATE XI. ALL NORTH AMERICA.

Years.	Exports.	Imports.
1700,	200,000	280,000
1710,	250,000	370,000
1720,	350,000	520,000
1730,	500,000	630,000
1740,	620,000	780,000
1750,	930,000	820,000
1760,	1,750,000	950,000
1770,	4,550,000	1,480,000
1780,	1,805,000	300,000

All North America

FIGURE 2.7 PG 84

U.S. Gun Killings in 2018

FIGURE 2.9 PG 86

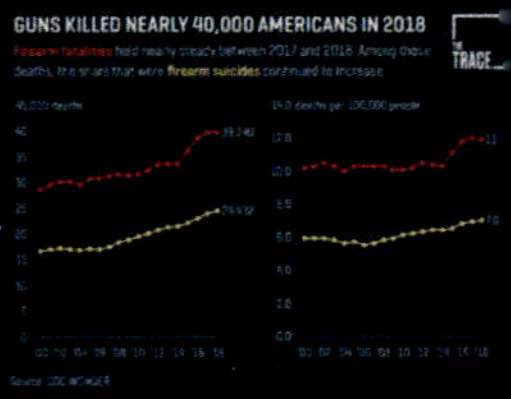

Guns Killed Nearly 40,000 Americans in 2018

FIGURE 2.10 PG 86

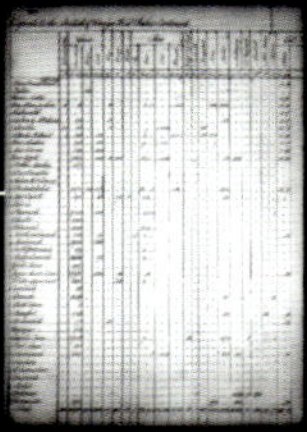

Exports to the British & Foreign West Indies

FIGURE 2.11 PG 90

Exports & Imports to and from the West Indies

FIGURE 2.12 PG 91

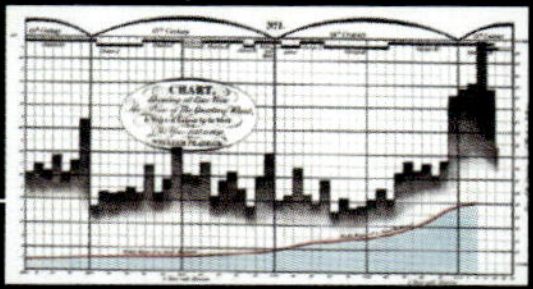

Chart, Shewing at One View the Price of the Quarter of Wheat, & Wages of Labour by the Week, from the Year 1656 to 1821

FIGURE 2.13 PG 92

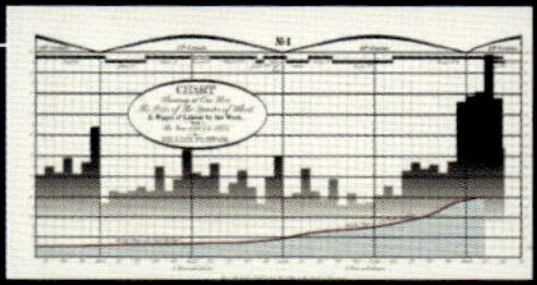

Recreation of "Chart, Shewing at One View the Price of the Quarter of Wheat, & Wages of Labour by the Week, from the Year 1656 to 1821"

FIGURE 2.14 PG 92

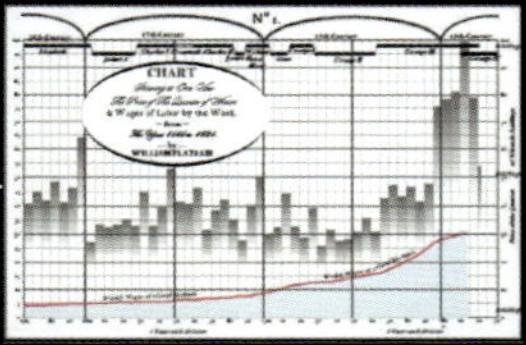

Recreation of "Chart, Shewing at One View the Price of the Quarter of Wheat, & Wages of Labour by the Week, from the Year 1656 to 1821"

FIGURE 2.15 PG 92

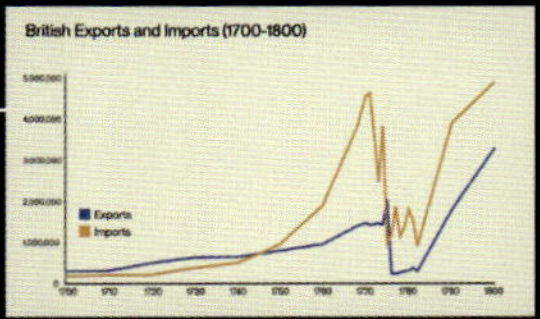

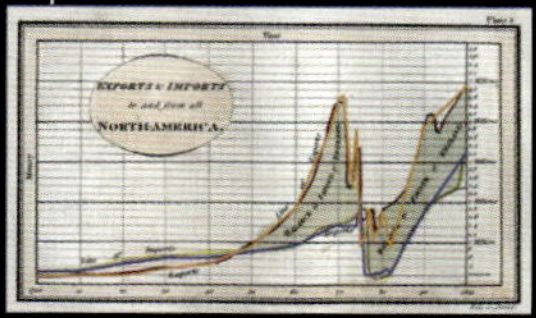

Comparison between Playfair's Data and His Chart

FIGURE 2.16B PG 95

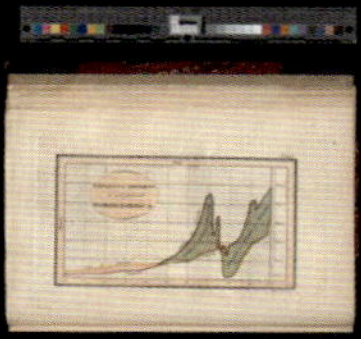

Uncropped Scan of Playfair's Chart

FIGURE 2.17A PG 96

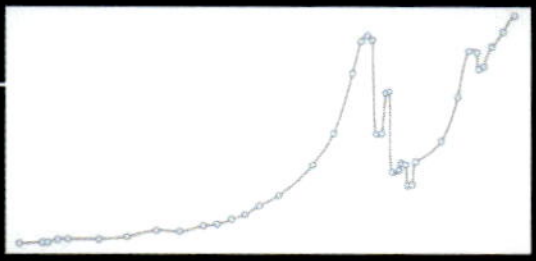

Path of Playfair's Data Line

FIGURE 2.17B PG 96

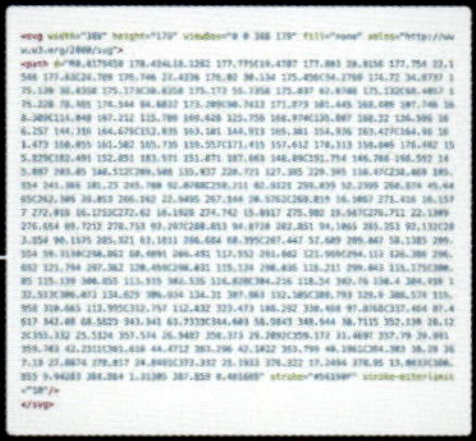

Vector Coordinates of Playfair's Data Line

FIGURE 2.17C PG 96

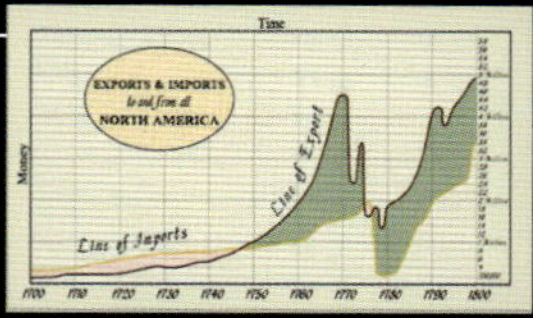

Recreation of "Exports & Imports to and from all North-America"

FIGURE 2.17D PG 97

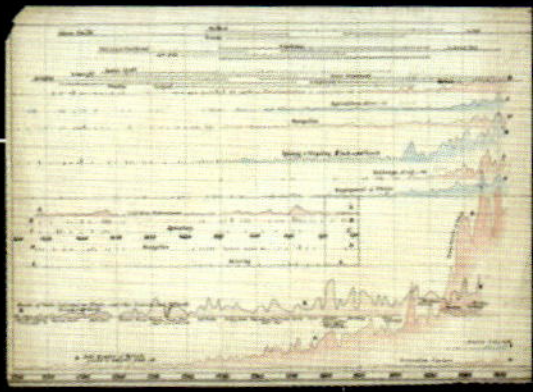

Chart by William Stanley Jevons

FIGURE 2.18 PG 99

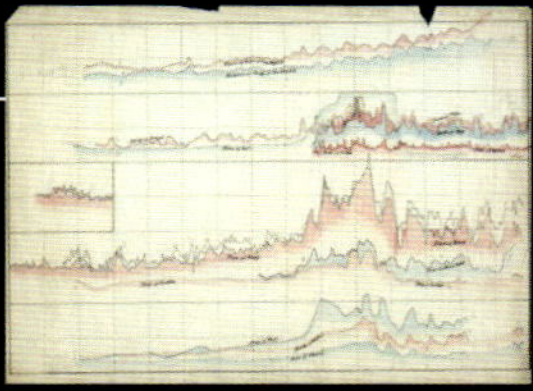

Chart by William Stanley Jevons

FIGURE 2.19 PG 99

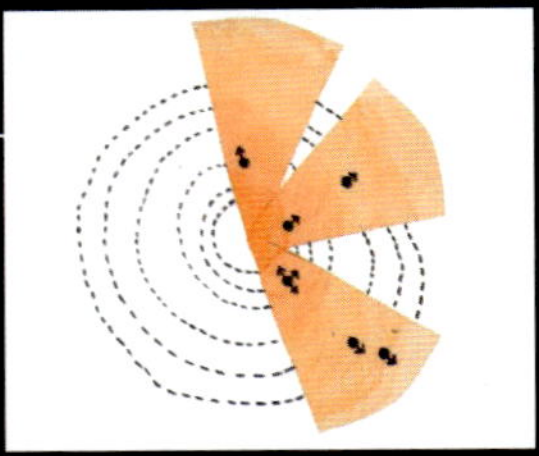

Diagram of Primary Sources
FIGURE 3.1 PG 106–107

The Kidnapping of Demasduit at Beothuk Lake, March 5, 1819 and Captain Buchan's Visit to Beothuk Lake in January 1811
FIGURE 3.2 PG 108–109

FIGURE 3.3 PG 110

FIGURE 3.4 PG 111

FIGURE 3.5 PG 111

FIGURE 3.6 PG 111

FIGURE 3.7 PG 112

FIGURE 3.8 PG 112

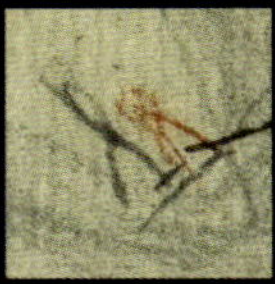

FIGURE 3.9 PG 113

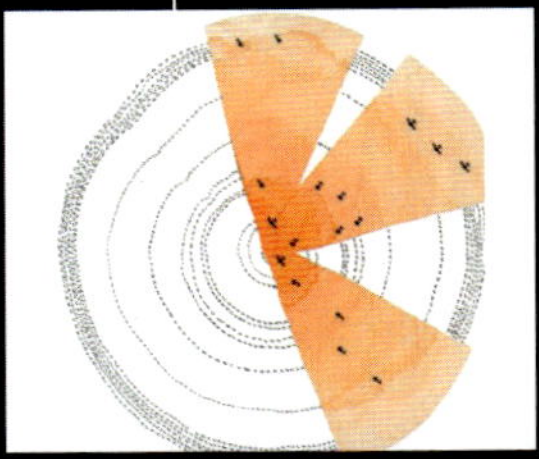

Diagram of People and Sources
FIGURE 3.10 PG 116–117

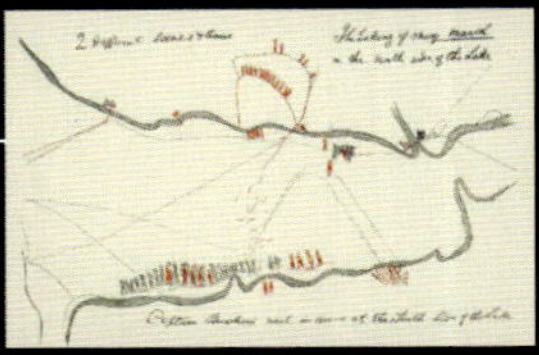

Howley's Recreation of Shanawdithit's map
FIGURE 3.11 PG 120

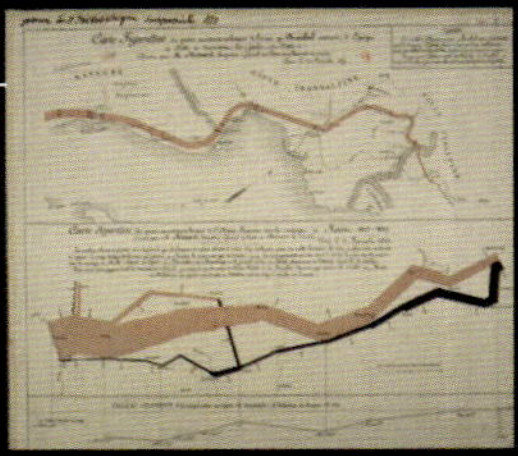

Carte Figurative des pertes successives en hommes de l'armée française dans la campagne de Russie 1812–1813 comparées à celle d'Hannibal durant la 2ème Guerre Punique
FIGURE 3.12 PG 121

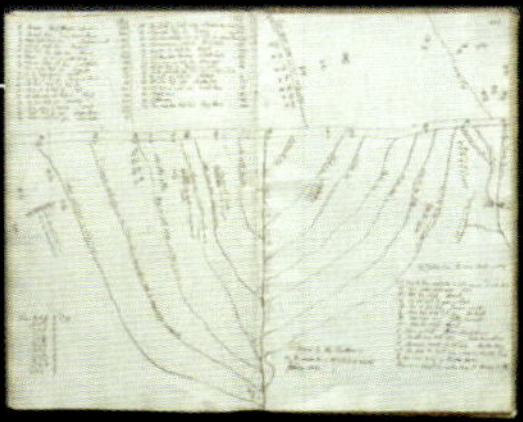

Map by Ac ko mok ki
FIGURE 3.13 PG 123

Introductory Map: Locations and Wanderings of the Aboriginal Tribes
FIGURE 3.14 PG 124

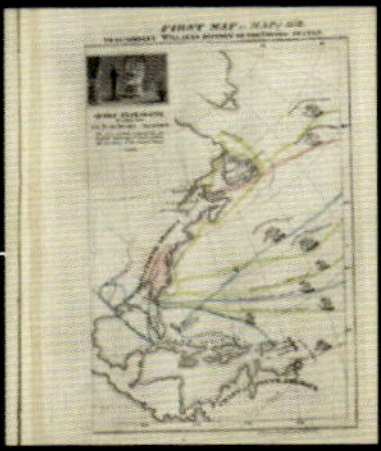

First Map or Map of 1578
FIGURE 3.15 PG 125

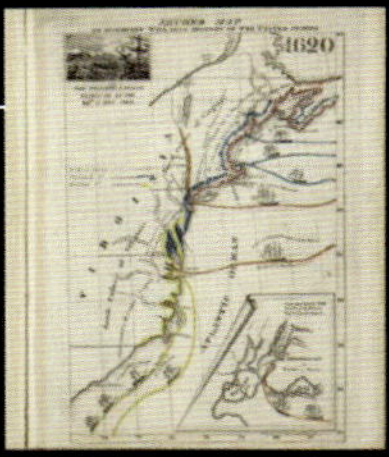

Second Map (1620)
FIGURE 3.16 PG 125

Third Map (1643)
FIGURE 3.17 PG 125

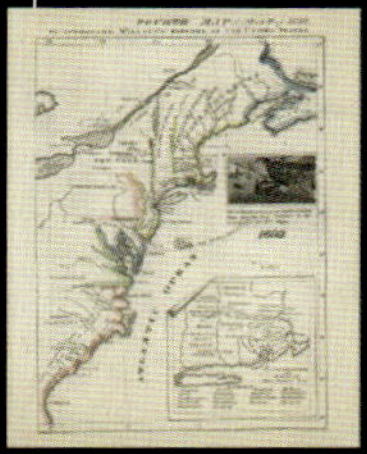

Fourth Map (1692)
FIGURE 3.18 PG 125

Fifth Map (1733)
FIGURE 3.19 PG 125

Sixth Map (1763)
FIGURE 3.20 PG 125

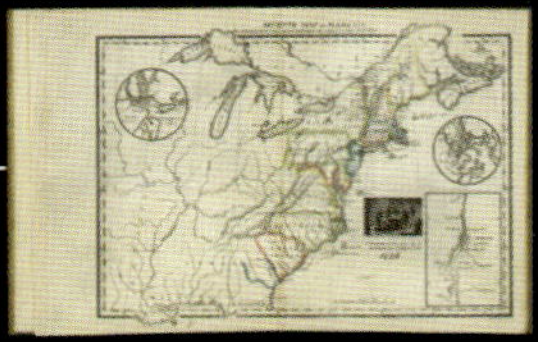

Seventh Map (1776)
FIGURE 3.21 PG 125

Eighth Map (1789)
FIGURE 3.22 PG 125

Ninth Map (1826)
FIGURE 3.23 PG 125

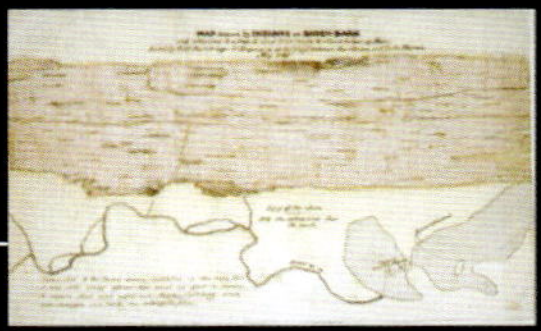

Map Drawn by Indians on Birch Bark and Attached to a Tree to Show Their Route to Others Following Them
FIGURE 3.24 PG 127

Birch Bark
FIGURE 3.25 PG 128

Lookout Tree at Red Indian Point
FIGURE 3.26 PG 130

View Across Red Indian Lake from Red Indian Point
FIGURE 3.27 PG 131

Diagram of People, Sources, and Tree
FIGURE 3.28 PG 132-133

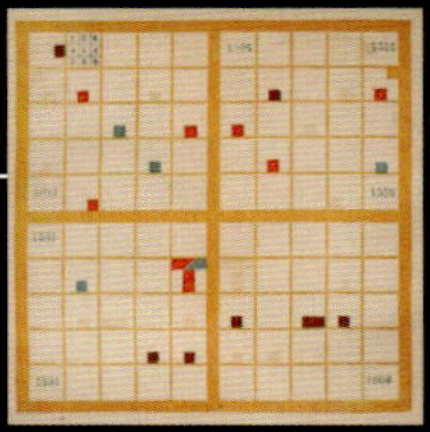

Chronological History of the Sixteenth Century
FIGURE 4.1 PG 138

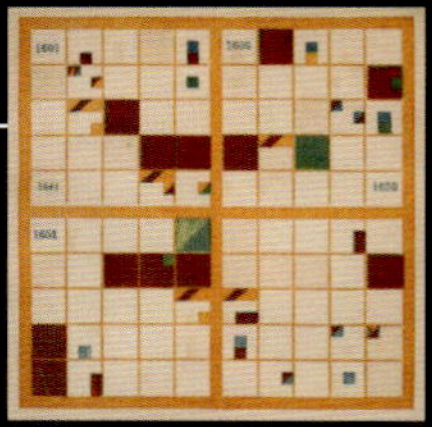

Chronological History of the Seventeenth Century
FIGURE 4.2 PG 138

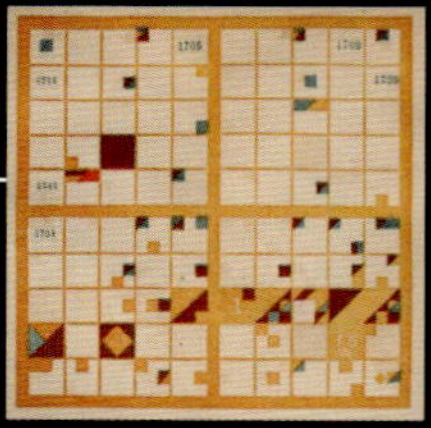

Chronological History of the Eighteenth Century
FIGURE 4.3 PG 138

Chronological History of the Nineteenth Century
FIGURE 4.4 PG 138

Chronological Exercise
FIGURE 4.5 PG 140

FIGURE 4.6 PG 141

FIGURE 4.7 PG 141

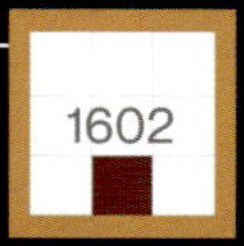

FIGURE 4.8 PG 142

FIGURE 4.9 PG 142

FIGURE 4.10 PG 142

FIGURE 4.11 PG 143

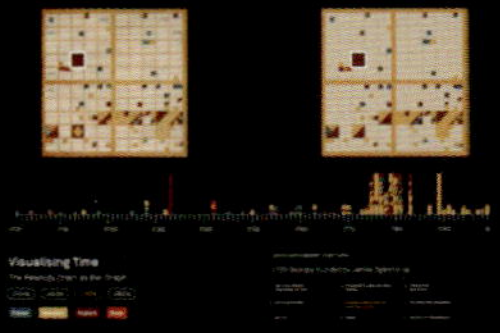

Visualizing Time
FIGURE 4.12 PG 144–145

A Blank Chart from Monographs of History
FIGURE 4.13 PG 147

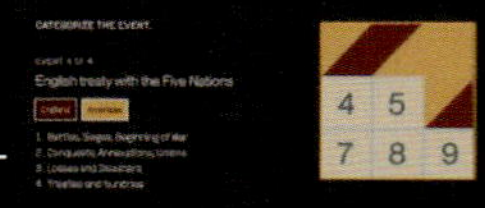

FIGURE 4.14 PG 148

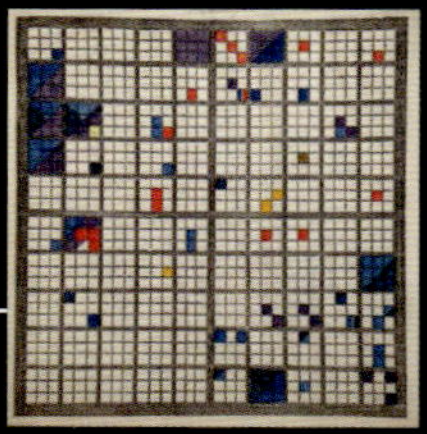

Student Chart
FIGURE 4.15 PG 150

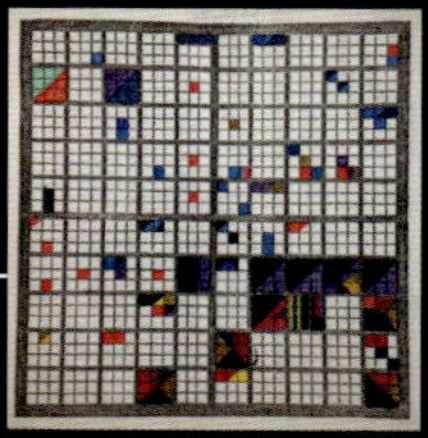

Student Chart
FIGURE 4.16 PG 150

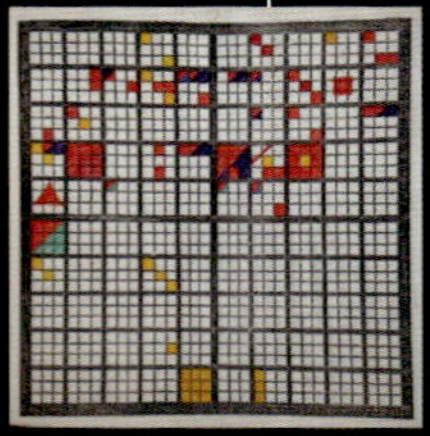

Student Chart
FIGURE 4.17 PG 150

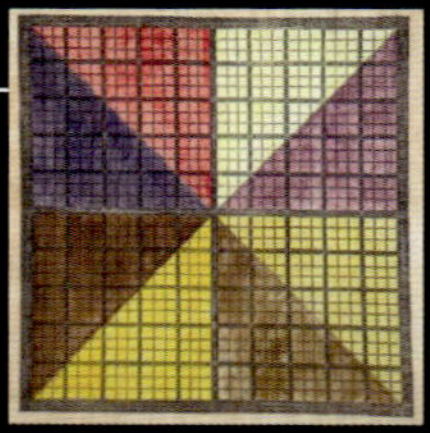

Student Chart
FIGURE 4.18 PG 150

Student Chart
FIGURE 4.19 PG 151

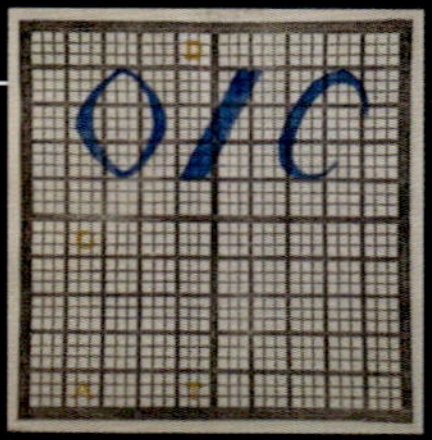

Student Chart
FIGURE 4.20 PG 151

Student Chart

FIGURE 4.21 PG 151

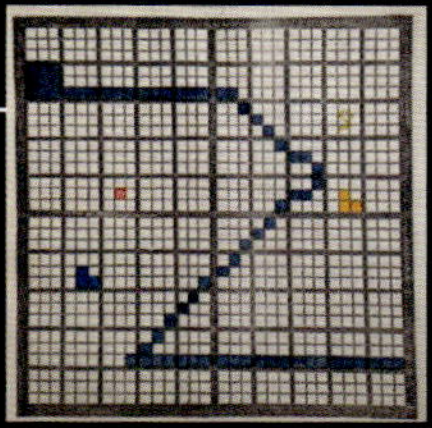

Student Chart

FIGURE 4.22 PG 151

Temple of Time

FIGURE 4.23 PG 153

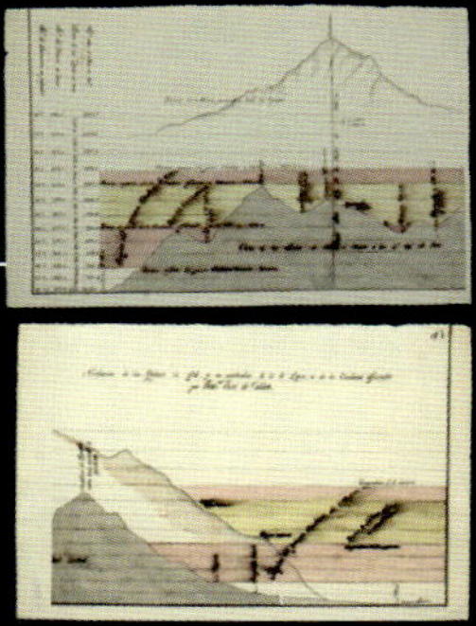

Elevation Profile of Chimborazo

FIGURE 4.24 PG 154–155

Diagram of the Vegetation of Mount Chimborazo

FIGURE 4.25 PG 156

Top of Chimborazo

FIGURE 4.26 PG 156

Picture of Nations or Perspective Sketch of the Course of Empire

FIGURE 4.27 PG 157

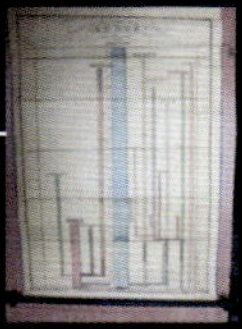

Clarke's Charts in her Family's Attic

FIGURE 4.28 PG 158

Clarke's Charts in her Family's Attic

FIGURE 4.29 PG 158

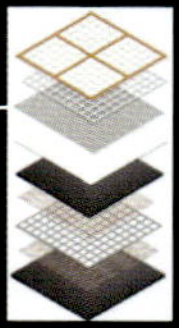

Diagram of the Floor Chart

FIGURE 4.30A PG 160

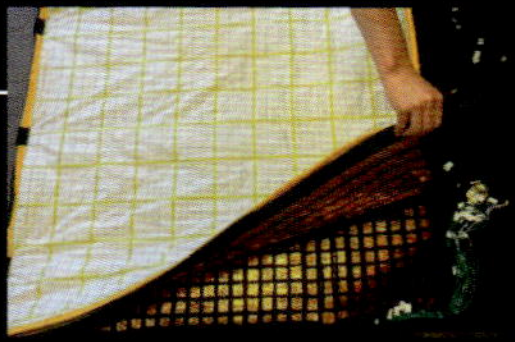

Floor Chart

FIGURE 4.30B PG 160

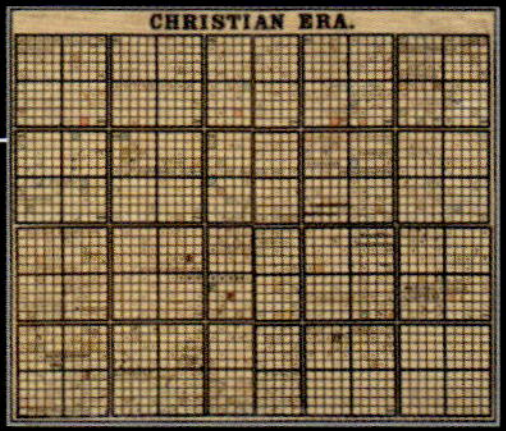

Christian Era

FIGURE 4.31 PG 162

Housetop—Sixteen-Block "Half-Log Cabin" Variation Sashed with Feed Sacks

FIGURE 4.32 PG 164

Housetop Variation

FIGURE 4.33 PG 164

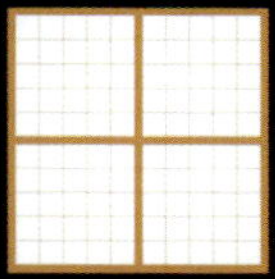

FIGURE 4.34 PG 166

Exhibit of American Negroes at the Paris Exposition
FIGURE 5.1 PG 170

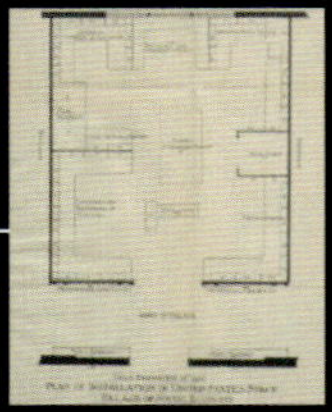

Floor Plan of the Palace of Social Economy and Congresses
FIGURE 5.2 PG 171

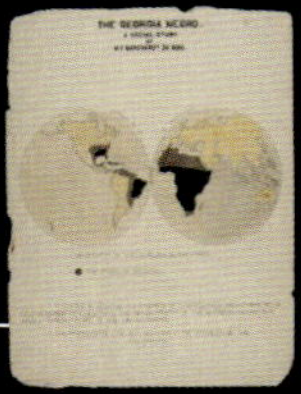

The Georgia Negro: A Social Study
FIGURE 5.3 PG 173

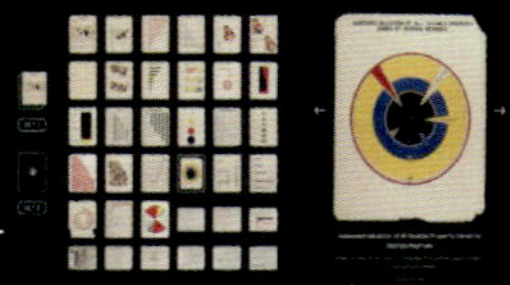

The 36 charts of The Georgia Negro: A Social Study
FIGURE 5.4 PG 174–175

The 27 extant charts in A Series of Statistical Charts
FIGURE 5.5 PG 176–177

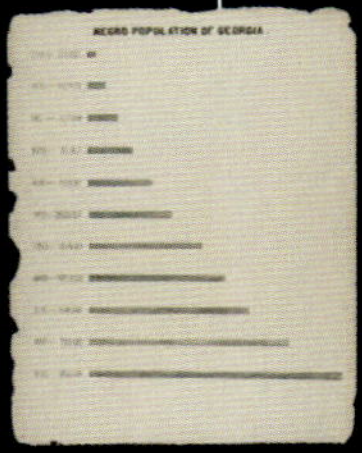

Negro Population of Georgia
FIGURE 5.6 PG 178

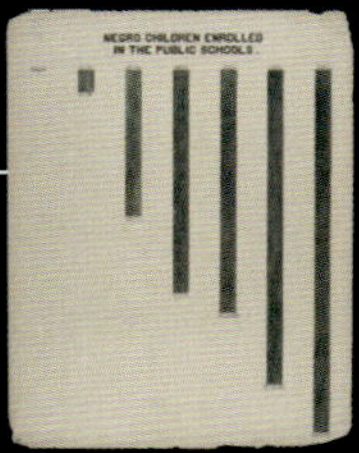

Negro Children Enrolled in the Public Schools
FIGURE 5.7 PG 178

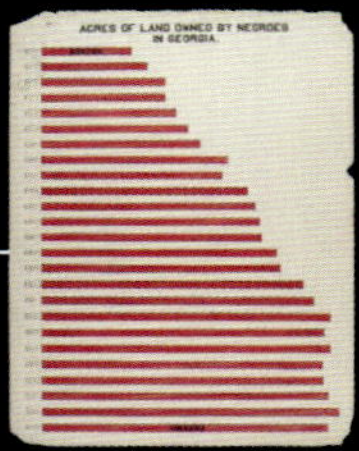

Acres of Land Owned by Negroes in Georgia
FIGURE 5.8 PG 178

Assessed Value of Household and Kitchen Furniture Owned by Georgia Negroes
FIGURE 5.9 PG 179

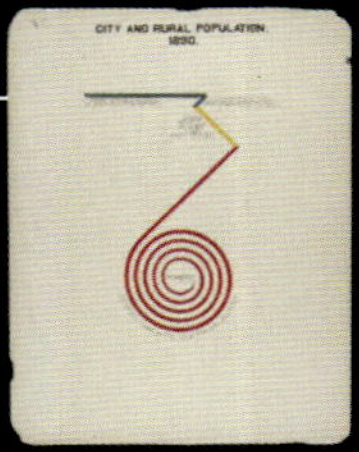

City and Rural Population, 1890
FIGURE 5.10 PG 179

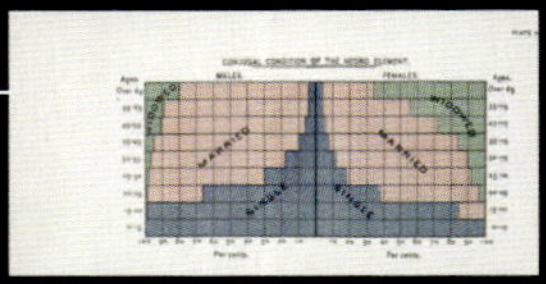

Conjugal Condition of the Negro Element
FIGURE 5.11 PG 180

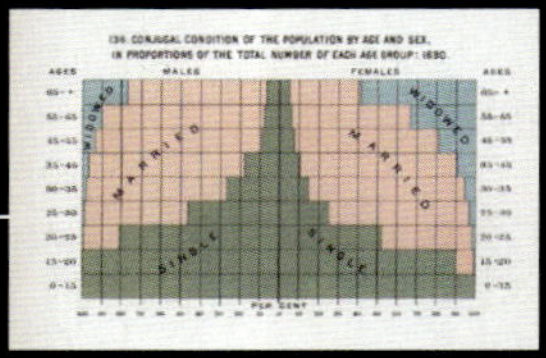

Conjugal Condition of the Population by Age and Sex, in Proportions of the Total Number of Each Age Group, 1890
FIGURE 5.12 PG 180

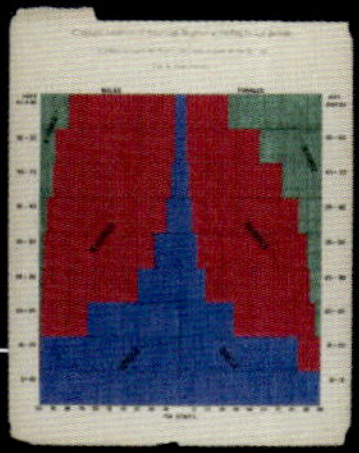

Conjugal Condition of American Negroes According to Age Periods
FIGURE 5.13 PG 181

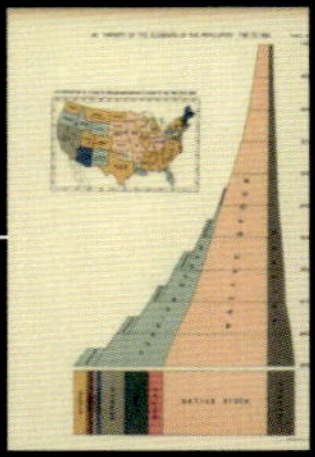

Growth of the Elements of the Population, 1790 to 1890
FIGURE 5.14 PG 182

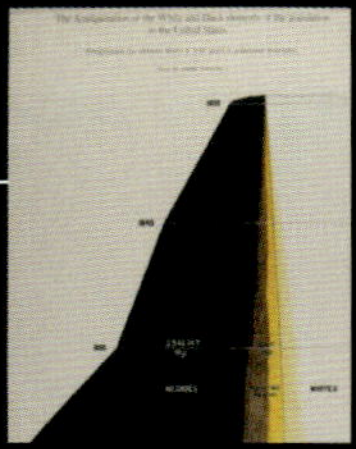

The Amalgamation of the White ar Black Elements of the Population the United States
FIGURE 5.15 PG 182

Four Atlanta University Women Students Posed on the Steps of a Classroom Building
FIGURE 5.16 PG 184

Perfectscope Showing Images of t Palace of Social Economy and Congresses
FIGURE 5.17 PG 185

The Alumni Listings in the Catalogue of the Officers and Students of Atlanta University, 1909–1910
FIGURE 5.18 PG 187

Occupations and Places of Residence of the Members of the Atlanta University Class of 1900
FIGURE 5.19 PG 188

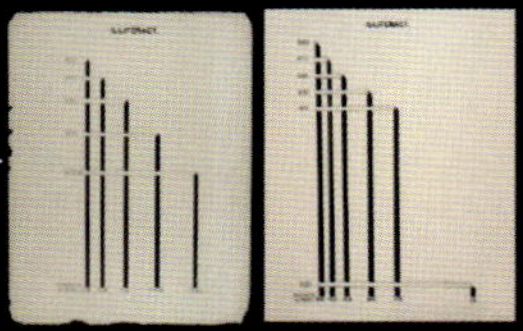

Illiteracy
FIGURE 5.20 PG 190

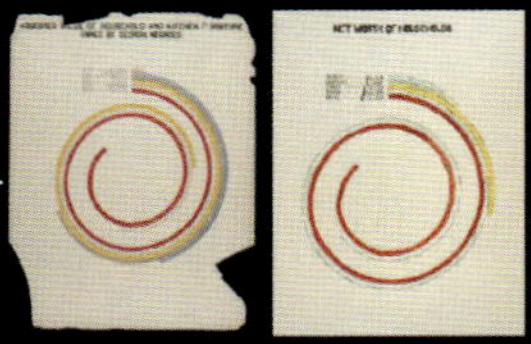

Net Worth of Households
FIGURE 5.21 PG 190

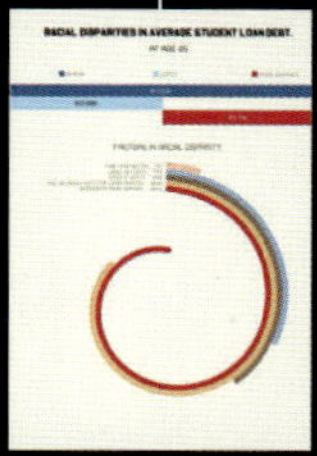

Factors in Racial Disparity
FIGURE 5.22 PG 191

In Absentia: This Land Is Your Land
FIGURE 5.23 PG 192

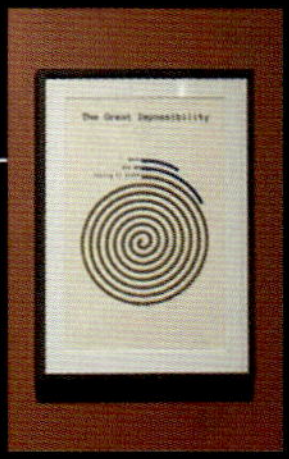

In Absentia: The Great Impossibility
FIGURE 5.24 PG 192

In Absentia: It Could Never Be Large Enough
FIGURE 5.25 PG 192

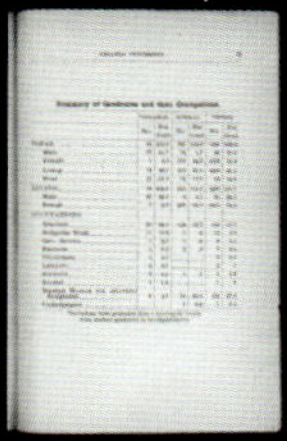

Summary of Graduates and Their Occupations
FIGURE 5.26 PG 194

FIGURE 5.27 PG 195

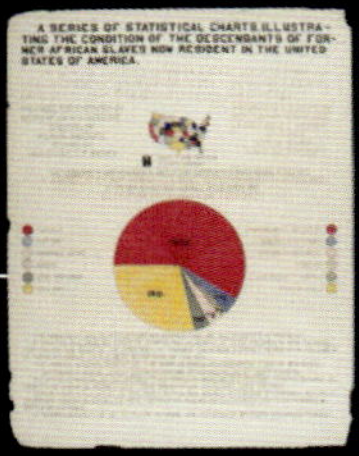

Title Image of A Series of Statistical Charts
FIGURE 5.28 PG 196

Center of Negro Population, Atlanta University
FIGURE 5.30 PG 198

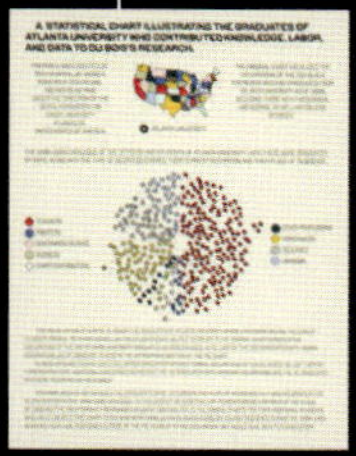

A Statistical Chart Illustrating the Graduates of Atlanta University Who Contributed Knowledge, Labor, and Data to Du Bois's Research
FIGURE 5.29 PG 197

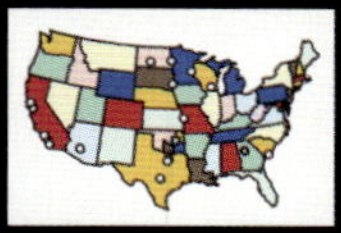

FIGURE 5.31 PG 198

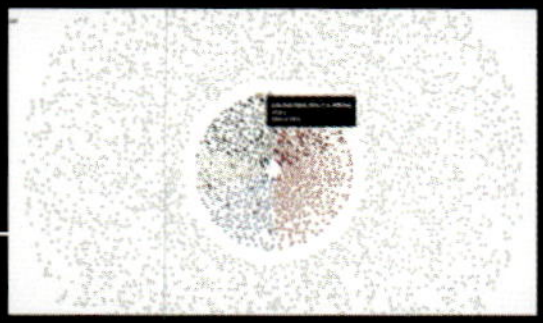

Students Known and Unknown
FIGURE 5.32 PG 200–201

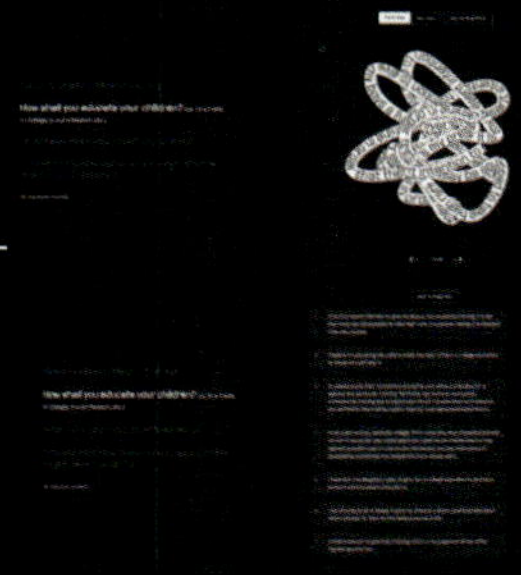

Entangled Data, Entangled Lives
FIGURE 5.33 PG 202

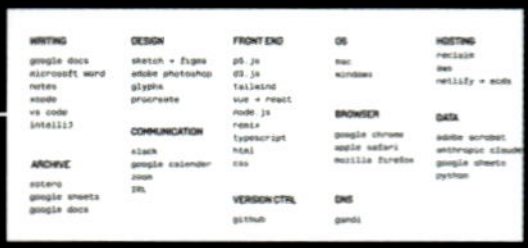

The Tech Stack of Data by Design

FIGURE 6.1 PG 209

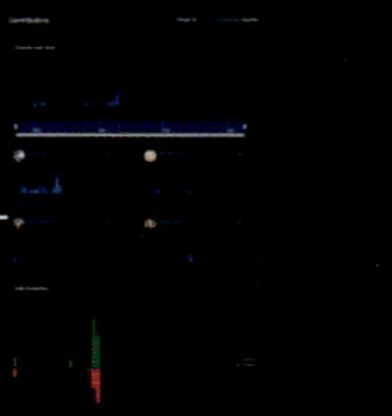

GitHub Insights

FIGURE 6.2 PG 211

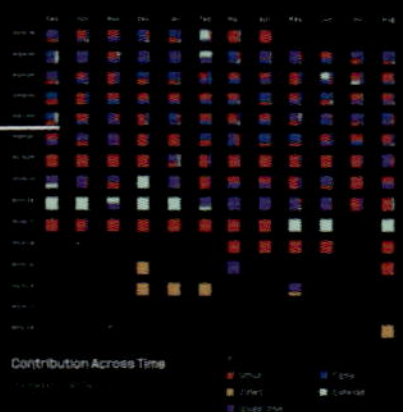

Quilt Chart

FIGURE 6.3 PG 212

FIGURE 6.4 PG 213

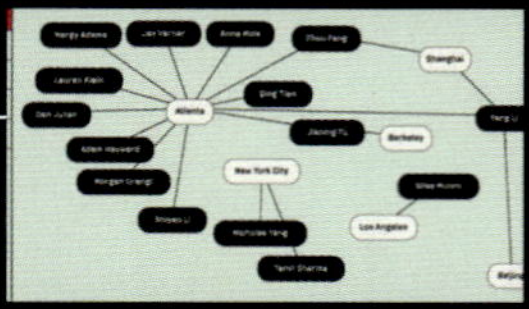

FIGURE 6.4 PG 215

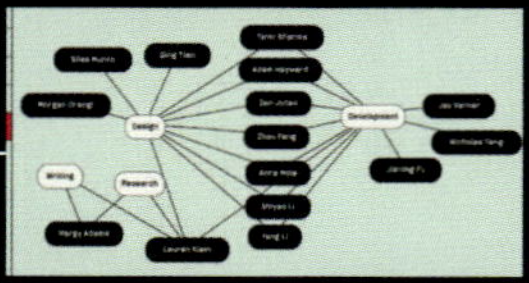

FIGURE 6.5 PG 215

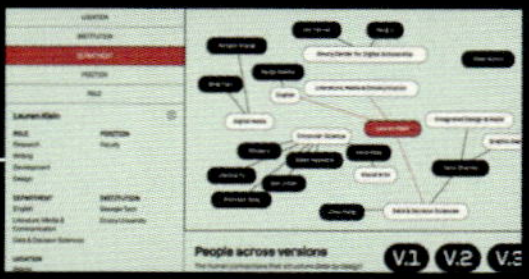

People Across Versions

FIGURE 6.6 PG 216–217

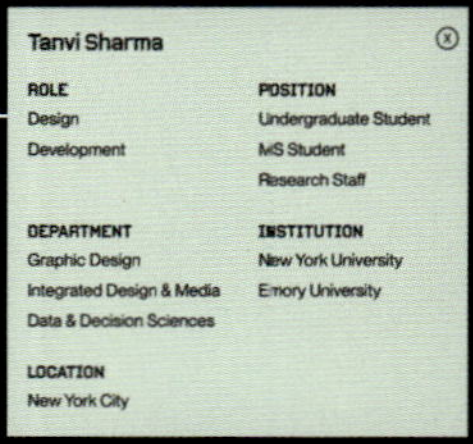

Tanvi Sharma

FIGURE 6.7 PG 219

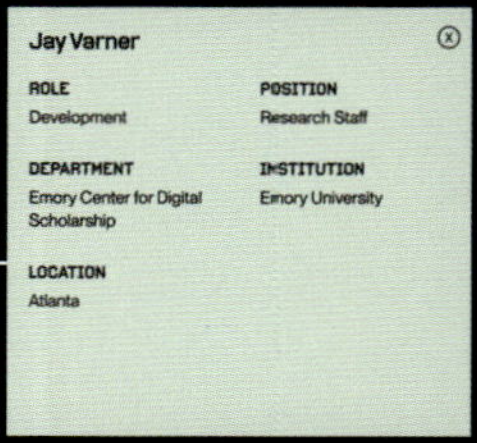

Jay Varner

FIGURE 6.8 PG 221

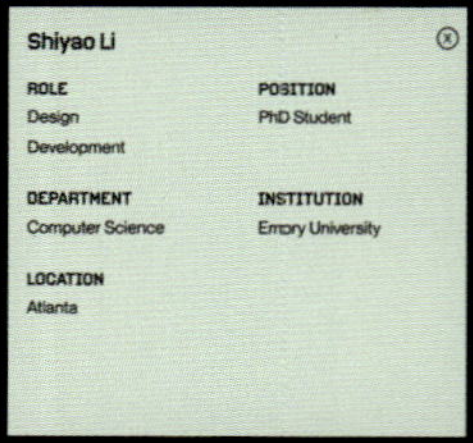

Shiyao Li

FIGURE 6.9 PG 222

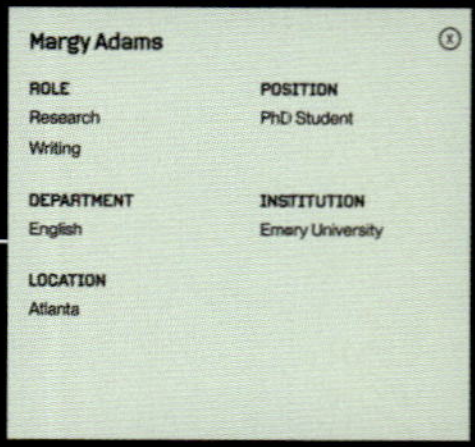

Margy Adams

FIGURE 6.10 PG 223

Price of Straw
SINGLE
SINGLE